Application of Artificial Neural Networks in Geoinformatics

Special Issue Editor
Saro Lee

MDPI • Basel • Beijing • Wuhan • Barcelona • Belgrade

MDPI

Special Issue Editor
Saro Lee
Korea University of Science and Technology
Korea Institute of Geoscience and Mineral Resources (KIGAM)
Korea

Editorial Office
MDPI AG
St. Alban-Anlage 66
Basel, Switzerland

This edition is a reprint of the Special Issue published online in the open access journal *Applied Sciences* (ISSN 2076-3417) from 2017–2018 (available at: http://www.mdpi.com/journal/applsci/special_issues/geoinformatics).

For citation purposes, cite each article independently as indicated on the article page online and as indicated below:

Lastname, F.M.; Lastname, F.M. Article title. *Journal Name* **Year**, *Article number, page range.*

First Edition 2018

ISBN 978-3-03842-742-1 (Pbk)
ISBN 978-3-03842-741-4 (PDF)

Table of Contents

About the Special Issue Editor

Saro Lee, PhD, received his B.Sc. degree in geology in 1991 and PhD. in landslide susceptibility mapping using GIS in 2000. He is currently a principal researcher at Geological Research Division, KIGAM. He is also a full professor at University of Science and Technology (UST). He started his professional career in 1995 as a researcher in the KIGAM. He carried out many International Cooperative Research Projects in the field of mineral potential and geological hazard. He also managed and had lectures in KOICA International Training Program (Mineral Exploration and GIS/RS). His research interests include geospatial predictive mapping with GIS and RS such as landslide susceptibility, ground subsidence hazard, groundwater potential, mineral potential and habitat mapping. He has published about 100 peer reviewed SCI(E) papers and he has a citation rate of about 9000 and h-index of 50 in Google. Also, he is associate editor and editorial board member of many international journals including "Landslide", "Arabian Journal of Geosciences" and "Geosciences Journal".

applied
sciences

MDPI

Editorial

Editorial for Special Issue: "Application of Artificial Neural Networks in Geoinformatics"

Saro Lee [1,2]

[1] Geological Research Division, Korea Institute of Geoscience and Mineral Resources (KIGAM), 124, Gwahak-ro Yuseong-gu, Daejeon 34132, Korea; leesaro@kigam.re.kr
[2] Department of Geophysical Exploration, Korea University of Science and Technology, 217 Gajeong-ro Yuseong-gu, Daejeon 34113, Korea

Received: 27 December 2017; Accepted: 28 December 2017; Published: 2 January 2018

1. Introduction

Recently, a need has arisen for prediction techniques that can address a variety of problems by combining methods from the rapidly developing field of machine learning with geoinformation technologies such as GIS, remote sensing, and GPS. As a result, over the last few decades, one particular machine learning technology known as artificial neural networks has been successfully applied to a wide range of fields in science and engineering. In addition, the development of computational and spatial technologies has led to the rapid growth of geoinformatics, which specializes in the analysis of spatial information. Thus, recently, artificial neural networks have been applied to geoinformatics and have produced valuable results in the fields of geoscience, environment, natural hazards, natural resources, and engineering. Hence, this special issue of the journal *Applied Sciences*, "Application of Artificial Neural Networks in Geoinformatics," was successfully planned, and we here publish many papers detailing novel contributions that are of relevance to these topics.

2. Applications of Artificial Neural Networks in Geoinformatics

In total, 23 papers were submitted to this special issue, 14 of which were accepted and published, constituting a 61% acceptance rate. The papers in this special issue cover various areas related to the application of artificial neural networks to GIS, remote sensing, and GPS, which are typical tools used by geoinformation researchers. These papers addressed problems such as the detection, assessment, and prediction of landslides, volcanos, forest, ozone, oil spills, buildings, ships, habitat, and traffic.

Four papers used GIS tools with artificial neural networks. The first and second papers, authored by Lee, S., Lee, M., Jung, H. [1] and Oh, H., Lee, S. [2], applied GIS and various machine learning algorithms such as artificial neural networks, support vector machines, and boosted tress to map landslide susceptibility. The third paper, authored by Lee, S., Lee, S., Song, W., Lee, M. [3], applied GIS with artificial neural networks to map potential marten and leopard habitats. The fourth paper, authored by Shah, S., Brijs, T., Ahmad, N., Pirdavani, A., Shen, Y., Basheer, M. [4], used data envelopment analysis in GIS with artificial neural networks to evaluate risks related to road safety.

Seven papers studied the applications of artificial neural networks to remote sensing. Among these, three papers used various image technologies and artificial neural networks for the detection of landslides, oil spills, and ships. Mezaal, M., Pradhan, B., Sameen, M., Mohd, S. H., Yusoff, Z. [5] used airborne laser scanning images to detect landslides. Chen, G., Li, Y., Sun, G., Zhang, Y. [6] used polarimetric synthetic aperture radar images to detect oil spills. Hwang, J., Chae, S., Kim, D., Jung, H. [7] used X-band Kompsat-5 images to detect ships. Additionally, Piscini, A., Romaniello, V., Bignami, C., Stramondo, S. [8] proposed a damage assessment method based on SAR and Sentinel-2 images, and Kadavi, P., Lee, W., Lee, C. [9] analyzed pyroclastic flow deposits using Landsat images. Kwon, S., Jung, H., Baek, W., Kim, D. [10] classified the vertical structures of forests using aerial orthophoto and Lidar images. Finally, Foody, G. [11] analyzed the impact of sample design on data validation using remote sensing data classified by feedforward neural networks, and then used a validation dataset to test the classification accuracy.

As the another geoinformation tool, GPS was used for real-time transportation mode identification with artificial neural networks by Byon, Y., Ha, J., Cho, C., Kim, T., Yeun, C. [12]. The paper authored by Sameen, M., Pradhan, B. [13] applied artificial neural networks to predict traffic accident recurrence, and Afonso, N., Pires, J. [14] applied artificial neural networks and genetic algorithms to characterize surface ozone behavior.

3. Future of Artificial Neural Networks in Geoinformatics

In this special issue, we only included papers on artificial neural networks and geoinformation technology. However, artificial neural networks are just one machine learning technique, albeit one of the most popular. Machine learning is a field of computer science that gives computers the ability to learn without being explicitly programmed. Machine learning explores the study and construction of algorithms that can learn from data and make data-driven predictions or decisions by building a model from sample inputs. There are numerous machine learning techniques, such as decision trees, support vector machines, naive Bayes classifier, clustering, inductive logic programming, and genetic algorithms. These machine learning techniques can be combined with geoinformation technologies, and further studies are required in this area.

As we enter the age of the fourth industrial revolution, artificial intelligence technologies have come to play a very important role in society. Machine learning technologies such as artificial neural networks are expected to play a key role in the fourth industrial revolution, especially in combination with geoinformation technology. However, these technologies do not come out of nowhere; they are developed by scientists. Therefore, many scientists will have to expand on the research presented in this special issue.

Acknowledgments: I would like to thank the authors of the papers submitted for this special issue, whether or not they were selected for publication. Also, I thank the reviewers, who are all experts on the theme, and the editorial team of *Applied Sciences*. This special issue was conducted by the Basic Research Project of the Korea Institute of Geoscience and Mineral Resources (KIGAM) funded by the Ministry of Science, ICT. This research (NRF-2016K1A3A1A09915721) was supported by Science and Technology Internationalization Project through National Research Foundation of Korea (NRF) grant funded by the Ministry of Science and ICT.

Conflicts of Interest: The authors declare no conflict of interest.

References

1. Lee, S.; Lee, M.; Jung, H. Data Mining Approaches for Landslide Susceptibility Mapping in Umyeonsan, Seoul, South Korea. *Appl. Sci.* **2017**, *7*, 683. [CrossRef]
2. Oh, H.; Lee, S. Shallow Landslide Susceptibility Modeling Using the Data Mining Models Artificial Neural Network and Boosted Tree. *Appl. Sci.* **2017**, *7*, 1000. [CrossRef]
3. Lee, S.; Lee, S.; Song, W.; Lee, M. Habitat Potential Mapping of Marten (*Martes flavigula*) and Leopard Cat (*Prionailurus bengalensis*) in South Korea Using Artificial Neural Network Machine Learning. *Appl. Sci.* **2017**, *7*, 912. [CrossRef]
4. Shah, S.; Brijs, T.; Ahmad, N.; Pirdavani, A.; Shen, Y.; Basheer, M. Road Safety Risk Evaluation Using GIS-Based Data Envelopment Analysis—Artificial Neural Networks Approach. *Appl. Sci.* **2017**, *7*, 886. [CrossRef]
5. Mezaal, M.; Pradhan, B.; Sameen, M.; Mohd Shafri, H.; Yusoff, Z. Optimized Neural Architecture for Automatic Landslide Detection from High-Resolution Airborne Laser Scanning Data. *Appl. Sci.* **2017**, *7*, 730. [CrossRef]
6. Chen, G.; Li, Y.; Sun, G.; Zhang, Y. Application of Deep Networks to Oil Spill Detection Using Polarimetric Synthetic Aperture Radar Images. *Appl. Sci.* **2017**, *7*, 968. [CrossRef]
7. Hwang, J.; Chae, S.; Kim, D.; Jung, H. Application of Artificial Neural Networks to Ship Detection from X-Band Kompsat-5 Imagery. *Appl. Sci.* **2017**, *7*, 961. [CrossRef]

8. Piscini, A.; Romaniello, V.; Bignami, C.; Stramondo, S. A New Damage Assessment Method by Means of Neural Network and Multi-Sensor Satellite Data. *Appl. Sci.* **2017**, *7*, 781. [CrossRef]
9. Kadavi, P.; Lee, W.; Lee, C. Analysis of the Pyroclastic Flow Deposits of Mount Sinabung and Merapi Using Landsat Imagery and the Artificial Neural Networks Approach. *Appl. Sci.* **2017**, *7*, 935. [CrossRef]
10. Kwon, S.; Jung, H.; Baek, W.; Kim, D. Classification of Forest Vertical Structure in South Korea from Aerial Orthophoto and Lidar Data Using an Artificial Neural Network. *Appl. Sci.* **2017**, *7*, 1046. [CrossRef]
11. Foody, G. Impacts of Sample Design for Validation Data on the Accuracy of Feedforward Neural Network Classification. *Appl. Sci.* **2017**, *7*, 888. [CrossRef]
12. Byon, Y.; Ha, J.; Cho, C.; Kim, T.; Yeun, C. Real-Time Transportation Mode Identification Using Artificial Neural Networks Enhanced with Mode Availability Layers: A Case Study in Dubai. *Appl. Sci.* **2017**, *7*, 923. [CrossRef]
13. Sameen, M.; Pradhan, B. Severity Prediction of Traffic Accidents with Recurrent Neural Networks. *Appl. Sci.* **2017**, *7*, 476. [CrossRef]
14. Afonso, N.; Pires, J. Characterization of Surface Ozone Behavior at Different Regimes. *Appl. Sci.* **2017**, *7*, 944. [CrossRef]

applied
sciences

MDPI

Article

Data Mining Approaches for Landslide Susceptibility Mapping in Umyeonsan, Seoul, South Korea

Sunmin Lee [1,2], Moung-Jin Lee [2,]* and Hyung-Sup Jung [1,]*

[1] Department of Geoinformatics, University of Seoul, 163 Seoulsiripdaero, Dongdaemun-gu, Seoul 02504, Korea; smlee@kei.re.kr
[2] Center for Environmental Assessment Monitoring, Environmental Assessment Group, Korea Environment Institute (KEI), Sejong-si 30147, Korea
* Correspondence: leemj@kei.re.kr (M.-J.L.); hsjung@uos.ac.kr (H.-S.J.); Tel.: +82-44-415-7314 (M.-J.L); +82-2-6490-2892 (H.-S.J.)

Academic Editor: Saro Lee
Received: 8 June 2017; Accepted: 27 June 2017; Published: 2 July 2017

Abstract: The application of data mining models has become increasingly popular in recent years in assessments of a variety of natural hazards such as landslides and floods. Data mining techniques are useful for understanding the relationships between events and their influencing variables. Because landslides are influenced by a combination of factors including geomorphological and meteorological factors, data mining techniques are helpful in elucidating the mechanisms by which these complex factors affect landslide events. In this study, spatial data mining approaches based on data on landslide locations in the geographic information system environment were investigated. The topographical factors of slope, aspect, curvature, topographic wetness index, stream power index, slope length factor, standardized height, valley depth, and downslope distance gradient were determined using topographical maps. Additional soil and forest variables using information obtained from national soil and forest maps were also investigated. A total of 17 variables affecting the frequency of landslide occurrence were selected to construct a spatial database, and support vector machine (SVM) and artificial neural network (ANN) models were applied to predict landslide susceptibility from the selected factors. In the SVM model, linear, polynomial, radial base function, and sigmoid kernels were applied in sequence; the model yielded 72.41%, 72.83%, 77.17% and 72.79% accuracy, respectively. The ANN model yielded a validity accuracy of 78.41%. The results of this study are useful in guiding effective strategies for the prevention and management of landslides in urban areas.

Keywords: spatial data mining; SVM; ANN; validation; ROC

1. Introduction

The rapid growth in data due to the development of information and communication technology (ICT) has spurred the demand for data mining, which is generally considered to be the most useful analytical tool in the analysis of large data sets. Data mining has been defined in various ways previously, and is commonly referred to as an exploratory analysis tool for large amounts of data [1]. Therefore, data mining is a process of finding useful information that is not easily exposed in a large amount of data, most recently known as "big data".

With the rapid growth of ICT since the 1980s, countries and corporations have made significant efforts to build databases to store and manage vast amounts of data. Rational and rapid decision-making is necessary to increase the utilization of large-scale databases. In this environment, it has become important to find meaningful new information that can support optimal decision-making. In this process, an efficient methodology for data mining that extracts not only known information from previously established databases but also information that was not previously known has gained popularity. Therefore, the data mining methodology makes it possible to effectively analyze large amounts of data in a database and the relationships between parameters.

In recent years, storms bringing localized heavy rains have been reported worldwide, and precipitation in the mid-latitude regions has increased since the 20th century [2]. In addition, rapid climate change and abnormal climate conditions have led to extreme weather phenomena, which provide the appropriate conditions for the creation of landslides, and the frequency of landslides has increased accordingly [3]. A landslide is the movement of soil down a slope when the stress exceeds the strength of the soil material due to rapid changes in the natural environment such as heavy rains or earthquakes [4]. Landslides are caused by various factors, such as topography, geology, soil characteristics, forest conditions, and climate variables [5–8]. The main purpose of this study was to generate and validate a landslide susceptibility map using data mining models for Umyeonsan, South Korea. Therefore, data mining techniques should be used appropriately for the analysis of the relationships between landslides and impact factors on landslides in landslide susceptibility analyses. Additionally, given that the data mining models derive their results from the database, the input data represent a key part in the model. Thus, the database should be constructed using reliable data, such as nationally distributed government data. Remote sensing data is one of the main sources to be used to generate the reliable data [9–11].

Due to the complex factors influencing landslides, it is important to perform an appropriate assessment of the potential for landslide susceptibility using data mining techniques. A susceptibility analysis based on data on previous damage should be performed to define landslide susceptibility in areas where landslides are anticipated so that the damage can be predicted in advance and appropriate mitigation measurements adopted. Over the past several decades, a number of data mining approaches have been developed for mapping landslide susceptibility [12,13]. The most commonly used methods are soft computing or statistical techniques such as artificial neural network (ANN) models [14,15], fuzzy logic methods [16,17] or logistic regression models [18,19]. Landslides are caused by multiple interactions between various factors, including geometry, geology, soil characteristics, and vegetation conditions [20,21], and are highly dependent on terrain features [22,23].

As geological hazards, landslides can limit the sustainable development of urban areas, causing environmental damage, serious casualties, and loss of property [24,25]. An appropriate assessment of the susceptibility for landslide damage is required to minimize the complex array of losses. In South Korea, according to an analysis of the scale of landslides by the Korea Forest Service (KFS), the average annual area affected by landslides increased markedly from 231 ha in the 1980s, to 349 ha in the 1990s, and 713 ha in 2000 [26]. Typhoons accompanied by intense storms caused by local climatic conditions often cause significant damage; and landslides are concentrated in the rainy season every summer. Because much of South Korea is mountainous, most studies on landslides in the country have focused on mountainous areas [27–29]; e.g., in Jangheung [23] or Yongjin [6,30]. Furthermore, due to the frequency of landslides in mountainous areas, most of the studies in Southeast Asia have been performed in non-urban areas [31–33]. Assessment of landslide susceptibility in urban areas could provide an important contribution to minimizing additional damage from these natural disasters; it could be also used to planning and multi hazard assessment [34–37].

Seoul, the capital of South Korea, also has urban areas occupying mountainous land. Unlike landslides that occur in sparsely populated mountainous areas, landslides that occur in mountainous areas in cities result in significant casualties/fatalities and structural damage. For example, landslides in Umyeonsan located in central Seoul in July 2011 caused unprecedented damage to the city. Following the 2011 Umyeonsan landslide, the downtown area, which had previously been considered safe, was no longer regarded as a landslide-safe zone. Various aspects of Korea's landslide policy have been re-evaluated and are expected to be revised in the near future. Landslides are more dangerous in urban areas than non-urban mountainous areas. Policies have shifted to preventative and resident-oriented initiatives, with a focus on minimizing injuries to citizens and restoration of their properties after landslides. Therefore, it is important to analyze potential landslide hazards in urban areas.

To apply the model for this type of data mining analysis, a spatial database should first be constructed containing information on the area of landslide damage. Because landslides are difficult to access, the collection of field survey data to build a landslide spatial database is costly and time-consuming. An effective alternative is to build a database using aerial photography and geographic information systems (GISs). Landslides leave traces that last for months or years; hence, aerial imagery can be used to extract data on areas damaged by landslides after the event. Using these data, the susceptibility for landslides can be analyzed by data mining modeling techniques.

2. Study Area

Seoul is the largest city in South Korea and is located in the center of the Korean Peninsula. The total urban area covered by Seoul (approximately 605 km^2) is less than 1% of the total land area of South Korea. However, Seoul is one of the most heavily populated cities in the world; over 10 million citizens were recorded in 2011. Thus, the city has a dense and complex transportation system. Natural disasters, including landslides, have the potential to cause serious damage in the city [2]. The study area, Umyeonsan, is an urbanized mountain rising 321.6 m above sea level in the southern sub-central area of Seoul (Figure 1). Due to the fact that it is not very high or steep, the mountain is a popular site for Seoul residents and it has a large transient population

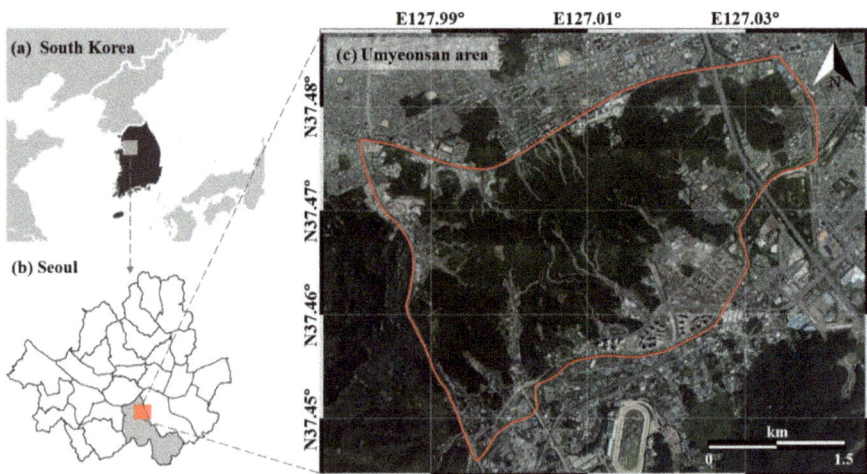

Figure 1. Map showing the location of the study area of Umyeonsan: (**a**) South Korea; (**b**) Seoul; and (**c**) Umyeonsan.

Umyeonsan is comprised of a low mountainous area with a slope of approximately 30° near the top of the mountain. The area is mainly comprised of gneissic rock formed by geomorphological movement and weathering. The composition of biotite gneiss, which is prone to weathering and fault formation, increases its susceptibility to landslides. The foliation structure of the gneiss is sparsely generated due to multiple flexures. The main soil type distributed in the study area is brown forest soil formed from the base materials of biotite gneiss and granite gneiss, which are metamorphic rocks, and the soil texture is sandy loam to loam, which has good drainage. Deep soils and high organic content brown forest soil is distributed in some valley areas in the study area, where the soil texture is silt loam in silt and sand. Due to the soil textures present, the Umyeonsan area is prone to severe weathering, causing the unstable condition of outcrops.

The main forest type in Umyeonsan is temperate deciduous broad-leaved forest. Broad-leaved forest, pine forest, poplar forest, and broad-leaved plantation forest are all present in the study area. The principal species in the forests are similar to those in the central Korean Peninsula: *Quercus mongolica, Robinia pseudoacacia, Quercus variabilis, oak trees* and *Asian black birch*.

On 27 July 2011, heavy rains occurred in central Korea and a large landslide occurred in Umyeonsan. The heavy rainfall lasted from 26–28 July. The cumulative precipitation in the Seoul metropolitan area was 587.5 mm, which is the largest three-day cumulative precipitation recorded since precipitation monitoring began in Seoul. Based on data from Seocho automatic weather system, the maximum precipitation per hour at the time of the landslide was 85.5 mm [38]. The landslide from the heavy rains affected an area of 73.23 ha. Sixteen people were killed, including 6 residents from the village of Namtaeryeong and 7 residents from the southern area of the southern ring roads; 2 people remained missing, and approximately 400 people were evacuated [39]. The water supply was cut off. Thousands of apartments near Gangnam and Umyeonsan were damaged; and over 20,000 households experienced water problems. Therefore, we chose this landslide to obtain data for our study (Figure 2d,e). The study area was determined by the road and river boundaries in the Umyeonsan area (Figure 2).

3. Data

Given that data mining models are a data-driven methodology, accurate data are essential to build a suitable database. Accurate data on the state of the earth's surface can be obtained rapidly using aerial photographs, which also effectively capture complicated urban structures. Moreover, despite the high cost of aerial photographs, it is cost-effective to acquire the images of and continuously monitor urban areas due to their dense nature. To acquire accurate data on the landslide occurrence, aerial photographs from before and after the Umyeonsan landslides were used (Figure 2a,b). Location data for the Umyeonsan landslide area were acquired through comparative analysis, as traces of landslides are visible for several years after the events.

In addition, a digital topographic map (1:5000) generated by the National Geographic Information Institute (NGII) was obtained, and geometric correction was performed. Overlay and comparative analysis of the aerial photographs and the 1:5000 digital topographic map indicated the landslide occurrence area (Figure 2c). The mapped landslide occurrence data were finally confirmed by comparison with data surveyed in the field by the Seoul metropolitan government [39].

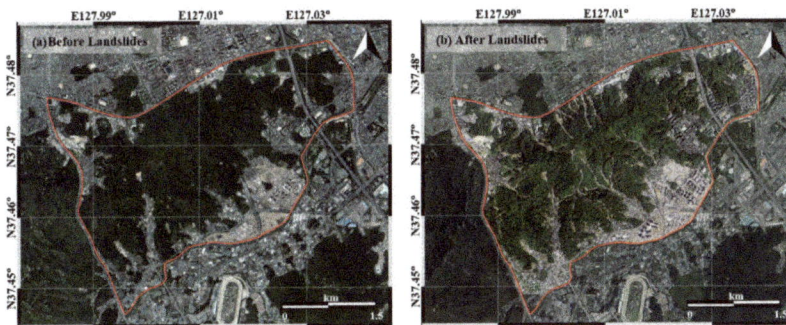

Figure 2. *Cont.*

Figure 2. Data of landslide occurrence locations from aerial photographs: (**a**) before landslides in 2011; (**b**) after landslides in 2011; (**c**) training and validation data; (**d**) damage situation in Southern Circular Road [40]; (**e**) Gyungnam apartment buildings in Seocho-gu [41].

Slopes are established by erosion and sedimentation of terrain surfaces and have a key effect on landslide occurrence as they affect the flow of water and the soil characteristics [42]. Therefore, among the factors influencing landslides, topographic factors were preferentially considered and were used as input data to identify the relationships with landslide occurrence (Figure 3). Prior to calculation of the topographic factors, a triangulated irregular network (TIN) was produced from a pre-generated topographic map, and a digital elevation model (DEM) was created using ArcGIS 10.1 (ESRI, Redlands, LA, USA).

All topographic factors were calculated from the DEM using ArcGIS software (ESRI, Redlands, LA, USA). The slope, indicated as the angle to the horizontal plane or the ratio of the vertical height to the horizontal distance as a percentage, was calculated using the SLOPE tool. The ASPECT tool was used to calculate the aspect, the compass direction of the slope faces. The curvature values of the terrain surface were calculated using the CURVATURE tool. The curvature represents the morphological characteristics of the study area; an upwardly convex surface possesses a positive value, while an upwardly concave surface possesses a negative value.

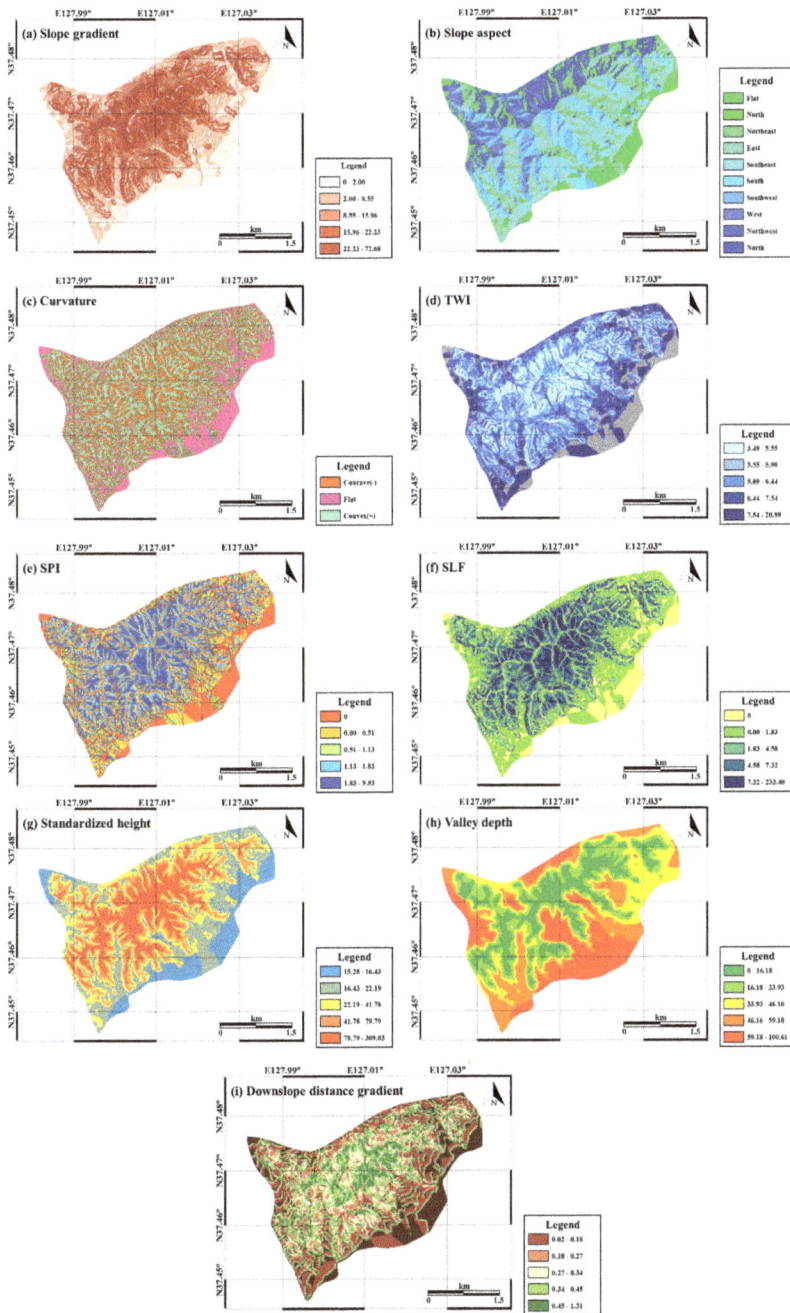

Figure 3. Topographical factors related to landslides used to construct a spatial database: (**a**) slope gradient; (**b**) slope aspect; (**c**) curvature; (**d**) topographic wetness index (TWI); (**e**) stream power index (SPI); (**f**) slope length factor (SLF); (**g**) standardized height; (**h**) valley depth; (**i**) downslope distance gradient.

The topographic wetness index (TWI) is defined for a steady state. The index is commonly used for hydrological processes to quantify the disposition of water flow and the impact on the topography [43]. The TWI is calculated by the run-off model as follows, which reflects the accumulation of water at a specific point in the catchment area and the force of gravity on the water in the area [43].

$$TWI = In\left(\frac{a}{\tan \beta}\right) \tag{1}$$

where α is the cumulative upslope area drained per unit contour length in a cell, and β is the local slope angle of the cell.

The stream power index (SPI) estimates the erosion power of a stream, which affects the stability of the area. Therefore, it is generally used to determine the locations for performing soil conservation measures to reduce the erosive effects of concentrated surface runoff. SPI can be calculated as follows [44]:

$$SPI = As \times \tan(\beta) \tag{2}$$

where As is the area of the target point of the study area, and β is the slope angle.

After calculating the slope, the slope length factor (SLF) for average erosion is calculated using the Revised Universal Soil Loss Equation (RUSLE) [45]. The SLF for slope of length λ is represented as:

$$SLF = \left(\frac{\lambda}{72.6}\right)^{m} \tag{3}$$

where the constant value 72.6 of the RUSLE is in feet, and the variable, m, is the slope-length exponent calculated from the local slope angle β [45]. The SLF is influenced by the ratio of rill erosion to interrill erosion, β. Rill erosion is caused by flow, while interrill erosion is caused by the effects from rainfall.

System for Automated Geoscientific Analyses (SAGA) GIS were used to calculate additional topographic factors [46]. The standardized height is the absolute height multiplied by the normalized height of the study area; normalized height is the normalization of the height of the study area between 0 and 1. The standardized height was calculated using linear regression to yield the suitable parameters for prediction of the soil attributes by simplifying the complex terrain states [47]. The valley depth is the vertical distance of the specific point from the base level of the channel network [46]. The elevation of the channel network base level is interpolated, and the valley depth is calculated by subtracting the base level of the channel network from the DEM [46]. The downslope distance gradient is a quantitative estimation of the hydraulic gradient. The downslope distance gradient is obtained by calculating the downhill distance when water loses a determined quantity of energy from precipitation [48]. $\tan a_d$ identifies the downslope distance gradient:

$$\tan a_d = \frac{d}{L_d} \tag{4}$$

where d and L_d are the elevation and the horizontal distance to the reference point, respectively.

The type and composition of the soil influence the occurrence of landslides as they determine the degree of erosion and saturation [49]. The slope stability can be strengthened by the roots of vegetation and the degree of strengthening depends on the specific attributes of the vegetation; hence, the impact of heavy rainfall on the slope can be mitigated by vegetation [50]. The vegetation map was provided by the Korea Forest Research Institute and was in polygon format with a scale of 1:25,000 (Figure 4). The attributes of timber diameter, type, density, and age were extracted from the vegetation map. In addition to the vegetation variables, the soil type and condition also influence the occurrence of landslides. The soil maps used in this study were provided by the National Academy of Agricultural Science (RDA) in polygon format with a scale of 1:25,000. Soil depth, drainage, topography, and texture were derived from the soil maps (Figure 5).

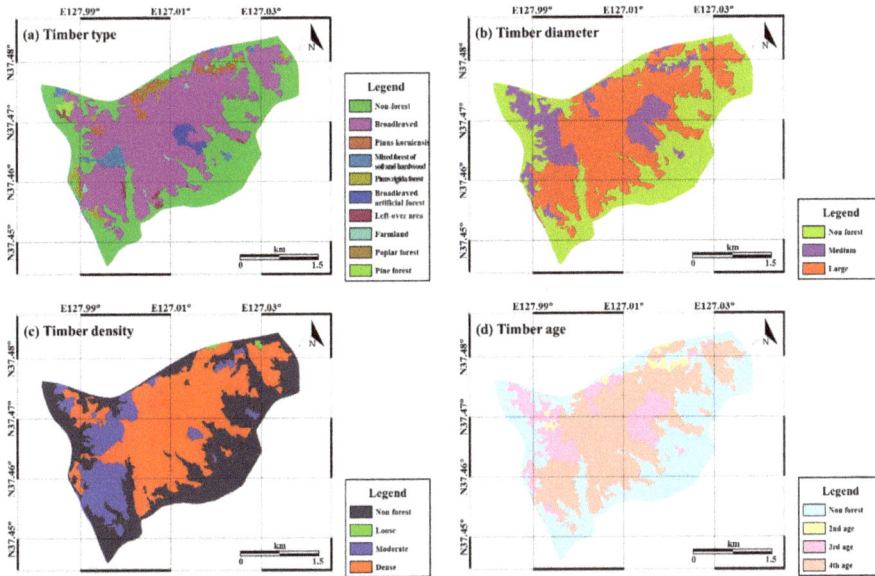

Figure 4. Landslide-related factors obtained from the vegetation map: (**a**) timber type; (**b**) timber diameter; (**c**) timber density; (**d**) timber age.

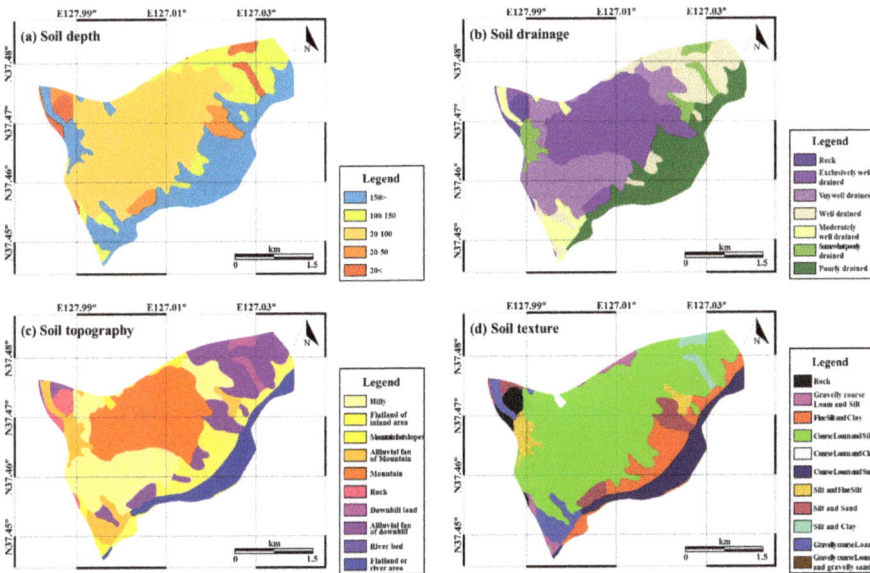

Figure 5. Landslide-related factors obtained from the soil map: (**a**) soil depth; (**b**) soil drainage; (**c**) soil topography; (**d**) soil texture.

All variables used in this study were resampled into a grid format of 5-m spatial resolution to create a spatial database. The total size of the spatial database was 1015 × 874, and the study area contained 457,133 cells. The 103 landslide occurrence locations in point format were also converted into the 5-m grid format to be included in the data in the spatial database. Half of the 103 landslide location data points (51 points) were used as training data, and the other half were used as validation data. All of the collected data were prepared in ASCII format from the grid. Finally, a spatial database was established to apply the model to (Table 1).

Table 1. Landslide-related factors used to construct spatial database.

Original Data	Factors	Data Type	Scale
Aerial photograph	Landslide location	Point	1:1000
Topographical map [a]	Slope gradient [°]	GRID	1:1000
	Slope aspect		
	Curvature		
	Topographic Wetness Index (TWI)		
	Stream Power Index (SPI)		
	Slope Length Factor (SLF)		
	Standardized height [m]		
	Valley depth [m]		
	Downslope distance gradient [rad]		
Forest map [b]	Timber diameter	Polygon	1:25,000
	Timber type		
	Timber density		
	Timber age		
Soil map [c]	Soil depth	Polygon	1:25,000
	Soil drainage		
	Soil Topography		
	Soil texture		

[a] Topographical factors were extracted from digital topographic map by National Geographic Information Institute; [b] The forest map produced by Korea Forest Service; [c] The detailed soil map produced by Rural Development Administration

4. Methodology

Landslides are affected by various factors and each element is simplified as input data in modeling studies. Given that natural phenomena such as landslides are composite events of each element, they should be analyzed using a modeling technique based on the relationships between data in a vast dataset such as a data mining model. Using data mining models, the relationships among all of the predisposing factors and the effects of all of the landslide-related factors on the model results can be analyzed. Therefore, in this study, the factors related to landslide susceptibility were used as the input data of the data mining models. The input variables were stored in a spatial database; 17 landslide-related variables were chosen from the spatial database, including half of the randomly selected landslide locations used as training data. Landslide susceptibility maps were created using two data mining models, specifically support vector machine (SVM) (Boulder, CO, USA) and ANN models, which are based on linear classifiers. Additionally, the weight of each factor was calculated in the data mining modeling process, and a validation was performed (Figure 6). A detailed description of the data mining models used in this study is given below.

Topographical map	Forest map	Landslide location data extraction from aerial photographs	
Slope Gradient	Timber diameter	Training data (Landslide area in 2011)	Validation data (Landslide area in 2011)
Slope Aspect	Timber type		
Curvature	Timber density	Applying the Data Mining Models	
Topographic Wetness Index	Timber age	SVM model	Applying the backpropagation ANN model
Stream Power Index	Soil map	linear / Radial basis function	
Slope Length Factor	Soil Depth	Polynomial / sigmoid	
Standard height	Soil Drainage	Generation of Landslide Susceptibility Maps	
Valley depth	Soil Topography	Validation	
Down-distance Gradient	Soil texture	Receiver Operating Characteristics (ROC) Analysis	Calculation of the Area Under the Curve (AUC)
Geospatial Database Construction			

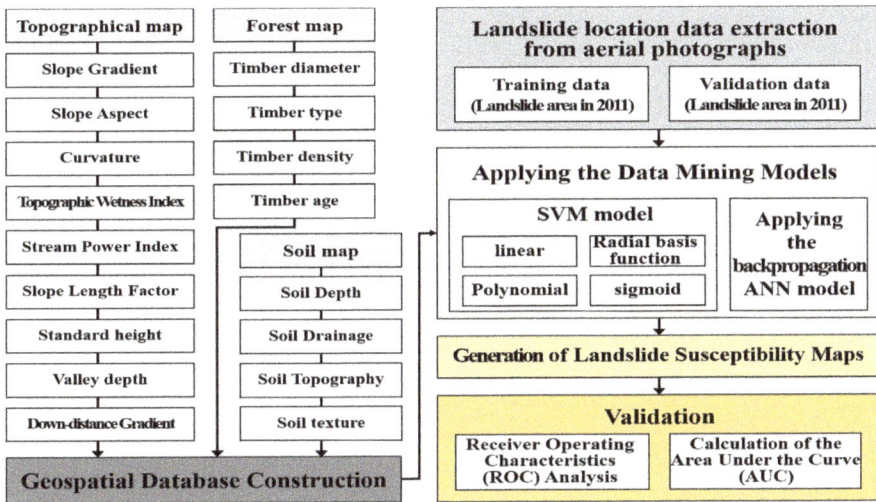

Figure 6. Flowchart of the spatial prediction of landslide susceptibility.

4.1. Support Vector Machine

4.1.1. Basic Principles of SVM

The SVM method is a training algorithm based on nonlinear transformations that uses a classification based on the principle of structural risk minimization, which has performed well in the test phase [51]. When training data are represented in multidimensional space, several hyperplanes can be used to distinguish the data into two classes; however, there is only one optimum linear separating hyperplane [52]. This optimal hyperplane should maximize the distance between the data closest to the hyperplane. Thus, the objective of the SVM method is to determine a multi-dimensional hyperplane that differentiates the two classes with the maximum margin. The SVM model performs this process according to three main concepts: margin, support vector, and kernel.

The margin is the distance between a specific data point and the hyperplane that separates the data patterns. It is the shortest distance from the data vector of each class to the given hyperplane. The larger the margin, the more accurately the SVM model performs the classification; hence, the optimal hyperplane maximizes the margin when $\vec{\omega}$ is the normal vector of the hyperplane, and is the norm of the normal of the hyperplane. $||\vec{\omega}||$ is the norm of the normal of the hyperplane.

$$margin = \frac{2}{||\vec{\omega}||} \qquad (5)$$

The hyperplane, which defines the two classes that maximize the margins when the data are linearly separable, can be defined by Equation (6). Under these conditions, the training data on these two hyperplanes constitute the support vector. Therefore, the support vector maintains the shortest distance to the hyperplane. The support vectors are dependable on the boundary of the hyperplane and they are time-efficient when a new dataset is tested with a support vector. The dataset of linear separable training vectors x_i ($i = 1, 2, \ldots, n$) and the training vectors consisting of two classes are represented by y_i as the value of ± 1 [53,54].

$$y_i\left((\vec{\omega} \cdot \vec{x}_i) + b\right) - 1 \geq 0 \qquad (6)$$

where b is the scalar base, and the scalar product operation is denoted by ($\cdot$).

In addition, because the two hyperplanes must maximize the margin between them, the optimization of objective Equation (7) with the constraints of Equation (6) yields the planes of the two classes:

$$\min(\frac{1}{2}||\vec{\omega}||^2) \tag{7}$$

However, most cases do not satisfy the constraint as the data are not linearly separable. A formula including the slack variable ξ_i indicates the distance from the hyperplane to the data located on the wrong side [52]; the penalty term C introduced to account for misclassification [55,56] was developed to address this problem.

$$y_i\left((\vec{\omega}\cdot\vec{x_i})+b\right) \geq 1-\xi_i \tag{8}$$

$$\min(\frac{1}{2}||\vec{\omega}||^2 + C\sum_{i=1}^{n}\xi_i) \tag{9}$$

4.1.2. SVM for Nonlinear Classification

In the preceding case of nonlinear boundary data, the SVM model is unable to generate an appropriate hyperplane. Therefore, the data can be classified into a linear problem as a hyperplane by mapping them to a high-dimensional space value from the low-dimensional data that cannot be linearly separated. To solve Equations (8) and (9) when the dataset is linearly separable, the training data are expressed as a vector inner product in the optimization process. This enables the inner product in the training algorithm to be expressed by the inner product of a function in a feature space when nonlinear data in the input space are mapped to a high-dimensional feature space by a specific function.

The hyperplane of the data with nonlinear data patterns can be derived by transforming the data space, but x cannot be computed directly; an arbitrary kernel function should be used to satisfy x [53]. The kernel functions, $K(x_i, x_j)$, convert the original data into linearly separable data in high-dimensional feature space [52]. The kernel function should satisfy Mercer's condition. The following four functions are the main kernel functions used in the SVM model:

$$\text{Linear}: K(x_i, y_i) = x_i^T \cdot x_j \tag{10}$$

$$\text{Polynomial}: K(x_i, y_i) = \left(\gamma\cdot\left(x_i^T\cdot x_j\right)+r\right)^d, \gamma > 0 \tag{11}$$

where γ and d are parameters of the kernel functions.

$$\text{Radial basis function}: K(x_i, y_i) = e^{-\gamma||x_i-x_j||^2}, \gamma > 0 \tag{12}$$

$$\text{Sigmond}: K(x_i, y_i) = \tanh\left(\gamma\cdot x_i^T\cdot x_j + r\right) \tag{13}$$

In radial basis function (RBF) kernel, $||x_i - x_j||$ is the squared Euclidean distance between the vectors, and γ is defined as follows:

$$\gamma = \frac{1}{2\sigma^2} \tag{14}$$

where σ is a freely adjustable parameter that controls the performance of the kernel, and r is a parameter of the sigmoid kernel.

In this study, the Environment for Visualizing Images (ENVI 5.0, Exelis Visual Information Solutions, Boulder, CO, USA) software program was used for the application of the SVM model (Boulder, CO, USA). The software provides four types of kernels in the SVM classifier as mentioned above: linear, polynomial, RBF, and sigmoid. Each kernel function has its own parameters, which should be adjusted appropriately according to the characteristics of the purpose. The default kernel of the ENVI software is the RBF kernel, which is known to provide more accurate prediction results in most classifications, particularly in nonlinear problems. After applying the kernel and classifying each input factor, the SVM model results were mapped using the process of overlapping and summation.

$$SVM_{all} = \sum_{i=1}^{n} SVM_i \ (n = \text{number of input factors}) \tag{15}$$

SVM_{all} is the sum of all classification results by the SVM model for all variables, and SVM_i is the single result of a given variable; in this study, the total number of input factors, n was 17. The results were mapped and validated using the landslide locations not used for training.

4.2. Artificial Neural Network

An ANN is a statistical learning algorithm that describes a neuronal signaling system. The human brain regulates the connection strength of synapses between neurons and determines whether signals are delivered or not; the ANN learns to derive the results with minimum error by adjusting the weight between input data and output data. The unit of the ANN is the perceptron, which is divided into an input layer and an output layer. The input layers are represented by nodes, which are linked to an edge having the weight of each node [57]. The model outputs 1 when the sum of the values of each layer multiplied by the weight is greater than or equal to the threshold; otherwise, the output is -1. This type of perceptron can only be applied to a linear classification and is not applicable in most cases.

In this study, multi-layered perceptrons allowing classification of nonlinear data were used with a hidden layer between the input and output layers. A feed-forward network was used for the map learning. The learning of neural networks progressed via a backpropagation algorithm that propagates the errors generated in the output layer to each layer and corrects the weights. In other words, the backpropagation algorithm learns the neural network by inverting the step, and the result most similar to the output layer value is derived based on the input layer. The process can be simplified as a combination of nodes and edges [58]. The backpropagation algorithm can be expressed by the following equation:

$$\omega_k^{up} = \omega_k - \eta * \frac{\partial E_{total}}{\omega_k} \tag{16}$$

where ω_k is the updated weight for connection k and η is the learning rate. The initial value of weight ω_0 is given randomly. E_{total} is the error of the final output result when the expected output value is assumed to be T, according to the input data [59].

$$E_{total} = \frac{1}{2} \sum_{j}^{n} \left(T_j - out_{o_j} \right)^2 \tag{17}$$

where j is the number assigned to the node in each layer, and n is the number of output layers. Because the backpropagation algorithm updates the weights while propagating the error generated from the output to the input direction, it updates the weights starting from the portion from the output. E_{total} is calculated repeatedly using the updated weight so as to be minimized; the differentiation of E_{total} is fundamental. Therefore, when the calculated value is passed to the next layer or is expected to be the final result, an activation function is used to convert it to a range of values. In this study, a unipolar sigmoid function was used. The equation of the unipolar sigmoid function is as follows:

$$f(x) = \frac{1}{1 + e^{-x}} \tag{18}$$

Using the activation function, the output value of each layer is adjusted from 0 to 1.

In this study, the ANN model was processed using MATLAB R2015a software (MathWorks, Massachusetts, MA, USA). Areas where the slope data were 0 were set as areas not prone to landslide occurrence and used for training data; landslide occurrence point data were also allocated to the training set. The backpropagation algorithm of the ANN was applied to estimate the weights between the input and the hidden layers, and between the hidden and the output layers by updating the hidden nodes and the learning rate. In this study, the input data were normalized to the range of 0.1–0.9 and a $17 \times 34 \times 1$ structure was selected for the networks. The learning rate was set to 0.01 with a randomly selected initial weight. The number of epochs was set to 2000, and the root mean square error value was set to 0.1, which was used for the stopping criterion.

5. Results

Figure 7a–d shows the landslide susceptibility maps produced by applying the SVM models with four kernels; and Figure 7e shows the map of the results from the ANN models. The values of the landslide susceptibility results were sorted in descending order and categorized into five classes for easy visual interpretation. For intuitive and efficient analysis, the results for the landslide susceptible areas were classified into five groups of approximately the same size: very high, high, medium, low, and very low susceptibility. The comparative analysis between the results and existing landslide occurrence locations was also made easier by representing the landslide susceptibility with the five categories. The spatial distribution of the landslide locations was concentrated in high-susceptibility areas.

In Figure 7, the results map of the linear, polynomial, and sigmoid kernels of the SVM models showed a similar pattern and represented the influence of the input factors of Figures 4 and 5. In addition, the RBF kernel showed a similar pattern to the ANN map in that the center of the map had relatively higher susceptibility. In general, highly susceptible areas were principally distributed in the northeast direction from the center of the study area. Areas with high susceptibility were located in dense forest areas of broadleaved forest, above-average soil depth, and steep slopes. The dominant rocks constituting the Umyeonsan area are gneissic rocks, which have high landslide susceptibility because they are heavily eroded; landslides occur in areas of high soil drainage.

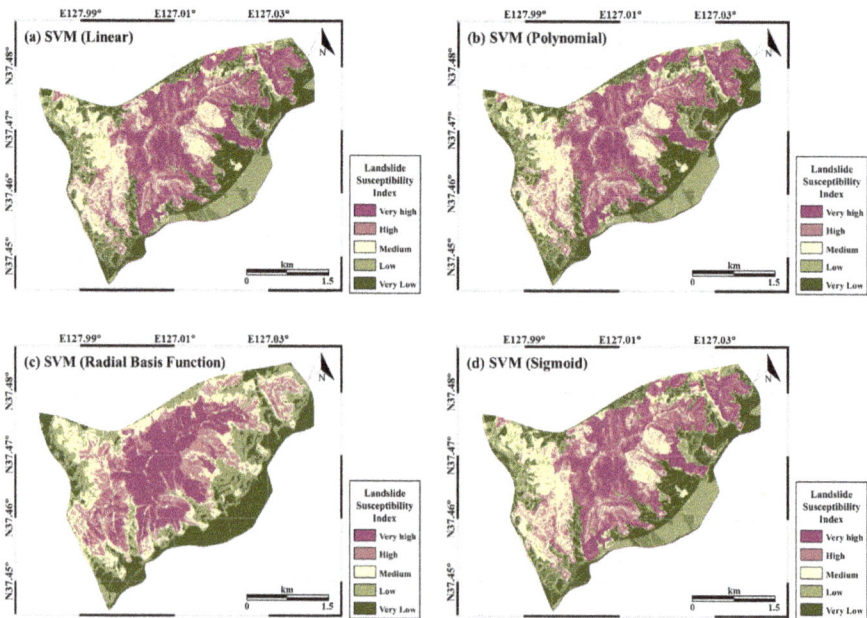

Figure 7. *Cont.*

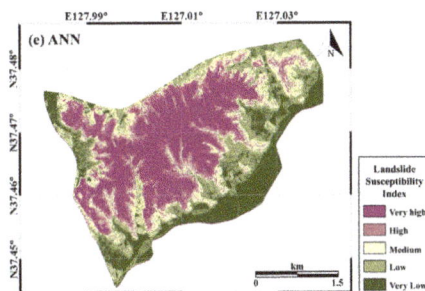

Figure 7. Landslide susceptibility maps generated using data mining models: Using an (**a**) support vector machine (SVM) model with linear kernel; (**b**) SVM model with polynomial kernel; (**c**) SVM model with radial basis function kernel; (**d**) SVM model with sigmoid kernel; (**e**) and the artificial neural network (ANN) model.

In Table 2, the area of each class is represented by percentage. The areas with very high indices in the SVM model were 2.2599, 2.2599, 2.2889 and 2.28369 km^2 for the linear, polynomial, RBF, and sigmoid kernels, respectively. Among the results from the SVM models, the RBF kernel accounted for a slightly higher percentage (20.03%) of the areas with a very high index compared with the linear, polynomial, and sigmoid kernels. The areas with a very high index in the ANN model covered 3.9262 km^2, which had the highest percentage (34.35%). In contrast, the areas with high index accounted for only a small proportion (8.65%). The area with a very low index accounted for similar percentages of 20.00, 20.00, 20.00 and 19.99% for the linear, polynomial, and sigmoid kernels, and ANN, respectively, while the RBF accounted for a lower percentage of 19.89%.

Table 3 shows the relative weights per variable, which were calculated from randomly selected training data for 20 epochs of 200 cycles. The weights between layers represent the contribution of each variable to the prediction of the landslide susceptibility. Timber age showed the lowest average value of weight, 0.04828. For the comparison analysis of the relativeness of the weights, the values were divided by the weight of the timber age and were normalized with respect to the average weight of the timber age. The highest average weight of 0.09615 for the standardized height was normalized to 2.00164, which was approximately twice the value of the weight of the timber age. The second major variable contributing to landslides was timber diameter (1.46886), followed by curvature (1.28335), soil depth (1.27746), and aspect (1.25804). The standard deviation ranged from 0.00022 to 0.01132.

To investigate the reliability of the proposed landslide susceptibility map using the previously described SVM and ANN data mining models, we performed a validation. The 2011 landslide occurrence data not used for the training were used for validation. After 51 points of training data had been extracted using random sampling from the landslide occurrence data from the 2011 Umyeonsan landslide and applied to the models, 52 points were used for the validation. Because the predicted landslide susceptibility map was an indicator of the potential possibility of landslides in the future, the data obtained after the occurrence of the landslides were a suitable choice for the validation data. However, an erosion control project implemented after the 2011 event has significantly reduced the possibility of landslides. Therefore, it is important to obtain information on landslide susceptibility in areas where landslides have not occurred. In addition, implementation of precautionary measures to prevent landslides from occurring in the future is necessary. Thus, half of the landslide occurrence points were used as training data, and the remaining points were used as validation data.

Table 2. Area and percentage of each category.

Landslide Susceptibility Index	SVM								Backpropagation ANN (km²)	Percentage (%)
	Linear (km²)	Percentage (%)	Polynomial (km²)	Percentage (%)	RBF (km²)	Percentage (%)	Sigmoid (km²)	Percentage (%)		
Very high	2.2599	19.77%	2.2599	19.77%	2.2889	20.03%	2.2836	19.98%	3.9262	34.35%
High	2.3081	20.20%	2.3081	20.20%	2.1835	19.11%	2.2866	20.01%	0.9881	8.65%
Moderate	2.2878	20.02%	2.2878	20.02%	2.2359	19.56%	2.2863	20.01%	1.9451	17.02%
Low	2.2866	20.01%	2.2866	20.01%	2.4474	21.41%	2.2859	20.00%	2.2849	19.99%
Very low	2.2858	20.00%	2.2858	20.00%	2.2727	19.89%	2.2859	20.00%	2.2841	19.99%
Total	11.4283	100.00	11.4283	100.00	11.4283	100.00	11.4283	100.00	11.4283	100.00

Table 3. Neural network weight between landslide and related factors.

200 Cycles for 1 epoch	1	2	3	4	5	6	7	8	9	10	Average	Standard Deviation	Normalized Weight with Respect to Timber Age
Slope gradient	0.0555	0.0548	0.0546	0.0546	0.0549	0.0552	0.0557	0.0561	0.0564	0.0567	0.05545	0.00072	1.14851
Slope aspect	0.0616	0.0611	0.0608	0.0607	0.0607	0.0606	0.0606	0.0605	0.0604	0.0603	0.06073	0.00036	1.25787
Curvature	0.0623	0.0615	0.0614	0.0615	0.0616	0.0618	0.0620	0.0623	0.0625	0.0629	0.06198	0.00048	1.28376
TWI	0.0512	0.0520	0.0512	0.0506	0.0500	0.0495	0.0490	0.0485	0.0480	0.0476	0.04976	0.00141	1.03065
SPI	0.0531	0.0520	0.0518	0.0517	0.0517	0.0517	0.0517	0.0517	0.0516	0.0516	0.05186	0.00043	1.07415
SLF	0.0579	0.0594	0.0590	0.0584	0.0578	0.0572	0.0567	0.0564	0.0561	0.0559	0.05748	0.00116	1.19056
Standardized height	0.0699	0.083	0.0909	0.0955	0.0987	0.1014	0.1036	0.1052	0.1063	0.1070	0.09615	0.01132	1.99151
Valley depth	0.0521	0.0517	0.0514	0.0512	0.0510	0.0509	0.0509	0.0508	0.0508	0.0509	0.05117	0.00042	1.05986
Downslope distance gradient	0.0626	0.0579	0.0559	0.0551	0.0547	0.0546	0.0548	0.0552	0.0557	0.0563	0.05628	0.00231	1.16570
Timber diameter	0.072	0.0717	0.0712	0.071	0.0708	0.0707	0.0705	0.0704	0.0704	0.0703	0.07090	0.00055	1.46852
Timber type	0.0546	0.0515	0.0514	0.0513	0.0512	0.0512	0.0513	0.0513	0.0512	0.0512	0.05162	0.00100	1.06918
Timber density	0.0607	0.0582	0.0577	0.0577	0.0579	0.0581	0.0583	0.0584	0.0584	0.0583	0.05837	0.00082	1.20899
Timber age	0.0489	0.0484	0.0482	0.0482	0.0481	0.0482	0.0482	0.0482	0.0482	0.0482	0.04828	0.00022	1.00000
Soil depth	0.0682	0.0686	0.0666	0.0646	0.0626	0.0606	0.0586	0.0566	0.0553	0.0543	0.06160	0.00506	1.27589
Soil drainage	0.0546	0.0544	0.0547	0.0554	0.0557	0.0560	0.0561	0.0562	0.0563	0.0564	0.05558	0.00072	1.15120
Soil Topography	0.0606	0.0607	0.0603	0.0600	0.0598	0.0597	0.0597	0.0596	0.0595	0.0594	0.05993	0.00043	1.24130
Soil texture	0.0542	0.0531	0.0529	0.0527	0.0526	0.0526	0.0526	0.0527	0.0528	0.0530	0.05292	0.00046	1.09611

The AUC of the receiver operating characteristic (ROC) curve was used for the validation of the models. The ROC consists of specificity, the *x*-axis, which shows the percentage of the areas where landslides are expected to occur but have not occurred, and sensitivity, the *y*-axis, which represents the predicted landslide areas according to training data of landslide occurrences. The ROC graph was drawn using the following procedure. To show the rank relative to the predicted results, the values of the results of landslide susceptibility from the models were sorted in descending order. The ranked results were represented in ROC curves using 100 classes; a single class comprised an area that was approximately 1% of the total study area. Figure 8a shows the AUC of the landslide susceptibility map for the SVM model; the linear, polynomial, RBF, and sigmoid kernel had 72.41 (0.7241), 72.83 (0.7283), 77.17 (0.7717) and 72.79% (0.7279) accuracy versus 78.41% (0.7841) accuracy in the ANN model (Figure 8b).

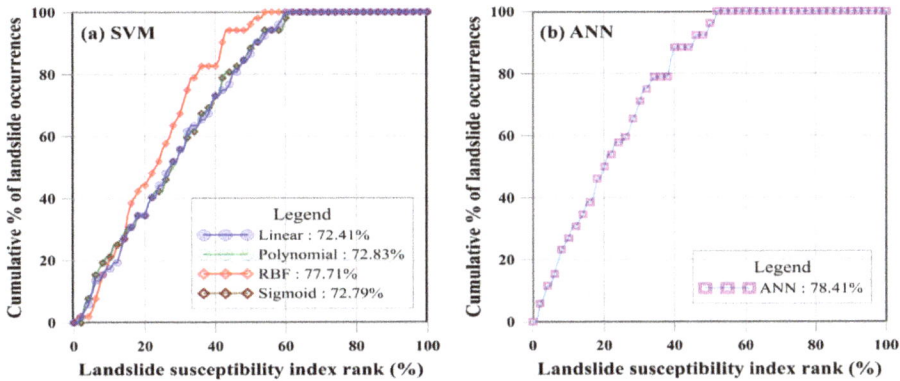

Figure 8. Validation results using data mining models: (**a**) SVM model; (**b**) ANN model.

A higher AUC value indicates better model performance, but it is important to perform additional empirical analyses based on input and ground truth data. Our validation results indicated that the five landslide sensitivity assessments performed well; the AUC values were all greater than 0.7 [60]. In the SVM model, the AUC value using the RBF kernel was 0.7717, which was superior to the other three kernels in the study area. The ANN model yielded an AUC value of 0.7841, which was approximately 6% higher than that of the SVM model with the linear kernel, and 0.7% higher than with the RBF kernel. The graphs of the validation results are depicted in Figure 8; in all of the models, 60% of the study area incorporated all of the landslide locations.

6. Discussion

In this study, two types of data mining models were used to create landslide susceptibility maps. Because the results of data mining models are derived from their data, the reliability of the input data is fundamental. Landslide location data were acquired from aerial photographs to provide accurate training datasets. The data mining models are suitable for observing the rapid changes in an urban area with its compactness and complex characteristics. In addition, other topographical-, forest-, and soil-related variables were obtained, and topographic-related factors were calculated from the topographical map produced from the aerial photographs.

The results from each model in Figure 7 were not statistically different in the spatial pattern of the landslide susceptibility area. This can be interpreted that all of the four prediction results from the SVM model were based on the same input factors, and all landslide predisposing factors were regarded as having equal importance so that the same weight was assigned to each predisposing factor. However, for the results from the RBF kernel, which had the highest accuracy among the SVM models, the susceptibility on the west, southwest, and southeast slopes was lower than that from the other kernels, which was similar to the trend for the ANN model. In general, the northern area of the mountain showed high susceptibility, and had the same pattern as the existing damaged area of the southern ring roads.

The percentages of the classes of five indices were also calculated. The linear, polynomial, and sigmoid kernels were distributed fairly evenly, while the RBF kernel showed a slightly higher percentage in the very high index class, and a lower percentage in the very low index class. This suggests that a high percentage in the high susceptibility index yields good validation results. Figure 8a shows that the RBF kernel had a percentage of 77.71%, which was approximately 5% higher than that of the other kernels. Similarly, the results from the ANN model yielded a high percentage (more than 30% of the total) in the very high class, which represented high risk areas, and this also led to a high validation percentage for the ANN model, 78.41%. In addition, the ANN model results in Table 3 show that the standardized height contributed most to landslide susceptibility with a normalized weight of greater than 2. This can be seen in Figure 7e, which shows the effect of standardized height with the high relative weight of the ANN model. Standardized height was the most important factor among all of the landslide-related factors. The factors related to landslide susceptibility should be managed based on their relative weight.

Because this study investigated the susceptibility of landslides in urban areas, the study area was fairly small, with an area of approximately 11.4 km^2 and a spatial resolution of 5 m along road and river boundaries rather than a rectangular boundary. It should be noted that the ROC was characterized using a percentage of the landslide susceptibility rates. Thus, in this study, the validation rate was relatively low compared with most previous studies, which were based on rectangles. Moreover, we could use the slope unit instead of common raster unit, which is closely correlated to the topography, for the further steps; the slope unit is defined as the unit between the valleys and ridges. Most of the mountains in the city are managed in the framework of city parks in which the various economic and social aspects of Umyeonsan are also considered, rather than the area being covered solely by forests that are more resistant to landslides. Given that landslide damage in urban areas can lead to human casualties, it is important to investigate the soil and topographical characteristics of the areas where landslides are likely to occur. Furthermore, landslide risk should be incorporated in city-level policies.

7. Conclusions

In this study, we applied a spatial data mining approach to identify areas where landslides are likely to occur using aerial photographs and GIS. Landslide-prone locations were determined using the interpretation of aerial photographs and field survey results to construct spatial datasets. The topographical predisposing factors were extracted from a digital topographical map constructed from aerial photographs, and the soil and forest predisposing factors were extracted from publicly available maps. A spatial database was constructed using the extracted and calculated factors including randomly extracted training data for the landslide area, and landslide susceptibility was mapped using SVM and ANN models. Finally, the results map was validated using the half of the landslide location data not used for training.

In urban areas such as Seoul, landslides have previously not been considered a serious concern, compared with floods and storms. Seoul has undergone rapid urbanization that has taken place with little consideration of natural disasters. In large cities such as Seoul, global warming, rapid urbanization, extreme precipitation, and population pressures could potentially lead to complex events, including landslides occurring in the city, which could have serious consequences. Therefore, it is important to create a scientifically valid landslide susceptibility map for urban areas such as Seoul.

The predicted rate of landslides was high on the steep mountain tops and similar to the general landslide pattern on the ridges. The map created in this study showed that susceptibility to landslides decreased with slope gradient. Because landslides are caused by differences in gravity due to terrain shape, it is logical that their occurrence is affected by the geographical structure of the area.

The relationship between landslide occurrence and forested land cover showed that broadleaved forests had a high susceptibility, and this type of forest occupied the largest proportion of the study area. In addition, large trunk-size, high-density forests, and middle-aged forests had a high risk of landslides. Among the soil factors, red and yellow soils had higher landslide susceptibility, as well as areas with large soil depth; the occurrence rate was higher in areas with good soil drainage. The performance of the SVM and ANN models was validated using AUC analysis. The SVM model yielded 72.41%, 72.83%, 77.17% and 72.79% accuracy for the linear, polynomial, RBF, and sigmoid kernels, respectively; while the ANN model had 78.41% accuracy. The results of this study are expected to support spatial decision-making from relevant agencies to formulate policies related to landslide disaster risk reduction in urban areas.

Acknowledgments: This research was conducted by Korea Environment Institute (KEI) with support of a grant (16CTAP-C114629-01) from Technology Advancement Research Program (TARP) funded by Ministry of Land, Infrastructure and Transport of Korean government. This work was supported by the Space Core Technology Development Program through the National Research Foundation of Korea, funded by the Ministry of Science, ICT and Future Planning, under Grant (NRF-2014M1A3A3A03034798).

Author Contributions: Sunmin Lee applied the SVM and ANN models and wrote the paper. Hyung-Sup Jung designed the experiments. Also, he interpreted the input data and the results. Moung-Jin Lee suggested the idea and organized the paperwork. In addition, he collected data and conducted pre-processing of the input data. All of authors contributed to the writing of each part.

Conflicts of Interest: The authors declare no conflict of interest.

References

1. Berry, M.J.; Linoff, G. *Data Mining Techniques: For Marketing, Sales, and Customer Support*; John Wiley & Sons, Inc.: Hoboken, NJ, USA, 1997.
2. Pachauri, R.K.; Allen, M.R.; Barros, V.R.; Broome, J.; Cramer, W.; Christ, R.; Church, J.A.; Clarke, L.; Dahe, Q.; Dasgupta, P. Contribution of working groups I, II and III to the fifth assessment report of the intergovernmental panel on climate change. In *Climate Change 2014: Synthesis Report*; IPCC: Geneva, Switzerland, 2014.
3. Centre for Research on the Epidemiology of Disasters, UN Office for Disaster Risk Reduction. *The Human Cost of Natural Disasters: A Global Perspective*; Centre for Research on the Epidemiology of Disaster (CRED): Brussels, Belgium, 2015.
4. Cruden, D.; Varnes, D. Landslide Types and Processes. In *Landslides: Investigation and Mitigation*; Turner, A.K., Schuster, R.L., Eds.; Special Report 247; Transportation Research Board: Washington, DC, USA, 1996; pp. 36–75.
5. Choi, J.; Oh, H.J.; Lee, H.J.; Lee, C.; Lee, S. Combining landslide susceptibility maps obtained from frequency ratio, logistic regression, and artificial neural network models using aster images and gis. *Eng. Geol.* **2012**, *124*, 12–23. [CrossRef]
6. Lee, S.; Ryu, J.H.; Won, J.S.; Park, H.J. Determination and application of the weights for landslide susceptibility mapping using an artificial neural network. *Eng. Geol.* **2004**, *71*, 289–302. [CrossRef]
7. Pradhan, B.; Lee, S. Landslide susceptibility assessment and factor effect analysis: Backpropagation artificial neural networks and their comparison with frequency ratio and bivariate logistic regression modelling. *Environ. Model. Softw.* **2010**, *25*, 747–759. [CrossRef]
8. Pradhan, B.; Lee, S. Utilization of optical remote sensing data and gis tools for regional landslide hazard analysis using an artificial neural network model. *Earth Sci. Front.* **2007**, *14*, 143–151.
9. Ahmed, B.; Dewan, A. Application of bivariate and multivariate statistical techniques in landslide susceptibility modeling in chittagong city corporation, bangladesh. *Remote Sens.* **2017**, *9*. [CrossRef]
10. Lee, S. Application of logistic regression model and its validation for landslide susceptibility mapping using gis and remote sensing data. *Int. J. Remote Sens.* **2005**, *26*, 1477–1491. [CrossRef]
11. Scaioni, M.; Longoni, L.; Melillo, V.; Papini, M. Remote sensing for landslide investigations: An overview of recent achievements and perspectives. *Remote Sens.* **2014**, *6*, 9600–9652. [CrossRef]

12. Akcay, O. Landslide fissure inference assessment by anfis and logistic regression using uas-based photogrammetry. *ISPRS Int. J. Geo-Inf.* **2015**, *4*, 2131–2158. [CrossRef]

13. Bui, D.T.; Pradhan, B.; Lofman, O.; Revhaug, I.; Dick, O.B. Landslide susceptibility assessment in the hoa binh province of vietnam: A comparison of the levenberg–marquardt and bayesian regularized neural networks. *Geomorphology* **2012**, *171*, 12–29.

14. Yilmaz, I. Landslide susceptibility mapping using frequency ratio, logistic regression, artificial neural networks and their comparison: A case study from kat landslides (Tokat—Turkey). *Comput. Geosci.* **2009**, *35*, 1125–1138. [CrossRef]

15. Ermini, L.; Catani, F.; Casagli, N. Artificial neural networks applied to landslide susceptibility assessment. *Geomorphology* **2005**, *66*, 327–343. [CrossRef]

16. Pourghasemi, H.R.; Pradhan, B.; Gokceoglu, C. Application of fuzzy logic and analytical hierarchy process (ahp) to landslide susceptibility mapping at haraz watershed, iran. *Nat. Hazards* **2012**, *63*, 965–996. [CrossRef]

17. Pradhan, B. Use of gis-based fuzzy logic relations and its cross application to produce landslide susceptibility maps in three test areas in malaysia. *Environ. Earth Sci.* **2011**, *63*, 329–349. [CrossRef]

18. Bai, S.; Lü, G.; Wang, J.; Zhou, P.; Ding, L. Gis-based rare events logistic regression for landslide-susceptibility mapping of lianyungang, china. *Environ. Earth Sci.* **2011**, *62*, 139–149. [CrossRef]

19. Gorsevski, P.V.; Gessler, P.E.; Foltz, R.B.; Elliot, W.J. Spatial prediction of landslide hazard using logistic regression and roc analysis. *Trans. GIS* **2006**, *10*, 395–415. [CrossRef]

20. Pradhan, B.; Mansor, S.; Pirasteh, S.; Buchroithner, M.F. Landslide hazard and risk analyses at a landslide prone catchment area using statistical based geospatial model. *Int. J. Remote Sens.* **2011**, *32*, 4075–4087. [CrossRef]

21. Rawat, M.; Joshi, V.; Rawat, B.; Kumar, K. Landslide movement monitoring using gps technology: A case study of bakthang landslide, gangtok, east sikkim, india. *J. Dev. Agric. Econ.* **2011**, *3*, 194–200.

22. Hess, D.M.; Leshchinsky, B.A.; Bunn, M.; Mason, H.B.; Olsen, M.J. A simplified three-dimensional shallow landslide susceptibility framework considering topography and seismicity. *Landslides* **2017**, *08*, 1–21. [CrossRef]

23. Lee, S.; Chwae, U.; Min, K. Landslide susceptibility mapping by correlation between topography and geological structure: The janghung area, korea. *Geomorphology* **2002**, *46*, 149–162. [CrossRef]

24. Zhou, S.; Chen, G.; Fang, L. Distribution pattern of landslides triggered by the 2014 ludian earthquake of china: Implications for regional threshold topography and the seismogenic fault identification. *ISPRS Int. J. Geo-Inf.* **2016**, *5*. [CrossRef]

25. Devkota, K.C.; Regmi, A.D.; Pourghasemi, H.R.; Yoshida, K.; Pradhan, B.; Ryu, I.C.; Dhital, M.R.; Althuwaynee, O.F. Landslide susceptibility mapping using certainty factor, index of entropy and logistic regression models in gis and their comparison at mugling–narayanghat road section in nepal himalaya. *Nat. Hazards* **2013**, *65*, 135–165. [CrossRef]

26. Korea Forest Service. *2013 Detailed Strategy for Primary Policy*; Korea Forest Service: Seoul, Korea, 2013; pp. 12–16.

27. Lee, M.; Park, I.; Won, J.; Lee, S. Landslide hazard mapping considering rainfall probability in inje, Korea. *Geomat. Nat. Hazards Risk* **2016**, *7*, 424–446. [CrossRef]

28. Lee, S.; Hong, S.M.; Jung, H.S. A support vector machine for landslide susceptibility mapping in Gangwon Province, Korea. *Sustainability* **2017**, *9*. [CrossRef]

29. Park, I.; Lee, S. Spatial prediction of landslide susceptibility using a decision tree approach: A case study of the pyeongchang area, Korea. *Int. J. Remote Sens.* **2014**, *35*, 6089–6112. [CrossRef]

30. Lee, S.; Min, K. Statistical analysis of landslide susceptibility at yongin, Korea. *Environ. Geol.* **2001**, *40*, 1095–1113. [CrossRef]

31. Lee, S.; Evangelista, D. *Landslide Susceptibility Mapping Using Probability and Statistics Models in Baguio City, Philippines*; Department of Environment and Natural Resources: Quezon City, Philippines, 2008.

32. Lee, S.; Pradhan, B. Landslide hazard assessment at cameron highland malaysia using frequency ratio and logistic regression models. *Geophys. Res. Abstr.* **2006**, *8*. SRef-ID: 1607-7962/gra/EGU06-A-003241.

33. Oh, H.-J.; Lee, S.; Soedradjat, G.M. Quantitative landslide susceptibility mapping at pemalang area, Indonesia. *Environ. Earth Sci.* **2010**, *60*, 1317–1328. [CrossRef]

34. Bathrellos, G.D.; Gaki-Papanastassiou, K.; Skilodimou, H.D.; Papanastassiou, D.; Chousianitis, K.G. Potential suitability for urban planning and industry development using natural hazard maps and geological–geomorphological parameters. *Environ. Earth Sci.* **2012**, *66*, 537–548. [CrossRef]

35. Bathrellos, G.D.; Skilodimou, H.D.; Chousianitis, K.; Youssef, A.M.; Pradhan, B. Suitability estimation for urban development using multi-hazard assessment map. *Sci. Total Environ.* **2017**, *575*, 119–134. [CrossRef] [PubMed]

36. Chousianitis, K.; Del Gaudio, V.; Sabatakakis, N.; Kavoura, K.; Drakatos, G.; Bathrellos, G.D.; Skilodimou, H.D. Assessment of earthquake-induced landslide hazard in greece: From arias intensity to spatial distribution of slope resistance demand. *Bull. Seismol. Soc. Am.* **2016**, *106*, 174–188. [CrossRef]

37. Papadopoulou-Vrynioti, K.; Bathrellos, G.D.; Skilodimou, H.D.; Kaviris, G.; Makropoulos, K. Karst collapse susceptibility mapping considering peak ground acceleration in a rapidly growing urban area. *Eng. Geol.* **2013**, *158*, 77–88. [CrossRef]

38. Korean Geotechnical Society. *Cause Investigation of Umyeonsan Landslide and Recovery Measure Establishment Service Final Report*; Korean Geotechnical Society: Seoul, Korea, 2011; p. 3.

39. Seoul Special City. *Supplemental Investigation Report of Causes of Umyeonsan Landslides*; Seoul Special City: Seoul, Korea, 2014; p. 447.

40. Nocutnews. A Part of Umyeonsan. Available online: http://news.Naver.Com/main/read.Nhn?Mode=lsd&mid=sec&sid1=102&oid=079&aid=0002273090 (accessed on 20 June 2017).

41. SBS. An Approval of a mAn-Made Disaster, Not a Natural Disaster, Umyeonsan Landslide. Available online: http://news.Sbs.Co.Kr/news/endpage.Do?News_id=n1002292872&plink=oldurl (accessed on 20 June 2017).

42. Jones, D.; Brunsden, D.; Goudie, A. A preliminary geomorphological assessment of part of the karakoram highway. *Q. J. Eng. Geol. Hydrogeol.* **1983**, *16*, 331–355. [CrossRef]

43. Beven, K.; Kirkby, M.J. A physically based, variable contributing area model of basin hydrology/un modèle à base physique de zone d'appel variable de l'hydrologie du bassin versant. *Hydrol. Sci. J.* **1979**, *24*, 43–69. [CrossRef]

44. Moore, I.D.; Grayson, R.; Ladson, A. Digital terrain modelling: A review of hydrological, geomorphological, and biological applications. *Hydrol. Process.* **1991**, *5*, 3–30. [CrossRef]

45. Wischmeier, W.H.; Smith, D.D. *Predicting Rainfall Erosion Losses—A Guide to Conservation Planning*; U.S. Department of Agriculture: Washington, DC, USA, 1978.

46. Conrad, O.; Bechtel, B.; Bock, M.; Dietrich, H.; Fischer, E.; Gerlitz, L.; Wehberg, J.; Wichmann, V.; Böhner, J. System for automated geoscientific analyses (saga) v. 2.1.4. *Geosci. Model Dev.* **2015**, *8*, 1991–2007. [CrossRef]

47. Böhner, J.; Selige, T. Spatial prediction of soil attributes using terrain analysis and climate regionalisation. *Gottinger Geographische Abhandlungen* **2006**, *115*, 13–28.

48. Hjerdt, K.; McDonnell, J.; Seibert, J.; Rodhe, A. A new topographic index to quantify downslope controls on local drainage. *Water Resour. Res.* **2004**, *40*, W05601–W05606. [CrossRef]

49. Doetterl, S.; Berhe, A.A.; Nadeu, E.; Wang, Z.; Sommer, M.; Fiener, P. Erosion, deposition and soil carbon: A review of process-level controls, experimental tools and models to address c cycling in dynamic landscapes. *Earth-Sci. Rev.* **2016**, *154*, 102–122. [CrossRef]

50. Anthes, R.A. Enhancement of convective precipitation by mesoscale variations in vegetative covering in semiarid regions. *J. Clim. Appl. Meteorol.* **1984**, *23*, 541–554. [CrossRef]

51. Choi, J.W.; Byun, Y.G.; Kim, Y.I.; Yu, K.Y. Support vector machine classification of hyperspectral image using spectral similarity kernel. *J. Korea Soc. Geospat. Inf. Sci.* **2006**, *14*, 71–77.

52. Cortes, C.; Vapnik, V. Support-vector networks. *Mach. Learn.* **1995**, *20*, 273–297. [CrossRef]

53. Yao, X.; Tham, L.; Dai, F. Landslide susceptibility mapping based on support vector machine: A case study on natural slopes of Hong Kong, China. *Geomorphology* **2008**, *101*, 572–582. [CrossRef]

54. Xu, C.; Dai, F.; Xu, X.; Lee, Y.H. Gis-based support vector machine modeling of earthquake-triggered landslide susceptibility in the jianjiang river watershed, China. *Geomorphology* **2012**, *145*, 70–80. [CrossRef]

55. Zhu, J.; Hastie, T. Kernel logistic regression and the import vector machine. *J. Comput. Graph. Stat.* **2005**, *14*, 185–205. [CrossRef]

56. Schölkopf, B.; Smola, A.J.; Williamson, R.C.; Bartlett, P.L. New support vector algorithms. *Neural Comput.* **2000**, *12*, 1207–1245. [CrossRef] [PubMed]

57. Atkinson, P.M.; Tatnall, A. Introduction neural networks in remote sensing. *Int. J. Remote Sens.* **1997**, *18*, 699–709. [CrossRef]

58. Magoulas, G.D.; Vrahatis, M.N.; Androulakis, G.S. Effective backpropagation training with variable stepsize. *Neural Netw.* **1997**, *10*, 69–82. [CrossRef]

59. Hines, J.; Tsoukalas, L.H.; Uhrig, R.E. *Matlab Supplement to Fuzzy and Neural Approaches in Engineering*; John Wiley & Sons, Inc.: Hoboken, NJ, USA, 1997.

60. Fawcett, T. An introduction to roc analysis. *Pattern Recognit. Lett.* **2006**, *27*, 861–874. [CrossRef]

applied
sciences

MDPI

Article

Shallow Landslide Susceptibility Modeling Using the Data Mining Models Artificial Neural Network and Boosted Tree

Hyun-Joo Oh [1] and Saro Lee [2,3,*]

[1] Department of Geological Hazards, Korea Institute of Geoscience and Mineral Resources (KIGAM), 124, Gwahang-ro, Yuseong-gu, Daejeon 34132, Korea; ohj@kigam.re.kr
[2] Geological Research Division, Korea Institute of Geoscience and Mineral Resources (KIGAM), 124, Gwahak-ro, Yuseong-gu, Daejeon 34132, Korea
[3] Korea University of Science and Technology, 217 Gajeong-ro Yuseong-gu, Daejeon 305-350, Korea
* Correspondence: leesaro@kigam.re.kr; Tel.: +82-42-868-3057

Received: 18 July 2017; Accepted: 21 September 2017; Published: 28 September 2017

Abstract: The main purpose of this paper is to present some potential applications of sophisticated data mining techniques, such as artificial neural network (ANN) and boosted tree (BT), for landslide susceptibility modeling in the Yongin area, Korea. Initially, landslide inventory was detected from visual interpretation using digital aerial photographic maps with a high resolution of 50 cm taken before and after the occurrence of landslides. The debris flows were randomly divided into two groups: training and validation sets with a 50:50 proportion. Additionally, 18 environmental factors related to landslide occurrence were derived from the topography, soil, and forest maps. Subsequently, the data mining techniques were applied to identify the influence of environmental factors on landslide occurrence of the training set and assess landslide susceptibility. Finally, the landslide susceptibility indexes from ANN and BT were compared with a validation set using a receiver operating characteristics curve. The slope gradient, topographic wetness index, and timber age appear to be important factors in landslide occurrence from both models. The validation result of ANN and BT showed 82.25% and 90.79%, which had reasonably good performance. The study shows the benefit of selecting optimal data mining techniques in landslide susceptibility modeling. This approach could be used as a guideline for choosing environmental factors on landslide occurrence and add influencing factors into landslide monitoring systems. Furthermore, this method can rank landslide susceptibility in urban areas, thus providing helpful information when selecting a landslide monitoring site and planning land-use.

Keywords: landslide susceptibility; artificial neural network; boosted tree; landslide inventory

1. Introduction

The mountainous area of Korea covers approximately 70% of the total land. Areas with landslide susceptibility in Korea have been reported in the steep slopes of mountainous areas consisting of granite or gneiss [1]. These conditions, in addition to low strengths of weathered soil and unstable slopes, are considered vulnerable to particularly shallow landslides when intense rainfall occurs during the summer rainy season. There is a tendency that suggests that the risk of landslides is increasing due to frequent localized heavy rain in Korea resulting from recent climate change [1,2]. In addition, earlier landslides that developed in the upper mountainous areas extend to debris flows in the valley area, and affect property damage and loss of human life in living areas which are developing and expanding [3]. Therefore, landslides are viewed as hazards for human life and artificial structures in Korea.

The damage caused by landslides is the same worldwide. To minimize the damage to people and property due to landslides, many efforts over the past few decades have been made to understand how to control landslides and predict their spatial and temporal distribution [4–13]. Most approaches have been applied on the Geographic Information System (GIS)-based landslide susceptibility assessment representing predicted landslide risks. The classification of these approaches (e.g., heuristic, statistical, probability, and deterministic approaches) are well documented in van Westen et al., 2006.

In particular, statistical and probability application models have been widely applied by several studies to predict landslide susceptibility using a past landslide inventory and their environmental factors. The models include frequency ratio [14,15], weight of evidence [16], logistic regression [17], and fuzzy logic [18]. Recently, data mining techniques have been developed and are extremely popular [19,20] when dealing with a variety of nonlinear issues. Techniques applied in landslide susceptibility modeling include: artificial neural network, decision tree, boosted tree, neuro fuzzy, Bayesian network, support vector machine, and random forest [21–30].

When using these approaches to predict landslide-susceptible areas, it is assumed that past landslide occurrence conditions are similar to the conditions for future landslide occurrence [12]. Therefore, it is necessary to train and explore the relationship between past landslide locations and environmental factors (e.g., topographic, hydrologic, soil, and forest data) when using these approaches to predict landslide-vulnerable areas. To do so, it is important to prepare accurate landslide maps and to select environmental variables that affect landslide occurrence to apply to models [31].

Although several different models have been compared in previous studies [22,26], this study analyzed landslide susceptibility based on artificial neural networks (ANN) and boosted tree (BT) models that have not been applied simultaneously in other studies. Furthermore, as various topographic and hydrologic factors have been calculated from a digital elevation model (DEM) using a System for Automated Geoscientific Analyses (SAGA) GIS Module, and landslide occurrences have accurately been detected from digital aerial photos, the contribution of these factors were evaluated from these models. Therefore, this research aimed to: (i) investigate and compare the performance of data mining-based ANN and BT models, (ii) prepare accurate landslide maps using digital aerial photographs with high resolution, and (iii) determine the contribution of the environmental factors.

The preparation of the landslide susceptibility model was accomplished in three major steps.

(1) Compilation of a spatial database. A total of 82 debris flows were detected by visual interpretation of aerial photographs with a 50 cm resolution before and after landslide events. The environmental factors were constructed into a spatial database including eight topographic factors: slope gradient, aspect, plan curvature, convexity, mid-slope position (MSP), terrain ruggedness index (TRI), topographic position index (TPI), and landforms; three hydrologic factors: slope length (SL), stream power index (SPI), and topographic wetness index (TWI); four soil factors: land-use, material, thickness, and topography; and three timber factors: age, density, and diameter.

(2) Processing the data from the database. The number of debris flows were randomly divided into training (50%) and validation (50%) data for landslide susceptibility analysis using ANN and BT models.

(3) The influence of environmental factors on landslide occurrences as the training set was calculated as the weight of the factor using both models.

(4) Mapping landslide susceptibility using ANN and BT, and assessing both maps using known landslide occurrences as a validation set.

2. Study Area and Materials

The study area, Yongin City, recorded over 350 mm of cumulative rainfall on 27 July 2011, and a shallow landslide occurred due to intense rainfall. Next, the debris flows collapsed houses and parts of buildings, and resulted in loss of life and property (Figure 1). In this paper, the landslide inventory was mapped from digital aerial photographs with a high resolution of 50 cm. The 82 landslides were detected from visual interpretation of before and after photos of landslide events in the study area. The altitude of the study area ranges from 47 m to 457 m with 140 m of average and 79 m of standard deviation. The landslide occurred at an altitude of 70 m~267 m. Specifically, 80% of total landslides occurred between 100 m and 200 m. Biotite gneiss and alluvium are composed of about 65% and 20% of the study area, respectively. Almost all landslides occurred in biotite gneiss. There are two fault lines from the geological map in the study areas.

Topographic and hydrologic factors were constructed from DEM using the terrain analysis of the SAGA GIS module (Table 1). The soil and timber factors were extracted from soil and forest maps. The locations of landslides and environmental factors were denoted by pixels of 5 m by 5 m, and the dimension of the study area had a total number of 1,918,400 cells with 1760 columns by 1090 rows.

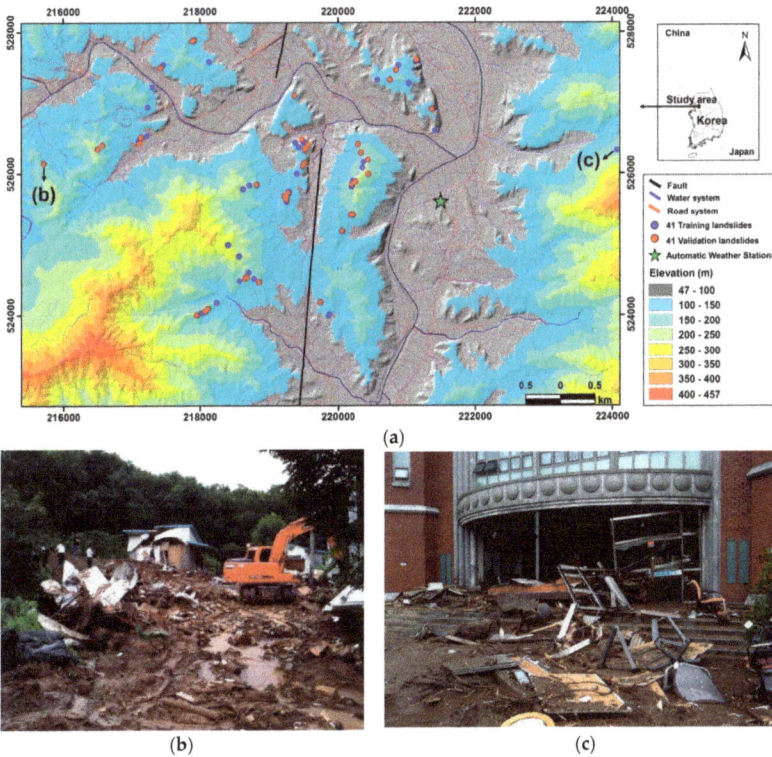

Figure 1. Digital elevation model (DEM) and landslide occurrences in the study area: (**a**) collapsed houses; (**b**) building; and (**c**) debris flows in Neungwonri and Hankuk University of Foreign Studies located in Figure 1a, respectively.

Table 1. Data layer related to the landslide of the study area.

Category		Factors	Data Type	Scale	Source
DEM	Topographic factors	Slope Aspect Plan curvature Convexity Mid-slope position (MSP) Terrain ruggedness index (TRI) Topographic position index (TPI) Landforms	Grid	1:5000	National Geographic Information Institute (NGII) in Korea
	Hydrologic factors	Slope length (SL) Stream power index (SPI) Topographic wetness index (TWI)			
Soil map		Land-use Material Thickness Topography	Polygon	1:5000	National Academy of Agricultural Science (NAAS) in Korea
Forest map		Timber age Timber density Timber diameter	Polygon	1:5000	Korea Forest Research Institute (KFRI)

2.1. Precipitation Characteristics

Rainfall affects the slope stability by means of its influence on run-off and pore water pressure [32]. Specifically, high intensity rainfall usually relates to a high concentration of landslide events in time and space [33]. In this study, rainfall characteristics were analyzed over the period 26–28 July 2011, which affected landslide occurrences. Hourly rainfall data were collected from one automatic weather station (AWS) in the study area (Figure 2). Figure 2 shows the amount of hourly rainfall and its accumulative rainfall. It was reported in articles that the landslides occurred around 1:00 p.m. on 27 July 2011. Before the landslide event, the rain fell for 10 h from 3:00 a.m. that day. The highest hourly rainfall recorded was 78 mm at 10:00 a.m. The second highest recorded hourly rainfall was 68 mm at 12:00 a.m. before the landslide events at 1:00 p.m. The volume of rainfall accumulation at the time of the landslide events was recorded at 385 mm in the study area. The different seven AWS sites outside the study area showed a value of 205 mm, 282 mm, 188 mm, 182 mm, 178 mm, 222 mm, and 130 mm. The study area has the highest volume of rainfall accumulation when compared to the accumulated volume of rainfall in the surrounding area.

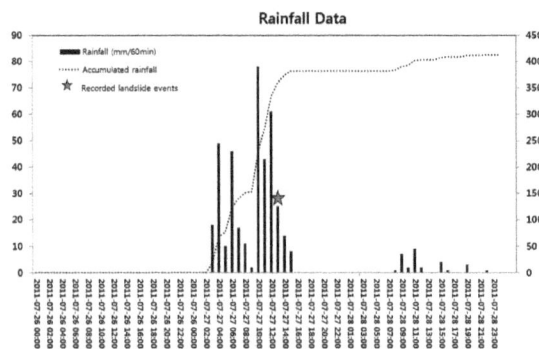

Figure 2. Hourly precipitation characteristics in the past led to landslide events in the study area.

2.2. Landslide Inventory

A landslide map is based on important information to determine the quantitative zoning of landslide susceptibility, hazards, and risk [34,35]. In this study, a visual interpretation of digital aerial photographs with a high resolution of 50 cm was used for accurate landslide mapping. Although visual interpretation is a classical method [34], it is very useful in detecting accurate landslide locations and scars using high resolution photographs, as landslide scars are similar to tombs and their surrounding area in the study area.

These types of photographs without ground control points (GCPs) can be freely obtained at portal sites such as DAUM [36] and Skymap [37] ("Skymap") in Korea [38]. The eighteen photos (taken before and after the landslide events) were selected from each region of landslide occurrences and five GCPs were applied to each photo from digital topographic features using ArcMap 10.2. Three out of the 82 landslides detected from the visual interpretation of the photos are shown in Figure 3. The photos taken before and after landslide occurrences are shown in Figure 3a–c, respectively. Figure 3c,d shows the blue plastic-covered area to prevent soil flow after landslides, and Figure 3d was taken by field survey. Most of the intensive rainfall-triggered debris flows were approximately 10–70 m in length, 3–20 m in width range, and less than one m in depth.

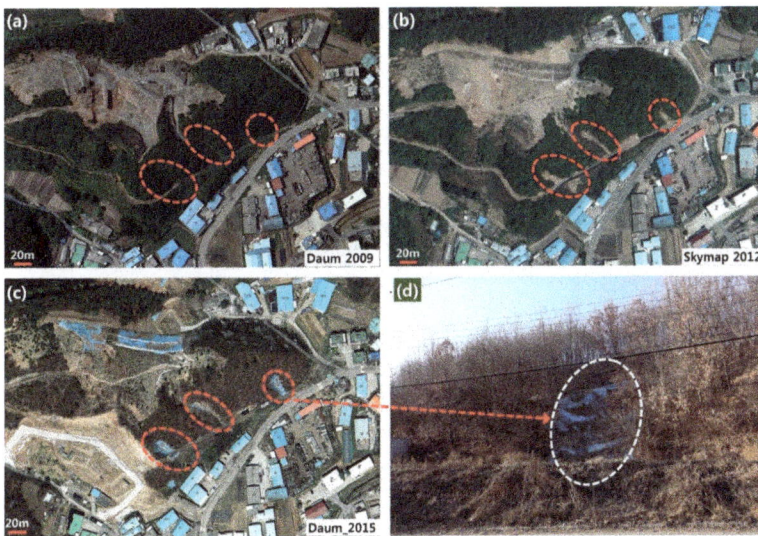

Figure 3. Digital aerial photographs of (**a**) pre-; (**b**,**c**) post-landslide occurrences; and (**d**) covered blue plastic at landslide scar after landslide occurrence in 2011.

2.3. Environmental Factors

Intensive rainfall-triggered debris flows are controlled by the interaction of various factors including topography, hydrology, soil, and forests [39]. Topography and hydrology influence debris flow initiation through the effect of gradient on slope stability with rainfall. These factors also determine the concentration and dispersion of the material and the material balance on the slope associated with the slope stability. In addition, soil and timber factors on the slope affect the spatial distribution of debris flows. These factors are significant controls, and can be represented as spatial distribution from digital elevation models (DEM) and soil and forest maps. Geology and faults were not considered as environmental factors in this study because shallow soil failure was mainly related to positive pore water pressure in saturated soils by intensive rainfall [39–41]. In this study, 18 environmental factors were considered for landslide susceptibility modeling based on ANN and BT (Table 1, Figure 4).

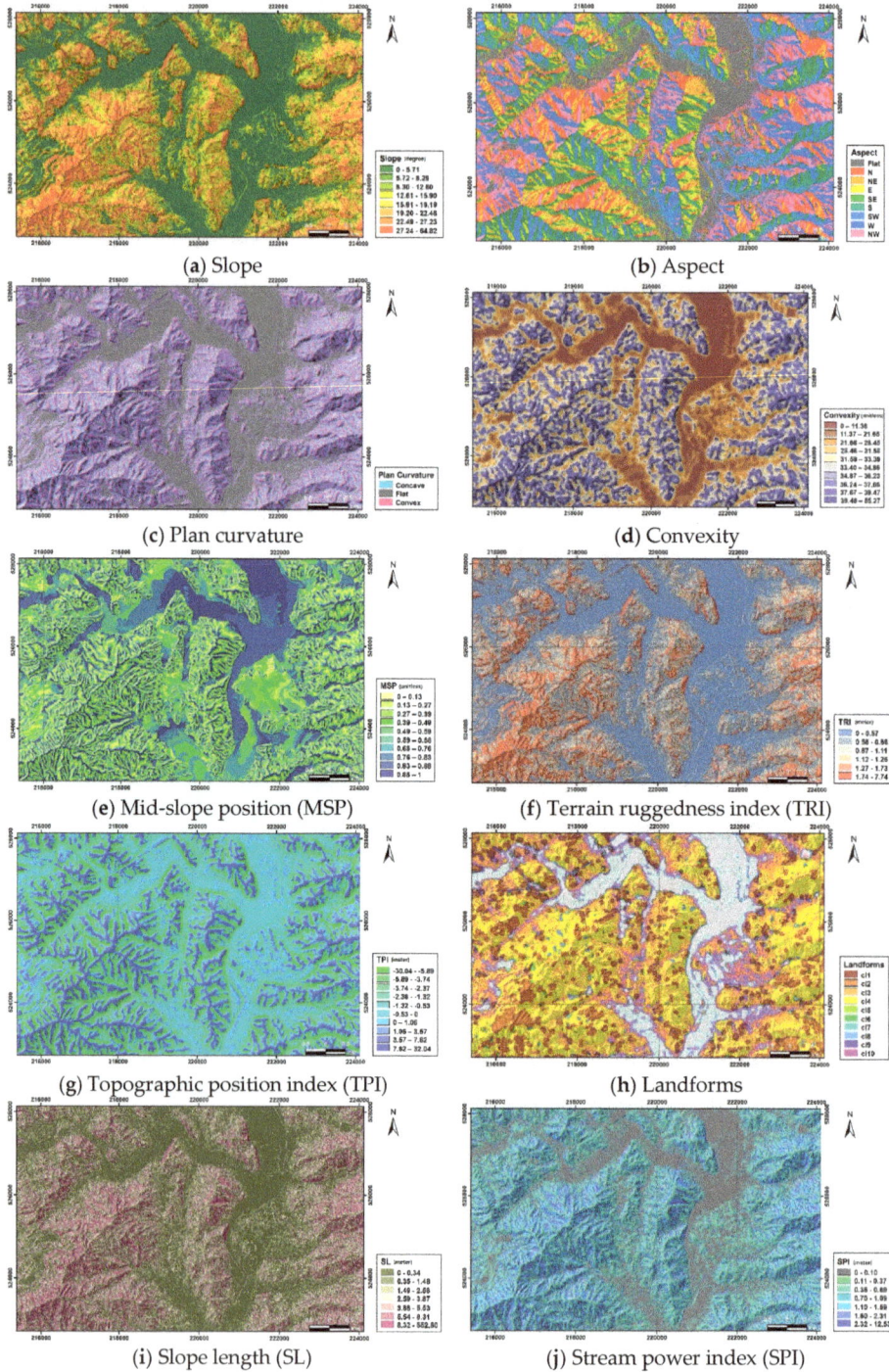

(a) Slope

(b) Aspect

(c) Plan curvature

(d) Convexity

(e) Mid-slope position (MSP)

(f) Terrain ruggedness index (TRI)

(g) Topographic position index (TPI)

(h) Landforms

(i) Slope length (SL)

(j) Stream power index (SPI)

Figure 4. *Cont.*

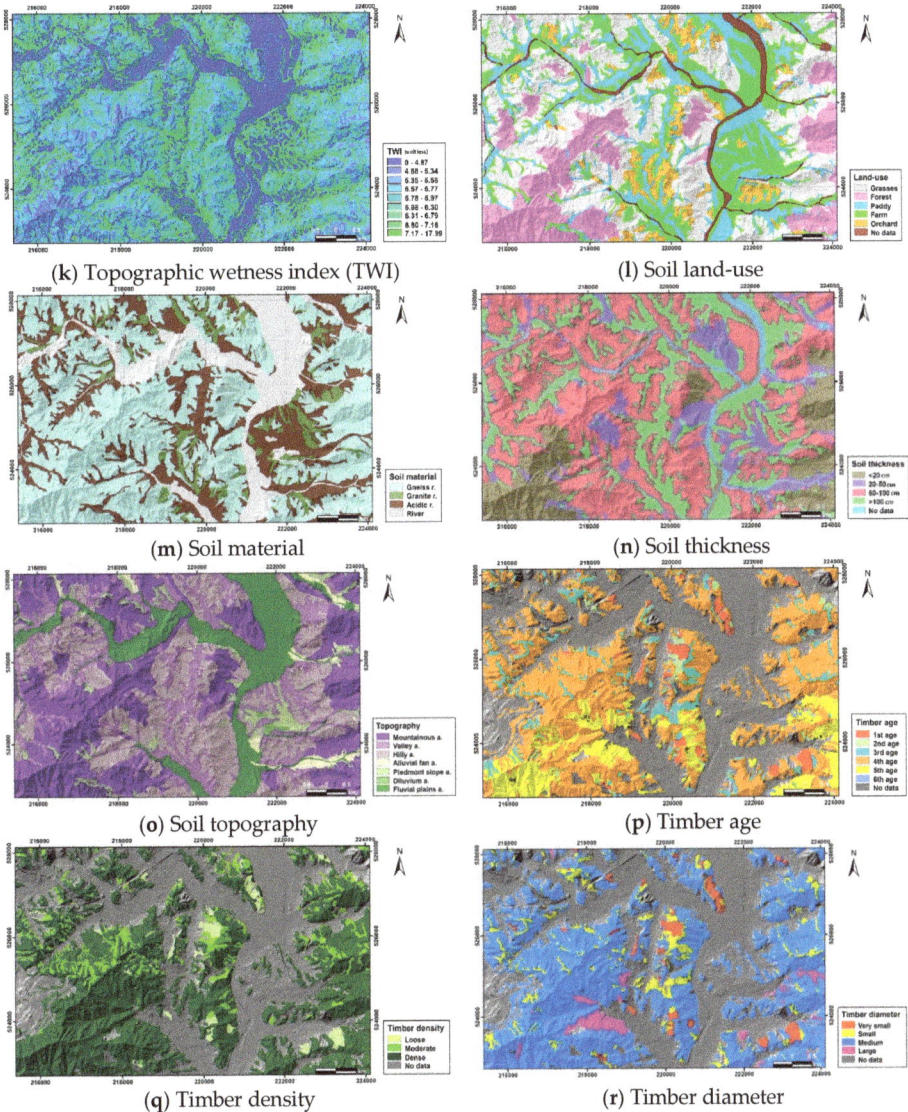

(k) Topographic wetness index (TWI)

(l) Soil land-use

(m) Soil material

(n) Soil thickness

(o) Soil topography

(p) Timber age

(q) Timber density

(r) Timber diameter

Figure 4. Spatial database of the landslide causative factors.

Topographic and hydrologic factors were extracted from the DEM for determining the relationship between these factors and debris flow using SAGA GIS modules [42]. A DEM with a 5 × 5 m grid format was generated from a triangulated irregular network (TIN) derived from a digital elevation contour with 5 m interval lines in ArcGIS 10.2. Soil and forest factors were also extracted from soil and forest maps with a scale of 1:5000.

The extracted topographic factors were slope, aspect, plan curvature, convexity, topographic position index (TPI), terrain ruggedness index (TRI), min-slope position (MSP), and landforms (Figure 4a–h). The considered hydrologic factors were slope length (SL), stream power index (SPI), and topographic wetness index (TWI) (Figure 4i–k). Slope indicated the steepness of a hill, and aspect was the steepest downhill direction. Plan curvature was perpendicular to the slope and affects the divergence and convergence of flow across the surface. Terrain surface convexity was described as positive surface curvature and represented the percentage of convex-upward cells [43]. TPI was the difference between the elevation of each cell and the mean elevation for a neighborhood of cells [44]. Negative values represented lower features than surrounding features, values near zero were flat areas, and positive values represented features typically higher. TRI were absolute values obtained by squaring the difference between the value of a cell and neighbor cells, and convex and concave areas could have similar values. MSP were assigned a 0 value, while maximum vertical distances to the mid-slope in crest or valley directions were assigned a 1 value. Landform classification (cl 1: deeply incised streams, cl 2: shallow valleys, cl 3: upland drainages, cl 4: U-shape valleys, cl 5: plains, cl 6: open slopes, cl 7: upper slopes, cl8: local ridges, cl 9: mid-slope ridges, and cl 10: high ridges) was derived by ranges of TPI values [45]. SL was based on specific catchment areas and slope, with the former used as a substitute for slope length. SPI represented the erosive power of a water flow [46]. TWI indicated the effect of topography on the location of the saturated area size of runoff generation [46]. In general, higher SL and SPI, and lower TPI, represented a higher landslide susceptibility.

The attribute columns in the digital soil map (Table 1) included land-use, material, thickness, and topographical values (Figure 4l–o). Land-use was classified into natural grasses, forests, paddy fields, and farm orchard areas. Soil material included three classes: gneiss, acidic residuum, and granite residuum. The class of soil thickness from the soil maps was divided into four classes: very shallow (<20 cm), shallow (20–50 cm), moderate (50–100 cm), and deep (>100 cm). Topography was classified into mountainous areas, fluvial plains, valley areas, hilly areas, alluvial fan areas, piedmont slope areas, and diluvium areas.

Timber factors from the digital forest map (Table 1) included timber age, density, and diameter (Figure 4p–r). Timber age was grouped into the 1st to 6th ages; over 50% of the timber in the study area belonged to the 1st age (less than 10 years), and the rest were classed as either the 2nd age (11–20 years), 3rd age (21–30 years), 4th age (31–40 years), 5th age (41–50 years), or 6th age (51–60 years). Timber density was divided into three classes: loose (less than 50% of a covered area), moderate (51–70%), and dense (over 71%). Timber diameter was divided into four classes: very small (over 51% of area with <6 cm), small (over 51% of area with <18 cm), medium (over 51% of area with <30 cm), and large (over 51% of area with >30 cm).

3. Application of Artificial Neural Network (ANN) and Boosted Tree (BT) Models for Landslide Susceptibility Mapping

3.1. Artificial Neural Network (ANN)

The ANN is an abstract mathematical model based on the knowledge of the human brain and its activities. The scope of possible applications of ANN is practically unlimited in fields such as pattern recognition (also known as classification), decision making, automatic control systems, and many others. Thus, ANN can be applied to the classification of landslide susceptibility by solving the non-linear relationship between landslides and their spatial environmental factors [47].

A feedforward ANN model called a multilayer perceptron (MLP) maps a set of input values onto a set of suitable outputs. The MLP comprises of an input and an output with one or more hidden layers of nonlinearly-activating nodes. Each node in one layer is connected with a certain weight to every node in the next layer. MLP utilizes a backpropagation algorithm for training the network. The algorithm trains the network until some goal minimal error is reached between the anticipated and actual output values of the network. At the end of this training step, the neural network produces a model that should be able to calculate a target value from a given input value [48].

It is important to select training data, such as landslide- and non-landslide locations, to be used as input to the ANN's learning algorithm [49]. In this study, areas with zero slope value were assigned as areas not prone to landslides, and areas with known landslides were assigned as areas prone to landslide in the training set. Both groups had 41 datasets. The values of the 18 landslide-related environmental factors were normalized to a range of 0.1–0.9 as input data. The backpropagation algorithm, as one of the most popular training algorithms, was used in this study. The three layered feed-forward network based on the framework provided by [50] was applied using the MATLAB software package as 18 (input layer) × 36 (hidden layer) × 2 (output layer). A log-sigmoid transfer function was used in the hidden layer and the output layer.

The flow of data processing was as follows. First, feedforward sent the input data to the neural network, and then the cost function with weight and bias were calculated. Many iterations of training satisfactorily minimized errors in updating optimized weight and bias for the training data. The relative influence indexes of the variables were calculated as the maximum repetitive number before reaching the targeted error of 2000, the learning rate of 0.01, and root mean square error (RMSE) of 0.001 using MATLAB. If the RMSE value of 0.001 was not achieved, then the maximum number of iterations was terminated at 2000 epochs. When the latter case occurred, then the maximum RMSE value was <0.1. As the calculated weights were granted to each factor (Table 2), landslide susceptibility for the whole study area was classified.

Table 2. Summary of the influence weights of predictor variables for Artificial Neural Network (ANN) and Boosted Tree (BT).

Normalized Weights Based on ANN		Normalized Weights Based on BT	
Soil thickness	0.00	Soil material	0.00
Plan curvature	0.05	Soil thickness	0.11
Aspect	0.14	Plan curvature	0.13
Slope length (SL)	0.19	Soil topography	0.18
Mid-slope position (MSP)	0.22	Landforms	0.21
Soil topography	0.24	Mid-slope position (MSP)	0.27
Topographic position index (TPI)	0.25	Stream power index (SPI)	0.33
Soil land-use	0.30	Soil land-use	0.34
Timber diameter	0.31	Convexity	0.36
Terrain ruggedness index (TRI)	0.35	Topographic position index (TPI)	0.42
Soil material	0.37	Timber density	0.43
Stream power index (SPI)	0.39	Aspect	0.45
Timber age	0.43	Slope length (SL)	0.65
Convexity	0.45	Slope gradient	0.66
Landforms	0.54	Topographic wetness index (TWI)	0.67
Timber density	0.58	Terrain ruggedness index (TRI)	0.71
Slope gradient	0.60	Timber diameter	0.73
Topographic wetness index (TWI)	1.00	Timber age	1.00

3.2. Boosted Tree (BT)

The boosted-tree technique has emerged as one of the most-influential methods for predictive data mining over the past few years. The boost-tree algorithm stems from one of the general computational approaches of stochastic-gradient boosting, also known as TreeNet (TM Salford Systems, Inc., 9685 Via Excelencia, Suite 208, San Diego, CA 92126, USA). These potent algorithms can effectively be used for regression as well as classification with continuous and categorical predictors. Boosted trees can ultimately produce a more-effective fit of the prediction values to the observation values, despite its complex relationship with the predictor and dependent variables, such as a nonlinear relationship; therefore, the boosted-tree algorithm can serve as a reliable machine-learning algorithm by fitting a weighted additive expansion of simple trees.

The training set in the BT model was the same as the training set in ANN. In the BT model in STATISTICA 10.0 [51], where the learning rate = 0.01, the tree complexity = 5, and the bag fraction = 0.5, the optimal number of trees was reached/selected at 262. The relative influence indexes of the variables were calculated summing the contribution of each variable (Table 2).

4. Landslide Susceptibility Mapping and Validation

The probability for landslide susceptibility was predicted by reflecting the relative influence indexes of predictor variables calculated in the ANN and BT models. The predicted landslide susceptibility index was classified into four classes based on area for simple and visual interpretation: very high, high, medium, and low index ranges in 5%, 10%, 15%, and 70% of the study area, respectively (Figure 5a,b).

Susceptibility maps were verified and compared by using known 41 actual landslide events as a validation set that were not used in the ANN and BT training to evaluate whether they could effectively reflect future landslide hazard areas. The landslide susceptibility indexes were sorted in descending order, and divided by 100 classes with cumulative 1% intervals. The cumulative distributions of landslide occurrence were compared with receiver operating characteristics (ROC) curves in 100 classes [52]. The ANN and BT models had a reasonable performance of 82.25% and 90.79% as percentage of area under ROC curves, respectively (Figure 6).

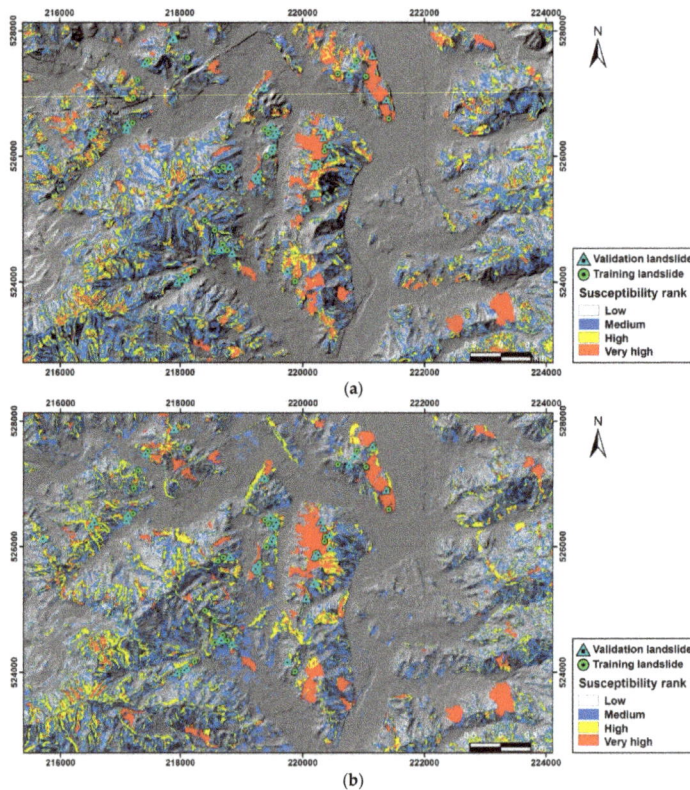

Figure 5. Landslide susceptibility maps based on (**a**) ANN; and (**b**) BT approaches. The rank was divided into four classes based on area: very high, high, medium, and low index ranges in 5%, 10%, 15%, and 70% of the study area, respectively.

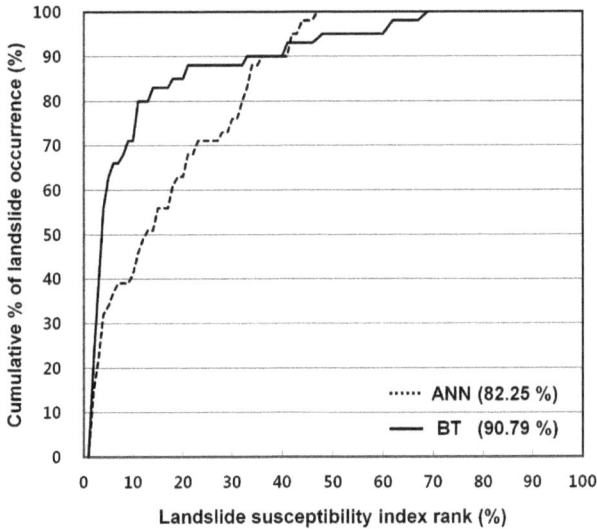

Figure 6. Percentage of area under curves (AUCs) of the landslide susceptibility maps based on ANN and BT models.

5. Discussion and Conclusions

Digital aerial photographs of high resolution are very useful in constructing detailed landslide inventory maps, as it is difficult to separate the similar shapes of landslide scar areas and surrounding tombs in the study area using satellite images or panchromatic aerial photographs. Therefore, both shapes could be easily interpreted visually in high-resolution aerial photographs taken in a high-vegetation season. Using aerial photographs could also save time and costs in field surveying to identify damage from natural disasters.

In Korea, debris flows occur randomly in several slope regions due to intensive rainfall per day. Therefore, it is necessary to select factors related to landslide occurrence and analyze the landslide susceptibility using pattern classification by looking at the relationship between the various factors and landslide location. However, it is not possible to know quantitatively how the environmental factors relate to the occurrence of landslides. ANN and BT models, which are used in many fields as sophisticated modeling techniques, were applied in identifying the influence of environmental factors to landslide occurrence and in mapping landslide susceptibility.

The training and validation sets, which were used in the ANN and BT models, were the same: 50% and 50% of a total of 82 landslide occurrences, respectively. ANN modeling was performed while changing the number of hidden layers (18 to 36), the value of learning rate (0.1 to 0.01), and RMSE (0.01~0.001). A sigmoid function for a backpropagation network was used as one of the more popular activation functions. The result of ANN modeling was best in 36 hidden layers, with 0.01 of learning rate and 0.001 of RMSE. BT modeling was performed at 0.01 of learning rate, 5 of tree complexity, and 0.5 of bag fraction. The optimal number of trees was reached at 262.

The weights of all factors from the ANN and BT models were normalized from 0 to 1. The factors' weights were divided into three groups: high, medium, and low influence groups from the ANN and BT models. The high group of ANN and BT included three factors: timber age, TWI, and slope gradient. The medium group of both models had three factors: TPI, soil land-use, and SPI. The low group had four factors: MSP, soil topography, plan curvature, and soil thickness. Although it was difficult to identify the influence ranking of the environmental factor (i.e., topographic, hydrologic, soil, forest, etc.) to landslide occurrence because of these intersections from intensive rainfall, the common factors in each ANN and BT group could be identified and can be used as a guideline for selecting environmental factors affecting landslide occurrence in other study areas in Korea.

ANNs are capable of handling complex and robustly nonlinear processes without previously assuming the relationships between the input and output variables (Lollino et al., 2014). Boosted tree has all of the strengths of decision trees, including the advantage of being able to handle both continuous and categorical variables (Krauss et al., 2017). In this study, the validation result of the ANN and BT models was 82.25% and 90.79%, respectively, which demonstrated reasonably good performance. In particular, the BT model had a higher accuracy (about 8%) than the ANN model. In other fields (using both models) [53,54], BT reported a better performance than ANN. The results of this study demonstrate the benefits of selecting optimal data mining techniques in landslide susceptibility modeling.

For future study, it is necessary to generalize the regularized relationships between the classes of each factor and landslide occurrence, and to combine the influence score of factors and factor classes. This approach could be used as a guideline to apply threshold values to a landslide monitoring system in the Korean rainy season. In addition, it could rank landslide susceptibility in urban areas, making this information helpful in selecting landslide monitoring sites and planning land-use.

Acknowledgments: This research was conducted by the Basic Research Project of the Korea Institute of Geoscience and Mineral Resources (KIGAM) funded by the Ministry of Science, ICT. This research (NRF- 2016K1A3A1A09915721) was supported by Science and Technology Internationalization Project through National Research Foundation of Korea (NRF) grant funded by the Ministry of Science and ICT.

Author Contributions: Hyun-Joo Oh organized the paperwork, constructed the input database, and performed the experiments; Saro Lee suggested the idea, collected the data, and performed the experiments. All authors contributed to the writing of each part.

Conflicts of Interest: The authors declare no conflict of interest.

References

1. Kim, K.S.; Song, Y.S. Geometrical and geotechnical characteristics of landslides in Korea under various geological conditions. *J. Mt. Sci.* **2015**, *12*, 1267–1280. [CrossRef]
2. Jeong, S.; Kim, Y.; Lee, J.K.; Kim, J. The 27 July 2011 debris flows at Umyeonsan, Seoul, Korea. *Landslides* **2015**, *12*, 799–813. [CrossRef]
3. Ro, K.S.; Jeon, B.J.; Jeon, K.W. Induction wall influence review by debris flow's impact force. *J. Korean Soc. Hazard Mitig.* **2015**, *15*, 159–164. [CrossRef]
4. Aleotti, P.; Chowdhury, R. Landslide hazard assessment: Summary review and new perspectives. *Bull. Eng. Geol. Environ.* **1999**, *58*, 21–44. [CrossRef]
5. Carrara, A.; Cardinali, M.; Guzzetti, F.; Reichenbach, P. GIS technology in mapping landslide hazard. In *Geographical Information Systems in Assessing Natural Hazards*; Springer: Dordrecht, The Netherlands, 1995; pp. 135–175.
6. Carrara, A.; Guzzetti, F.; Cardinali, M.; Reichenbach, P. Use of GIS technology in the prediction and monitoring of landslide hazard. *Nat. Hazards* **1999**, *20*, 117–135. [CrossRef]
7. Carrara, A.; Pugliese-Carratelli, E.; Merenda, L. Computer-based data bank and statistical analysis of slope instability phenomena. *Z. Geomorphol. N. F.* **1977**, *21*, 187–222.
8. Guzzetti, F.; Carrara, A.; Cardinali, M.; Reichenbach, P. Landslide hazard evaluation: A review of current techniques and their application in a multi-scale study, Central Italy. *Geomorphology* **1999**, *31*, 181–216. [CrossRef]
9. Soeters, R.; Van Westen, C.J. Slope instability recognition, analysis, and zonation. In *Landslides: Investigation and Mitigation*; National Academy Press: Washington, DC, USA, 1996; pp. 129–177.
10. Van Westen, C.J.; Castellanos, E.; Kuriakose, S.L. Spatial data for landslide susceptibility, hazard, and vulnerability assessment: An overview. *Eng. Geol.* **2008**, *102*, 112–131. [CrossRef]
11. Van Westen, C.J.; Soeters, R.; Sijmons, K. Digital geomorphological landslide hazard mapping of the Alpago area, Italy. *ITC J.* **2000**, *2*, 51–60. [CrossRef]

12. Van Westen, C.J.; van Asch, T.W.J.; Soeters, R. Landslide hazard and risk zonation—Why is it still so difficult? *Bull. Eng. Geol. Environ.* **2006**, *65*, 167–184. [CrossRef]
13. Varnes, D.J. *Landslide Hazard Zonation: A Review of Principles and Practice*; United Nations: New York, NY, USA, 1984; p. 63.
14. Choi, J.; Oh, H.-J.; Lee, H.-J.; Lee, C.; Lee, S. Combining landslide susceptibility maps obtained from frequency ratio, logistic regression, and artificial neural network models using aster images and GIS. *Eng. Geol.* **2012**, *124*, 12–23. [CrossRef]
15. Lee, M.-J.; Park, I.; Lee, S. Forecasting and validation of landslide susceptibility using an integration of frequency ratio and neuro-fuzzy models: A case study of Seorak mountain area in Korea. *Environ. Earth Sci.* **2015**, *74*, 413–429. [CrossRef]
16. Armaş, I. Weights of evidence method for landslide susceptibility mapping. Prahova subcarpathians, romania. *Nat. Hazards* **2012**, *60*, 937–950. [CrossRef]
17. Wang, L.-J.; Sawada, K.; Moriguchi, S. Landslide susceptibility analysis with logistic regression model based on FCM sampling strategy. *Comput. Geosci.* **2013**, *57*, 81–92. [CrossRef]
18. Pradhan, B. Landslide susceptibility mapping of a catchment area using frequency ratio, fuzzy logic and multivariate logistic regression approaches. *J. Indian Soc. Remote Sens.* **2010**, *38*, 301–320. [CrossRef]
19. Tsai, F.; Lai, J.-S.; Chen, W.W.; Lin, T.-H. Analysis of topographic and vegetative factors with data mining for landslide verification. *Ecol. Eng.* **2013**, *61*, 669–677. [CrossRef]
20. Nefeslioglu, H.A.; Sezer, E.; Gokceoglu, C.; Bozkir, A.S.; Duman, T.Y. Assessment of landslide susceptibility by decision trees in the metropolitan area of Istanbul, Turkey. *Math. Probl. Eng.* **2010**, *2010*. [CrossRef]
21. Kim, J.-C.; Lee, S.; Jung, H.-S.; Lee, S. Landslide susceptibility mapping using random forest and boosted tree models in Pyeong-Chang, Korea. *Geocarto Int.* **2017**, 1–16. [CrossRef]
22. Youssef, A.M.; Pourghasemi, H.R.; Pourtaghi, Z.S.; Al-Katheeri, M.M. Landslide susceptibility mapping using random forest, boosted regression tree, classification and regression tree, and general linear models and comparison of their performance at Wadi Tayyah Basin, Asir Region, Saudi Arabia. *Landslides* **2017**, *13*, 839–856. [CrossRef]
23. Huang, F.; Yin, K.; Huang, J.; Gui, L.; Wang, P. Landslide susceptibility mapping based on self-organizing-map network and extreme learning machine. *Eng. Geol.* **2017**, *223*, 11–22. [CrossRef]
24. Park, I.; Lee, S. Spatial prediction of landslide susceptibility using a decision tree approach: A case study of the Pyeongchang area, Korea. *Int. J. Remote Sens.* **2014**, *35*, 6089–6112. [CrossRef]
25. Oh, H.-J.; Pradhan, B. Application of a neuro-fuzzy model to landslide-susceptibility mapping for shallow landslides in a tropical hilly area. *Comput. Geosci.* **2011**, *37*, 1264–1276. [CrossRef]
26. Pradhan, B. A comparative study on the predictive ability of the decision tree, support vector machine and neuro-fuzzy models in landslide susceptibility mapping using GIS. *Comput. Geosci.* **2013**, *51*, 350–365. [CrossRef]
27. Liang, W.-J.; Zhuang, D.-F.; Jiang, D.; Pan, J.-J.; Ren, H.-Y. Assessment of debris flow hazards using a bayesian network. *Geomorphology* **2012**, *171–172*, 94–100. [CrossRef]
28. Song, Y.; Gong, J.; Gao, S.; Wang, D.; Cui, T.; Li, Y.; Wei, B. Susceptibility assessment of earthquake-induced landslides using bayesian network: A case study in Beichuan, China. *Comput. Geosci.* **2012**, *42*, 189–199. [CrossRef]
29. Ballabio, C.; Sterlacchini, S. Support vector machines for landslide susceptibility mapping: The Staffora River Basin case study, Italy. *Math. Geosci.* **2012**, *44*, 47–70. [CrossRef]
30. Lee, S.; Ryu, J.-H.; Kim, I.-S. Landslide susceptibility analysis and its verification using likelihood ratio, logistic regression, and artificial neural network models: Case study of Youngin, Korea. *Landslides* **2007**, *4*, 327–338. [CrossRef]
31. Samodra, G.; Chen, G.; Sartohadi, J.; Kasama, K. Comparing data-driven landslide susceptibility models based on participatory landslide inventory mapping in Purwosari area, Yogyakarta, Java. *Environ. Earth Sci.* **2017**, *76*, 184. [CrossRef]
32. Tsukamoto, Y.; Ohta, T. Runoff process on a steep forested slope. *J. Hydrol.* **1988**, *102*, 165–178. [CrossRef]
33. Bui, D.T.; Lofman, O.; Revhaug, I.; Dick, O. Landslide susceptibility analysis in the Hoa Binh province of Vietnam using statistical index and logistic regression. *Nat. Hazards* **2011**, *59*, 1413. [CrossRef]
34. Guzzetti, F.; Mondini, A.C.; Cardinali, M.; Fiorucci, F.; Santangelo, M.; Chang, K.-T. Landslide inventory maps: New tools for an old problem. *Earth Sci. Rev.* **2012**, *112*, 42–66. [CrossRef]

35. Hervas, J.; Van Den Eeckhaut, M.; Legorreta, G.; Trigila, A. *Landslide Science and Practice Volume 1: Landslide Inventory and Susceptibility and Hazard Zoning*; Introduction; Springer: Berlin, Germany, 2013.
36. Daum. Available online: http://map.daum.net/ (accessed on 2 March 2017).
37. Skymap. Available online: http://www.skymaps.co.kr/ (accessed on 2 March 2017).
38. Lee, S.; Song, K.-Y.; Oh, H.-J.; Choi, J. Detection of landslides using web-based aerial photographs and landslide susceptibility mapping using geospatial analysis. *Int. J. Remote Sens.* **2012**, *33*, 4937–4966. [CrossRef]
39. Montgomery, D.R.; Dietrich, W.E. A physically based model for the topographic control on shallow landsliding. *Water Resour. Res.* **1994**, *30*, 1153–1171. [CrossRef]
40. Baum, R.L.; Savage, W.Z.; Godt, J.W. *TRIGRS-A Fortran Program for Transient Rainfall Infiltration and Grid-Based Regional Slope-Stability Analysis*; US Geological Survey: Reston, VA, USA, 2002.
41. Montrasio, L.; Valentino, R. A model for triggering mechanisms of shallow landslides. *Nat. Hazards Earth Syst. Sci.* **2008**, *8*, 1149–1159. [CrossRef]
42. Conrad, O.; Bechtel, B.; Bock, M.; Dietrich, H.; Fischer, E.; Gerlitz, L.; Wehberg, J.; Wichmann, V.; Böhner, J. System for automated geoscientific analyses (SAGA) v. 2.1.4. *Geosci. Model Dev.* **2015**, *8*, 1991–2007. [CrossRef]
43. Iwahashi, J.; Pike, R.J. Automated classifications of topography from DEMs by an unsupervised nested-means algorithm and a three-part geometric signature. *Geomorphology* **2007**, *86*, 409–440. [CrossRef]
44. Guisan, A.; Weiss, S.B.; Weiss, A.D. Glm versus cca spatial modeling of plant species distribution. *Plant Ecol.* **1999**, *143*, 107–122. [CrossRef]
45. Weiss, A.D. Topographic position and landforms analysis. In Proceedings of the Poster Presentation, ESRI User Conference, San Diego, CA, USA, 9–13 July 2001.
46. Moore, I.D.; Grayson, R.B.; Ladson, A.R. Digital terrain modelling: A review of hydrological, geomorphological, and biological applications. *Hydrol. Process.* **1991**, *5*, 3–30. [CrossRef]
47. Gómez, H.; Kavzoglu, T. Assessment of shallow landslide susceptibility using artificial neural networks in Jabonosa River Basin, Venezuela. *Eng. Geol.* **2005**, *78*, 11–27. [CrossRef]
48. Pradhan, B.; Lee, S. Landslide susceptibility assessment and factor effect analysis: Backpropagation artificial neural networks and their comparison with frequency ratio and bivariate logistic regression modelling. *Environ. Model. Softw.* **2010**, *25*, 747–759. [CrossRef]
49. Paola, J.D.; Schowengerdt, R.A. A detailed comparison of backpropagation neural network and maximum-likelihood classifiers for urban land use classification. *IEEE Trans. Geosci. Remote Sens.* **1995**, *33*, 981–996. [CrossRef]
50. Hines, J.W.; Tsoukalas, L.H.; Uhrig, R.E. *Matlab Supplement to Fuzzy and Neural Approaches in Engineering*; John Wiley & Sons, Inc.: Hoboken, NJ, USA, 1997; p. 210.
51. STATISTICA. Boosted Trees for Regression and Classification Overview (Stochastic Gradient Boosting)—Basic Ideas. Available online: http://documentation.statsoft.com/STATISTICAHelp.aspx?path=Gxx/Boosting/BoostingTreesforRegressionandClassificationOverviewStochasticGradientBoostingBasicIdeas (accessed on 1 July 2017).
52. Deng, X.; Li, L.; Tan, Y. Validation of spatial prediction models for landslide susceptibility mapping by considering structural similarity. *ISPRS Int. J. Geo-Inf.* **2017**, *6*, 103. [CrossRef]
53. Suleiman, A.; Tight, M.R.; Quinn, A.D. Hybrid neural networks and boosted regression tree models for predicting roadside particulate matter. *Environ. Model. Assess.* **2016**, *21*, 731–750. [CrossRef]
54. Roe, B.P.; Yang, H.-J.; Zhu, J.; Liu, Y.; Stancu, I.; McGregor, G. Boosted decision trees as an alternative to artificial neural networks for particle identification. *Nucl. Instrum. Methods Phys. Res. Sect. A* **2005**, *543*, 577–584. [CrossRef]

applied
sciences

MDPI

Article

Habitat Potential Mapping of Marten (*Martes flavigula*) and Leopard Cat (*Prionailurus bengalensis*) in South Korea Using Artificial Neural Network Machine Learning

Saro Lee [1,2] , Sunmin Lee [3,4] , Wonkyong Song [5] and Moung-Jin Lee [4,*]

[1]   Geological Research Division, Korea Institute of Geoscience and Mineral Resources (KIGAM), 124, Gwahak-ro Yuseong-gu, Daejeon 34132, Korea; leesaro@kigam.re.kr
[2]   Korea University of Science and Technology, 217 Gajeong-ro Yuseong-gu, Daejeon 34113, Korea
[3]   Department of Geoinformatics, University of Seoul, 163 Seoulsiripdaero, Dongdaemun-gu, Seoul 02504, Korea; smlee@kei.re.kr
[4]   Center for Environmental Assessment Monitoring, Environmental Assessment Group, Korea Environment Institute (KEI), 370 Sicheong-daero, Sejong 30147, Korea
[5]   Department of Landscape Architecture, Dankook University, 119 Dandae-ro, Dongnam-gu, Cheonan-si, Chungnam 31116, Korea; wksong@dankook.ac.kr
*   Correspondence: leemj@kei.re.kr; Tel.: +82-44-415-7314

Received: 15 July 2017; Accepted: 31 August 2017; Published: 5 September 2017

Abstract: This study developed habitat potential maps for the marten (*Martes flavigula*) and leopard cat (*Prionailurus bengalensis*) in South Korea. Both species are registered on the Red List of the International Union for Conservation of Nature, which means that they need to be managed properly. Various factors influencing the habitat distributions of the marten and leopard were identified to create habitat potential maps, including elevation, slope, timber type and age, land cover, and distances from a forest stand, road, or drainage. A spatial database for each species was constructed by preprocessing Geographic Information System (GIS) data, and the spatial relationship between the distribution of leopard cats and environmental factors was analyzed using an artificial neural network (ANN) model. This process used half of the existing habitat location data for the marten and leopard cat for training. Habitat potential maps were then created considering the relationships. Using the remaining half of the habitat location data for each species, the model was validated. The results of the model were relatively successful, predicting approximately 85% for the marten and approximately 87% for the leopard cat. Therefore, the habitat potential maps can be used for monitoring the habitats of both species and managing these habitats effectively.

Keywords: habitat mapping; marten; leopard cat; ANN; South Korea

---

1. Introduction

Medium-sized predators control not only herbivorous animals but also intermediate predators, and have a combined effect on the overall food chain. Apex predators generally have low population densities and wide ranges of activities, and are vulnerable to local extinction due to habitat damage and disconnection [1]. In South Korea, members of the order Carnivora, such as tigers, leopards, and wolves, have become extinct or lost their ecological functions in natural ecosystems over the past century. As a result, the importance of the remaining carnivorous animals, such as martens, leopard cats, and otters, is increasing.

Martens (*Martes* sp.) are mammals with wide distribution worldwide, from subtropical to sub-Antarctic habitats. They are generally characterized by a larger sphere of activity than other mammals [2], and have roles as forest ecosystem indicator species sensitive to habitat disturbances and habitat fragmentation [3]. Therefore, the marten in Korea, the yellow-throated marten (*Martes flavigula*), is considered a second-grade endangered species of wild fauna and flora by the Ministry of Environment, South Korea, and belongs to Annex II of the Convention on International Trade in Endangered Species of Wild Fauna and Flora (CITES). Studies on marten habitats have been conducted mainly in tropical and subtropical areas such as Thailand [4], and there is insufficient information on marten habitats on the Korean Peninsula.

The leopard cat (*Prionailurus bengalensis*) is the only wild feline carnivore in South Korea, which also belongs to the second-grade endangered species of wild fauna and flora of the Ministry of Environment and the CITES Annex II list. Leopard cats are distributed within a wide range, from deep forests to rural and coastal areas in Korea. Moreover, they maintain relatively stable populations, due to their adaptability to inhabiting various habitats. It has been reported that leopard cats are distributed widely in forests and rural areas [5], but there have been few quantitative studies on the environmental spatial characteristics of their habitats. In particular, there is a lack of research on their distribution throughout South Korea and analysis of their habitat characteristics.

Machine learning is designed to process new data and predict results by learning patterns through training based on consistent patterns among variables in a dataset [6,7]. Recently, the amount of plant- and animal-related data has rapidly increased, and many studies using machine learning models have been conducted [8,9]. Depending on the approach applied, machine learning can use various algorithms, such as decision tree, neural network, and support vector machine, and it can be divided into supervised learning and unsupervised learning depending on the presence or absence of training data. Thus, many studies have been conducted to predict the distribution of plants and animals using machine learning techniques [10,11]. In addition, Geographic Information System (GIS) platforms have been used as a useful tool to model the spatial relationships between specific events and related factors [12–14].

Thus, recent studies have used GIS to indicate the distribution of habitats of various species. Studies on mapping and quantifying mammalian habitats have been conducted through GIS-based models [15,16]. Among mammals, studies on martens and leopard cats have been conducted mainly in Asia. Studies of the genetic structure and mitochondrial genome of the marten and the leopard cat have also been performed [17,18]. In particular, the complete mitochondrial genome has been analyzed for martens in Korea [19]. Meanwhile, studies on the movements and activity patterns of leopard cats have been conducted in China [20,21]. One study predicted the distribution of leopard cats in Borneo [22], and another similar study predicting the habitat potential of leopard cats was conducted in South Korea using GIS [23]. Habitat use and activity patterns of martens were also analyzed in central and northern Thailand [4]. Typically, the habitat distributions of badgers [24,25], leopard cats [26], and bears [27] have been mapped using logistic regression models. In recent years, research has been conducted on the creation of habitat distribution maps of leopard cats in South Korea through probabilistic and statistical models [23]. Supporting this, other probabilistic models have been used to map European bison habitats [28], and statistical models have been applied to the habitat distribution of bats [29,30]. Moreover, habitat mapping of water birds for conservation plans was conducted in the Hamun wetland [31]. In Canada, a high-resolution habitat map of Atlantic wolffish was also created [32]. However, artificial neural network (ANN) modeling has not been applied to analyzing the habitat of martens or leopard cats in South Korea. Therefore, the purposes of this study were to map the distribution of martens and leopard cats in South Korea using ANN modeling and to clarify the relationship between the habitat distributions and various environmental factors.

South Korea is located in East Asia between latitudes 33° and 39° N and longitudes 124° and 130° E, including all of its islands, and occupies the southern part of the Korean Peninsula (Figure 1); the total area of the country is approximately 100,032 km^2 [33]. The boundary of the study area in this study is marked with a red line in the Figure 1. South Korea can be divided into four regions: a western region of broad coastal plains, an eastern region of high mountain ranges, a southwestern mountainous region, and a southeastern region of a broad basin. Mountains cover about 70% of South Korea, and the country is surrounded by the Yellow Sea to the west, the East Sea to the east, and the Korea Strait and the East China Sea to the south. The north of the country is bordered by the Democratic People's Republic of Korea, via the demilitarized zone; therefore, it is difficult to support the return of extirpated terrestrial animals in South Korea, since their path is blocked in all directions. To support conservation efforts, habitat potential maps should be created to manage habitats before extirpation, including the selection of development areas. Therefore, in this study, ANN modeling was applied to generate habitat potential maps for martens and leopard cats. The weights of factors related to the habitat potentials were calculated with the model, and the results were validated to ensure the reliability of the maps.

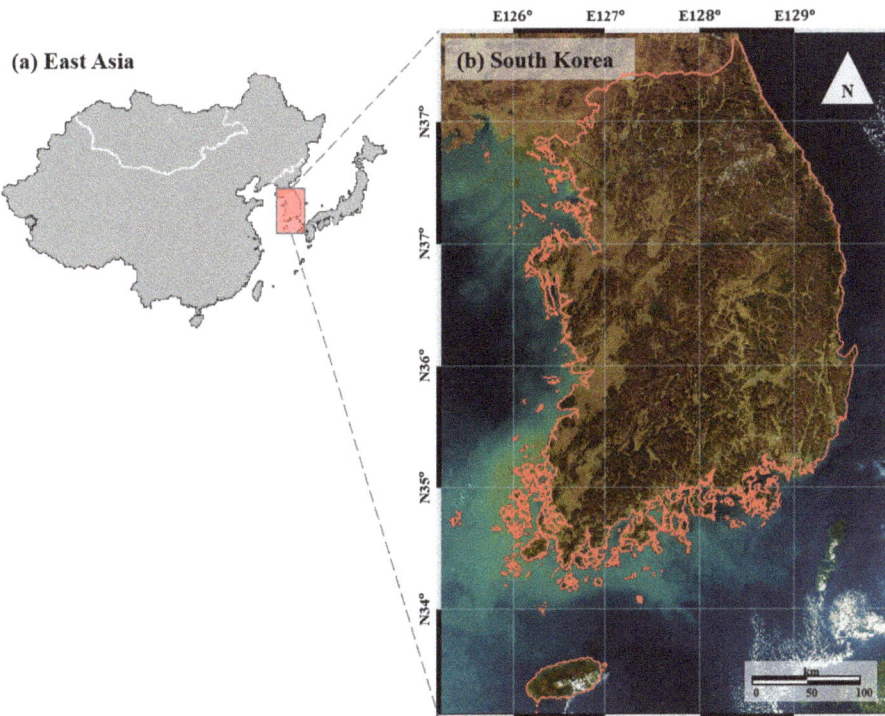

Figure 1. Study area: (**a**) South Korea in East Asia; and (**b**) South Korea.

2. Data

2.1. Habitat Survey

This study used data from the Second National Survey on Natural Environment in South Korea, which included the identification of habitats of endangered species of wild animals and plants. From 1997 to 2003, the National Institute for Environmental Studies conducted a survey of species appearances and spatial distributions of wild animals by experts from various research institutes. Surveys were performed from February to October every year for mammals to consider changes in seasonal patterns. The survey methods were based on field observations, including direct observations, community surveys, tracking, feces, and footprints, to investigate the species occurrence and spatial distribution of wild animals. The locations observed or detected by the traces were geocoded via Global Positioning System and were composed of GIS data.

The marten has the second largest sphere of action among mammals in the Korean Peninsula after the Asiatic black bear. Martens prefer broadleaf forests and mainly inhabit mature forests, such as fourth-grade forests that have ≥50% occupancy rates of 31–40-year-old trees with diameters of ≥30 cm. Martens inhabiting the Jirisan area occur at a density of 1–1.6 per 10 km^2, and are diurnal animals with a wide range of behaviors, moving an average of 11.2 ± 5.4 km per day. Martens mainly hunt Eurasian red squirrels, rats, and other rodents as food, but they also hunt hares, young roe deer, and wild boars. They also consume tree fruit, such as those from the lotus persimmon tree. In winter, they often hunt for food in shrubs on the edge of forests and frequently cross two-lane roads in their radius of action [34].

Leopard cats have activity areas of 3.69 ± 1.34 km^2, and the core space of the species is estimated to be 0.64 ± 0.47 km^2 [5]. Leopard cats prefer forests, as well as adjacent grassland and agricultural land, as habitats. They have a high preference for inland wetlands, such as wild grassland along riverbanks. Owing to their behavioral radius and nocturnality of movement in a variety of regions, they suffer a high frequency of road kills on roads adjacent to forests and rivers. Rodents, birds, and small mammals are their main sources of food, and forest ridges and valleys are used as main transport routes by leopard cats.

In the Second National Survey on Natural Environment, martens were found at 156 points, and leopard cats were found at 630 points in the study area (Figure 2). In this study, the observed data were divided randomly in half for model training and validation. This confirmed that the distribution of martens is mainly limited to large forest areas, including the Mt. Baekdu range. Meanwhile, leopard cats inhabit a wide range of emergence sites, from major forests to rivers and agricultural areas.

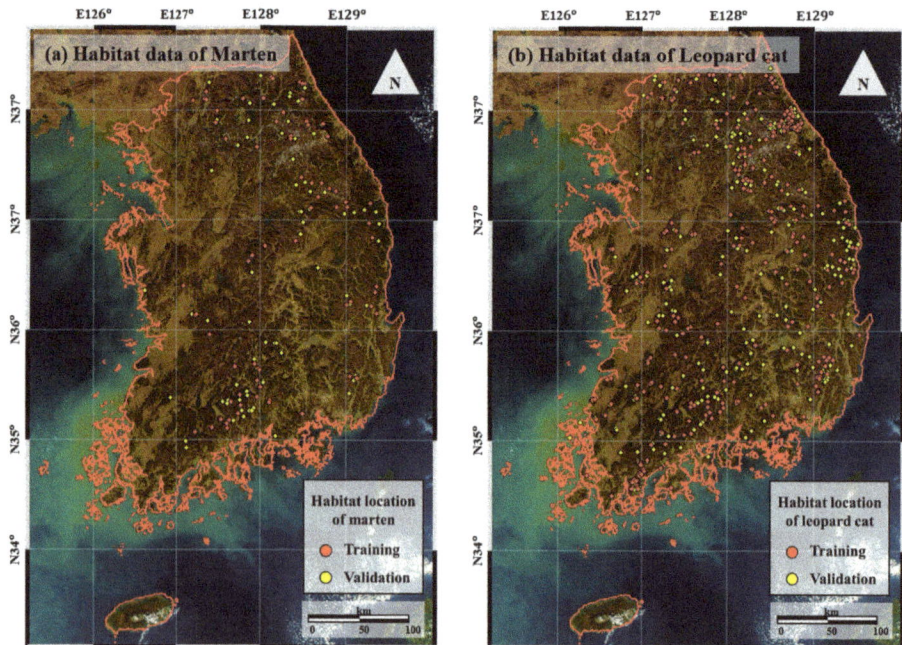

Figure 2. Data obtained from field observations of (**a**) Martens; and (**b**) Leopard cats.

2.2. Habitat-Related Factors

Habitat distributions of wild animals are influenced by the combined impacts of various factors. Environmental factors such as topographic characteristics, forest properties [22], and distance from essential factors to support life, such as water, influence the habitat distributions of mammals, including martens and leopard cats [22]. In this study, ground elevation, slope gradient, slope aspect, timber type, timber age, land cover, and distance from road, water, and forest were selected as factors influencing the habitat distributions of marten and leopard cat (Table 1 and Figure 3). The attribute information of the timber type of forest map is shown separately in Table 2, due to the limitation of the figure size. The determination and collection of factors were a fundamental component of mapping the potential habitats of these species. The factors were collected from nationally generated thematic maps and field investigations, such as those described in Section 2.1.

Table 1. Data layers related to marten and leopard cat habitat.

Original Data	Factors	Data Type	Scale
Habitat	Marten Leopard cat	Point	-
Topographical map [a]	Ground elevation (m) Slope gradient (°) Slope aspect	GRID	1:5000
Forest map [b]	Timber type Timber age	Polygon	1:25,000
Land cover map [c]	Land cover Distance from road (m) Distance from water (m) Distance from forest (m)	Polygon	1:25,000

[a] The digital topographic map by National Geographic Information Institute (NGII); [b] The forest map published by Korea Forest Service (KFS); [c] The land use map offered by the Korea Ministry of Environment.

A digital elevation model (DEM) was produced by generating a triangulated irregular network and digitizing the contours in 100-m intervals from topographical maps published by the National Geographic Information Institute. The slope gradient and slope aspect calculations were performed in ArcGIS 10.3. Thematic maps of forest and land cover were prepared at a 1:25,000 scale in a polygon format. The forest map was provided by the Korea Forest Service, and the timber type and age were prepared from this map. The land cover map was generated and published by the Ministry of Environment, South Korea. Land cover, distance from road, water, and forest were also calculated using ArcGIS.

All factors were converted into a 100 m × 100 m grid format to apply the ANN model. The row and column sizes of the study area were 6056 and 3533 cells, and the study area consisted of a total of 9,952,165 cells, except in the no data region, based on the DEM. The marten and leopard cat habitat data were converted from point formats into grid formats. Half of both the marten and the leopard cat habitat data were selected via random sampling as input data. The other half of the data were used for validation. Ultimately, all factors were collated into one input dataset with randomly extracted habitat sample data for both marten and leopard cat.

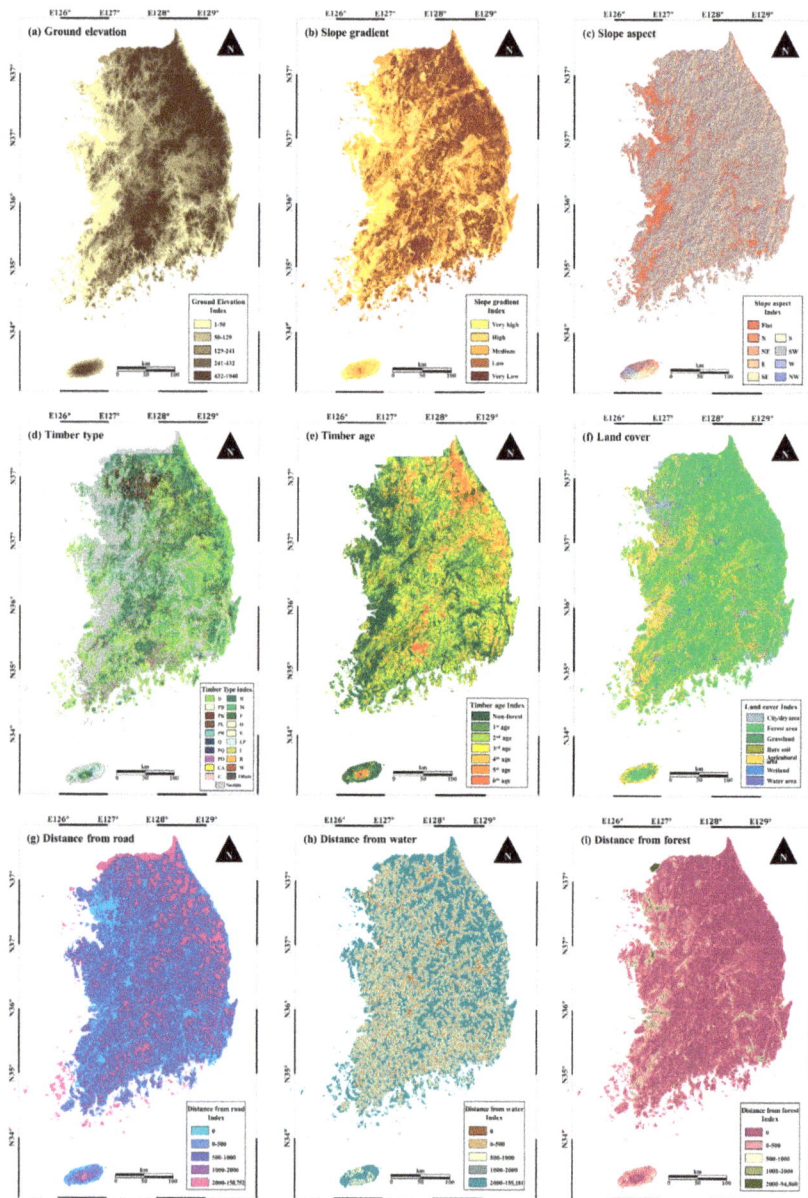

Figure 3. Factors related to the habitat potentials of marten and leopard cat: (**a**) ground elevation, (**b**) slope gradient, (**c**) slope aspect, (**d**) timber type, (**e**) timber age, (**f**) land cover, (**g**) distance from road, (**h**) distance from water, and (**i**) distance from forest.

Table 2. Attribute information of timber type from forest map.

	Code	Forest Type		Code	Forest Type
				C	Conifer mixed forest
	D	Pinus densiflora Forests	Forest physiognomy	H	Broadleaved forest
	PD	Pinus densiflora artificial forest		M	Mixed forest of soft and hardwood
	PK	Pinus koraiensis forest		F	Cut-over area
	PL	Larch		O	Non-stocked forest land
Forest species	PR	Pinus rigida forest	Dentuded area	E	Dentuded land
	Q	Oak forest		LP	Grassland
	PQ	Oak artificial forest		L	Farmland
	PO	Poplar forest		R	Left-over area
	CA	Chestnut artificial forest	Left-over area	W	Water

3. Methods

The machine learning technique determines the algorithm patterns using the sampled training data and derives the results through learned patterns. In other words, it is possible to learn from results that are already known according to information from the samples. Habitats are influenced by various environmental factors based on the nature of animals; therefore, even if there are no animals at the present time, the possibility of migration into an area with a similar environment exists at any time. Therefore, the habitat pattern can be determined by training based on habitat data using the machine learning method, and the habitat potential of an entire study area can be created. Among various machine learning techniques, habitat potential mapping of martens and leopard cats was conducted using an ANN in this study.

Interest in ANNs has increased recently. ANN is a highly sophisticated modeling method that can model complex functions such as environmental problems or social issues with large numbers of variables. Mimicking the human biological neural system, an ANN adjusts the weight from the basic unit of perceptron between the input and output data, minimizing the error of the result. In addition to these strengths, ANNs have been used for prediction or classification in a remarkable range of fields, such as engineering, physics, geology, and environmental science. In particular, neural networks can be applied to nonlinear problems of dimensionality and can be used to identify spatial relationships between observation location data and influencing factors.

Multi-layer perceptron is one of the most popular ANN models, and was created by [35]. This network, which allows for the regression of nonlinear data, was used in this study with a hidden layer between the input layers of habitat-related factors and the output layer. In this study, a back-propagation algorithm was used for neural network training. In back propagation, the algorithm calculates the gradient vector of the error surface and distance from the current point. The error can be decreased when the point moves a short distance from the output layer to other layers. The process iterates through a number of epochs, submitting the training data and calculating the error by comparing the target and output. The weights can be corrected through the error from the iteration and surface gradient. The initial network is randomly set, and the stopping point is determined by a set number of epochs or a user-selected stopping point. The following equation indicates the weight-updating architecture of the back-propagation algorithm.

$$\omega_k^{up} = \omega_k - \eta * \frac{\partial E_{total}}{\omega_k} \tag{1}$$

where ω_{k_0} is the initial value of the weight ω_k, which is given randomly, and ω_k is the updated weight for connection k. η is the learning rate that determines the step size and is typically chosen experimentally. E_{total} is the error of the output.

$$E_{total} = \frac{1}{2} \sum_{j}^{n} \left(T_j - out_{o_j} \right)^2 \tag{2}$$

T is the expected output value according to the input data [36] and j is the number assigned to the node in each layer. n is the number of output layers. E_{total} is minimized through the updated weight from Equation (16). To detect the minimum point, differentiation of E_{total} is necessary. Thus, an activation function is needed for conversion from the result of real number x to a range of values between 0 and 1 when the value is passed through the next layer or when the final result is expected. The unipolar sigmoid function was used for the activation function in this study. The equation of the unipolar sigmoid function is as follows.

$$f(x) = \frac{1}{1 + e^{-x}} \tag{3}$$

In this study, the ANN was supported by MATLAB software. Using the training data of randomly extracted habitat sample data, the slope data of the flat area were used as non-habitat area for training, and the input data of the study area were rearranged between 0.1 and 0.9. The weights of the input layers were calculated using the back-propagation algorithm of the ANN with a $9 \times 18 \times 1$ structure of the networks. The initial weight was set randomly as described previously, and the learning rate and number of epochs were set to 0.01 and 1000, respectively. The root-mean-square error, the criterion of the stopping point of decreasing error, was set to 0.01.

The process of potential mapping for martens and leopard cats can be explained in three steps, as shown in Figure 4. First, a spatial database was constructed with nine potential habitat-influencing factors, including the observation points of martens and leopard cats. Half of the marten and leopard distribution data selected by random sampling were used as training sets. The locations of observations of martens and leopard cats were confirmed in field surveys for the application and validation of habitat potential models described in Section 2.2 (Figure 2). The observation location data of martens and leopard cats were set as dependent variables. Second, the ANN model was applied to a spatial database to map the habitat potentials. Nine factors that were considered to affect the marten and leopard cat habitats were set as independent variables. Elevation, slope, and aspect derived from topographic maps, timber type and timber age from forest maps, and land cover, distance from road, distance from water, and distance from forest from land cover map were included in the spatial database with half of the randomly sampled marten and leopard cat habitat data. The ANN model was applied after constructing the spatial database. To apply the model, MATLAB programming was used with the GIS data. The resulting maps of the habitat potentials were validated using a receiver operating characteristic (ROC) curve. Finally, validation of the marten and leopard cat habitat potential maps was performed using the remaining marten and leopard cat distribution data that were not included in the spatial database.

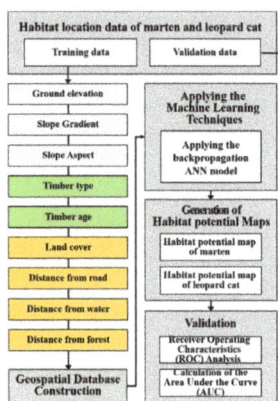

Figure 4. Flow chart of the steps used in this study.

4. Results

4.1. Weight of Related Factors and Habitat Potential Mapping

The weights of the related factors between the layers acquired during the ANN model training process are shown in Table 3. The weight values show the contribution of each weight to the ANN model. Since the weights were initialized during the first training, the results could differ. This study attempted to obtain similar results by performing the calculation several times. The weight calculation was repeated for 10 epochs, with 100 cycles for each epoch, to identify the influence of the randomly extracted sample for a city/dry area (public area), non-habitat area, since most wild animals do not live in populated urban areas. No leopard cats were observed in public areas in the results of the frequency ratio analysis in a previous study [23]. In the iterations of each epoch, the weights were updated via back propagation. The standard deviation was 0.001–0.018 for martens and 0.002–0.021 for leopard cats. The ranges of standard deviations indicated that there were minimal effects of random sampling of non-habitat area on the results.

Table 3. Neural network weight between martens and leopard cats, and habitat-related factors.

	Marten			Leopard Cat		
	Average	Standard Deviation	Normalized Weight with Respect to Land Cover	Average	Standard Deviation	Normalized Weight with Respect to Land Cover
DEM	0.1657	0.0175	2.3877	0.1728	0.0208	2.3883
Slope gradient	0.1335	0.0102	1.9242	0.1939	0.0103	2.6790
Slope aspect	0.0901	0.0019	1.2986	0.0768	0.0019	1.0619
Timber type	0.1453	0.0107	2.0945	0.1159	0.0185	1.6014
Timber age	0.0704	0.0104	1.0146	0.1053	0.0059	1.4545
Land cover	0.0694	0.0015	1.0000	0.0724	0.0030	1.0000
Distance from road	0.1092	0.0023	1.5741	0.0817	0.0054	1.1285
Distance from water	0.1210	0.0032	1.7439	0.0854	0.0037	1.1803
Distance from forest	0.0953	0.0008	1.3735	0.0959	0.0044	1.3259

Regarding the average weights of the related factors for the habitat potential mapping of martens, DEM had the highest value (0.166), followed by timber type (0.145), while land cover showed the lowest value (0.069). The factor weights for leopard cats were similar. Slope gradient had the highest value (0.194), followed by DEM (0.173), and land cover had the lowest value (0.072).

However, for a comparison of the same standards between the two analyses, all average weight values were normalized by dividing by the smallest average weight value, land cover, for each factor for martens and leopard cats. The lowest values of 0.069 and 0.072 were normalized to 1.000. DEM showed the highest normalized weight value of 2.388 for martens and timber type was the second highest normalized factor at 2.095. Likewise, the weight analysis showed that the slope gradient and DEM presented the highest and second highest normalized weight values of 2.679 and 2.388 for leopard cats, respectively. The least important factors, land cover for both martens and leopard cats, were used as references for the normalized weights. Timber age (1.015) and slope aspect (1.299) for martens and slope aspect (1.062) and distance from road (1.129) for the leopard cats followed.

The weights calculated via iteration, shown above, were applied throughout the study area, and the final weights determined for martens and leopard cats were applied to the corresponding factor in each dataset. Figure 5 shows the habitat potential maps for martens and leopard cats. To simplify the interpretation of the habitat potential maps, the results were classified into five classes of potentials: very high (10%), high (10%), medium (20%), low (20%), and very low (40%). Based on the habitat potential maps of both martens and leopard cats, the eastern region of South Korea appeared to have high potential.

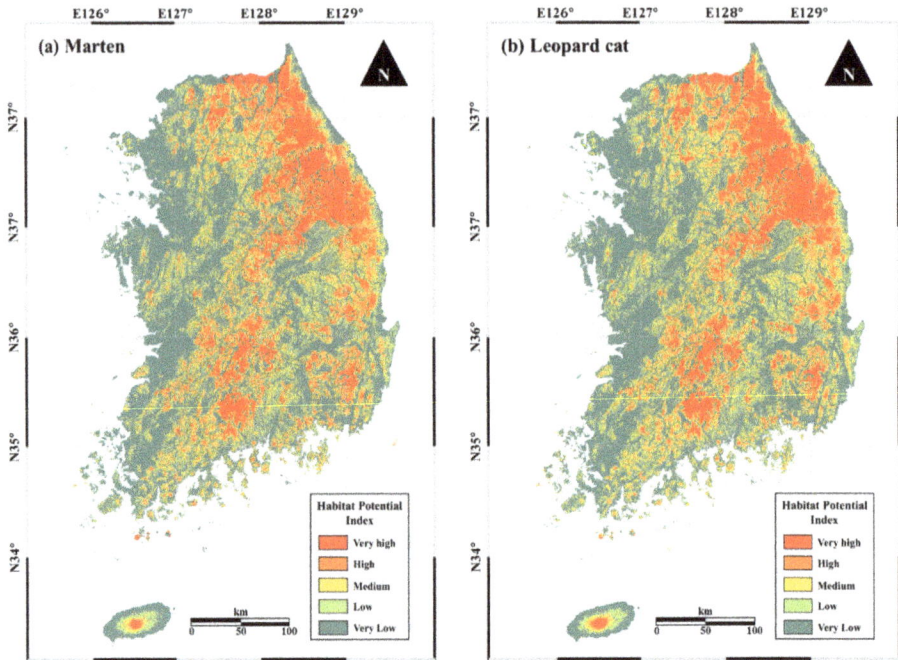

Figure 5. Habitat potential maps generated using ANN model of (**a**) martens; and (**b**) leopard cats.

4.2. Validation

Habitat data can be used to validate whether the habitat potential maps were effectively predicted. Therefore, the halves of marten and leopard cat habitat data not used for training were used as the validation data. ROC curves were used to validate the habitat potential maps of martens and leopard cats generated using the ANN model. The ROC curves were generated for the marten and leopard cat habitat potential maps to compare the habitat locations of martens and leopard cats, respectively.

The ROC curve is a graphical plot showing the diagnostic capabilities of classification models with various thresholds. It is a widely used method for validation [37–40]. ROC curves can show sensitivity and specificity on the *x*-axis and *y*-axis, respectively. In this study, specificity represented the percentage of area that martens and leopard cats could inhabit, while sensitivity represented the predicted potential marten and leopard cat habitat locations. For the ROC graph, the predicted habitat potential values after applying the ANN model were sorted in descending order. The ranks of the habitat potential values were identified and the values were equally classified into 100 classes of study area.

Figure 6 shows the ROC curves of the marten and leopard cat habitat potential maps; martens showed 85.01% (0.8501) accuracy, versus 87.03% (0.8703) accuracy for leopard cats. These results are depicted as a graph of the validation rate (Figure 6), where 65% of the study area included 100% of all habitat locations for both species.

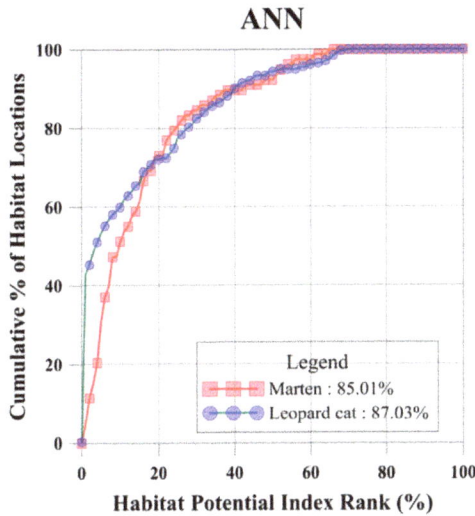

Figure 6. Validation results of the habitat potential maps for martens and leopard cats using the ANN model.

5. Discussion

Machine learning focuses on processing new data and predicting outcomes using patterns learned through training [12,13]. In other words, the most noticeable advantage of ANN modeling is that the algorithm itself can learn to be more accurate, although it requires sufficient data to support this. In South Korea, a government-affiliated organization constructed detailed country-scale thematic maps. These previously constructed data were sufficient for generalizing and correctly processing new incoming untrained data. Therefore, a spatial database was constructed based on nationally constructed data, including topographical, forest, and land use maps. The spatial database built in this process could be used in conjunction with other methods. MATLAB was used to process large amounts of data quickly, and the spatial distribution of martens and leopard cats could be confirmed through a GIS analysis. The results for leopard cats using the ANN showed approximately 5% higher accuracy than those obtained using the frequency ratio (82.15%) and logistic model (81.48%) in a previous study [23], even though the input data differed. This confirmed that the ANN model had been applied appropriately to habitat potential mapping.

The results of machine learning models, such as the ANN model applied in this study, are strongly influenced by the input data. Therefore, in this study, an input spatial database was constructed with marten and leopard cat habitat point data based on the Second National Survey on Natural Environment, and data extracted and calculated from national theme maps, which reflected regional characteristics. The impacts of factors besides the input data, such as climate or sudden natural disasters, on martens and leopard cats were assumed to be negligible. In the case of the habitat data, the generation of errors may depend on the survey method. However, the behavior radii of martens and leopard cats were within 10 km, and the size of the entire study area was considerably larger by comparison; therefore, all data were considered to be reliable.

In both species, DEM, slope gradient and timber type were the three most important factors related to habitat. The relative weight of slope gradient was higher for leopard cats than for martens, indicating a greater influence of slope gradient in the result map for leopard cats (Figure 5). The smaller radius of action of leopard cats compared to martens may have influenced this result. Martens mainly inhabit forests; therefore, the habitat potential of martens was high in relatively dense forests. In contrast, the habitat potential of leopard cats tended to be evenly distributed in relatively low areas compared to martens, which was particularly evident on Jeju Island.

Considering the results of the habitat potential map of this study, elevation, slope gradient and timber type information of mountain areas should be considered as indices for prioritizing the protection and management of marten and leopard cat habitats. Moreover, the index should be managed with weighted indices for each factor. In the future, national natural environment survey data that are constructed and accumulated in a timely manner will enable the continuous construction and management of habitat maps for not only marten and leopard cat, but also other endangered species.

6. Conclusions

In this study, ANN modeling was used to predict the habitat potentials for martens and leopard cats. The factors affecting the habitat distribution of each species were selected and a spatial database was constructed with the collected data. Using the collected data and half of the field-surveyed habitat data, the habitat potentials of martens and leopard cats were predicted using the ANN model. The predicted maps were validated with the other half of the habitat data not used for training.

The first- to third-highest normalized weights with respect to the lowest average weight for both martens and leopard cats (land cover) were DEM, slope gradient and timber type. This indicates that factors related to DEM are positively correlated with the habitat location of martens and leopard cats. In contrast, land cover showed the lowest normalized weights for both martens and leopard cats. Subsequently, martens had lower weights for timber age and slope aspect, whereas leopard cats had lower weights for slope aspect and distance from road. Since both species are mammals with large activity radii, and steep slopes are considered an important factor, the weight value for distance from mountains was relatively low.

In South Korea, martens and leopard cats exhibited similar habitat patterns, with high habitat potentials in eastern mountainous areas, except along beach lines and southern areas. Likewise, coastal areas beyond the eastern range and western plains of the study area showed very low habitat potentials for both species. Nearly all areas with low habitat potentials were lowlands, coastal areas, and non-forest areas. These trends are likely similar even though the habitat characteristics of martens and leopard cats differ because of the large size of the study area.

The results of the marten and leopard cat habitat potential maps were validated with the half of the marten and leopard cat habitat data not used for modeling. The results of the validation showed accuracy of 85.01% for martens and 87.03% for leopard cats, both of which were satisfactory (>85%). These results could lead to more informed decisions for wildlife management planning and land development planning in areas with high habitat potentials.

Owing to the characteristics of the machine learning model, input data accuracy is very important. However, it is difficult to determine the exact habitat location of mammals with large activity radii, such as martens and leopard cats. Since inaccurate location data could result in difficulties in performing spatial analyses, reliable habitat data should be used when research is conducted for smaller administrative units. Quantitative evaluation of the ecological consequences of a wide range of spatial data could be performed using GIS. Integrating several characteristics representing the habitat potential of mammals is an important aspect of ecological management research.

This study identified habitat-related factors for martens and leopard cats. The methodology applied in this study could also be used to generate a time-series of habitat maps, and marten and leopard cat habitat data could be used, since the habitat-related data will be additionally and continuously constructed during the next national natural environment survey. In addition to martens and leopard cats, this method can also be applied to habitat mapping of other mammalian species examined in national natural environment surveys. Moreover, the habitat potential mapping results can be used as a basis for determining the locations of monitoring sites or creating protection plans for mammalian species. However, more studies are essential to generalize the habitat-related factors that affect the habitat characteristics for individual mammalian species.

Acknowledgments: This research was conducted by Korea Environment Institute (KEI) with support of a grant (16CTAP-C114629-01) from Technology Advancement Research Program (TARP) funded by Ministry of Land, Infrastructure and Transport of Korean government. This research was supported by the Basic Research Project of the Korea Institute of Geoscience and Mineral Resources (KIGAM) funded by the Ministry of Science and ICT. This research (NRF-2016K1A3A1A09915721) was supported by Science and Technology Internationalization Project through National Research Foundation of Korea (NRF) grant funded by the Ministry of Education, Science and Technology (MEST).

Author Contributions: Saro Lee organized the paperwork and constructed the input database. Sunmin Lee designed and performed the experiments. Wonkyong Song interpreted the input data and this resulted in the habitat aspect. Moung-Jin Lee suggested the idea and collected the data. All authors contributed to the writing of each part.

Conflicts of Interest: The authors declare no conflict of interest.

References

1. Ripple, W.J.; Estes, J.A.; Beschta, R.L.; Wilmers, C.C.; Ritchie, E.G.; Hebblewhite, M.; Berger, J.; Elmhagen, B.; Letnic, M.; Nelson, M.P. Status and ecological effects of the world's largest carnivores. *Science* **2014**, *343*, 1241484. [CrossRef] [PubMed]

2. Taylor, S.L.; Buskirk, S.W. Forest microenvironments and resting energetics of the American marten *Martes americana*. *Ecography* **1994**, *17*, 249–256. [CrossRef]

3. Harrison, D.J.; Fuller, A.K.; Proulx, G. *Martens and Fishers (Martes) in Human-Altered Environments: An International Perspective*; Springer Science & Business Media: Berlin, Germany, 2004.

4. Grassman, L.I., Jr.; Tewes, M.E.; Silvy, N.J. Ranging, habitat use and activity patterns of binturong arctictis binturong and yellow-throated marten *Martes flavigula* in north-central Thailand. *Wildl. Biol.* **2005**, *11*, 49–57. [CrossRef]

5. Park, C.-H.; Choi, T.-Y.; Kwon, H.-S.; Woo, D.-G. Habitat selection and management of the leopard cat (*Prionailurus bengalensis*) in a rural area of Korea. *Korean J. Environ. Ecol.* **2012**, *26*, 322–332.

6. Ramcharan, A.; Hengl, T.; Nauman, T.; Brungard, C.; Waltman, S.; Wills, S.; Thompson, J. Soil property and class maps of the conterminous us at 100 meter spatial resolution based on a compilation of national soil point observations and machine learning. *arXiv*, **2017**, arXiv:1705.08323.

7. Veronesi, F.; Korfiati, A.; Buffat, R.; Raubal, M. Assessing accuracy and geographical transferability of machine learning algorithms for wind speed modelling. In Proceedings of the International Conference on Geographic Information Science, Wageningen, The Netherlands, 9–12 May 2017; Springer: Berlin, Germany, 2017; pp. 297–310.

8. Folmer, E.O.; van Beusekom, J.E.; Dolch, T.; Gräwe, U.; Katwijk, M.M.; Kolbe, K.; Philippart, C.J. Consensus forecasting of intertidal seagrass habitat in the wadden sea. *J. Appl. Ecol.* **2016**, *53*, 1800–1813. [CrossRef]

9. Sahragard, H.P.; Chahouki, M.A.Z. Comparison of logistic regression and machine learning techniques in prediction of habitat distribution of plant species. *Range Manag. Agrofor.* **2016**, *37*, 21–26.

10. Fox, C.; Huettmann, F.; Harvey, G.; Morgan, K.; Robinson, J.; Williams, R.; Paquet, P. Predictions from machine learning ensembles: Marine bird distribution and density on Canada's pacific coast. *Mar. Ecol. Prog. Ser.* **2017**, *566*, 199–216. [CrossRef]

11. Wieland, R.; Kerkow, A.; Früh, L.; Kampen, H.; Walther, D. Automated feature selection for a machine learning approach toward modeling a mosquito distribution. *Ecol. Model.* **2017**, *352*, 108–112. [CrossRef]

12. Anderson, D.J.; Rojas, L.F.; Watson, S.; Knelson, L.P.; Pruitt, S.; Lewis, S.S.; Moehring, R.W.; Bennett, E.E.S.; Weber, D.J.; Chen, L.F. Identification of novel risk factors for community-acquired clostridium difficile infection using spatial statistics and geographic information system analyses. *PLoS ONE* **2017**, *12*, e0176285. [CrossRef] [PubMed]

13. Bui, D.T.; Bui, Q.-T.; Nguyen, Q.-P.; Pradhan, B.; Nampak, H.; Trinh, P.T. A hybrid artificial intelligence approach using gis-based neural-fuzzy inference system and particle swarm optimization for forest fire susceptibility modeling at a tropical area. *Agric. For. Meteorol.* **2017**, *233*, 32–44.

14. Rahimi, S.; Martin, M.J.; Obeysekere, E.; Hellmann, D.; Liu, X.; Andris, C. A geographic information system (gis)-based analysis of social capital data: Landscape factors that correlate with trust. *Sustainability* **2017**, *9*, 365. [CrossRef]

15. Ottaviani, D.; Panzacchi, M.; Lasinio, G.J.; Genovesi, P.; Boitani, L. Modelling semi-aquatic vertebrates' distribution at the drainage basin scale: The case of the otter lutra lutra in Italy. *Ecol. Model.* **2009**, *220*, 111–121. [CrossRef]
16. Poplar-Jeffers, I.O.; Petty, J.T.; Anderson, J.T.; Kite, S.J.; Strager, M.P.; Fortney, R.H. Culvert replacement and stream habitat restoration: Implications from brook trout management in an appalachian watershed, USA. *Restor. Ecol.* **2009**, *17*, 404–413. [CrossRef]
17. Patel, R.P.; Wutke, S.; Lenz, D.; Mukherjee, S.; Ramakrishnan, U.; Veron, G.; Fickel, J.; Wilting, A.; Förster, D.W. Genetic structure and phylogeography of the leopard cat (*Prionailurus bengalensis*) inferred from mitochondrial genomes. *J. Hered.* **2017**, *108*, 349–360. [CrossRef] [PubMed]
18. Tan, S.; Xu, J.-T.; Zou, F.-D.; Peng, Q.-K.; Peng, R. The complete mitochondrial genome of leopard cat, *Prionailurus bengalensis* chinensis (carnivora: Felidae). *Mitochondrial DNA A* **2016**, *27*, 3073–3074.
19. Jang, K.H.; Hwang, U.W. Complete mitochondrial genome of korean yellow-throated marten, *Martes flavigula* (carnivora, mustelidae). *Mitochondrial DNA A* **2016**, *27*, 1785–1786. [CrossRef] [PubMed]
20. Chen, M.-T.; Liang, Y.-J.; Kuo, C.-C.; Pei, K.J.-C. Home ranges, movements and activity patterns of leopard cats (*Prionailurus bengalensis*) and threats to them in Taiwan. *Mamm. Study* **2016**, *41*, 77–86. [CrossRef]
21. Vigne, J.-D.; Evin, A.; Cucchi, T.; Dai, L.; Yu, C.; Hu, S.; Soulages, N.; Wang, W.; Sun, Z.; Gao, J. Earliest "domestic" cats in China identified as leopard cat (*Prionailurus bengalensis*). *PLoS ONE* **2016**, *11*, e0147295. [CrossRef] [PubMed]
22. Mohamed, A.; Ross, J.; Hearn, A.J.; Cheyne, S.M.; Alfred, R.; Bernard, H.; Boonratana, R.; Samejima, H.; Heydon, M.; Augeri, D.M.; et al. Predicted distribution of the leopard cat *Prionailurus bengalensis* (mammalia: Carnivora: Felidae) on borneo. *Raffles Bull. Zool.* **2016**, *33*, 180–185.
23. Lee, M.-J.; Song, W.; Lee, S. Habitat mapping of the leopard cat (*Prionailurus bengalensis*) in South Korea using gis. *Sustainability* **2015**, *7*, 4668–4688. [CrossRef]
24. Bui, D.T.; Lofman, O.; Revhaug, I.; Dick, O. Landslide susceptibility analysis in the hoa binh province of vietnam using statistical index and logistic regression. *Nat. Hazards* **2011**, *59*, 1413. [CrossRef]
25. Huck, M.; Davison, J.; Roper, T.J. Predicting European badger meles meles sett distribution in urban environments. *Wildl. Biol.* **2008**, *14*, 188–198. [CrossRef]
26. Gavashelishvili, A.; Lukarevskiy, V. Modelling the habitat requirements of leopard panthera pardus in west and Central Asia. *J. Appl. Ecol.* **2008**, *45*, 579–588. [CrossRef]
27. Northrup, J.; Stenhouse, G.; Boyce, M. Agricultural lands as ecological traps for grizzly bears. *Anim. Conserv.* **2012**, *15*, 369–377. [CrossRef]
28. Kuemmerle, T.; Perzanowski, K.; Chaskovskyy, O.; Ostapowicz, K.; Halada, L.; Bashta, A.-T.; Kruhlov, I.; Hostert, P.; Waller, D.M.; Radeloff, V.C. European bison habitat in the Carpathian mountains. *Biol. Conserv.* **2010**, *143*, 908–916. [CrossRef]
29. Clement, M.J.; Castleberry, S.B. Estimating density of a forest-dwelling bat: A predictive model for rafinesque's big-eared bat. *Popul. Ecol.* **2013**, *55*, 205–215. [CrossRef]
30. Greaves, G.J.; Mathieu, R.; Seddon, P.J. Predictive modelling and ground validation of the spatial distribution of the New Zealand long-tailed bat (*Chalinolobus tuberculatus*). *Biol. Conserv.* **2006**, *132*, 211–221. [CrossRef]
31. Maleki, S.; Soffianian, A.R.; Koupaei, S.S.; Saatchi, S.; Pourmanafi, S.; Sheikholeslam, F. Habitat mapping as a tool for water birds conservation planning in an arid zone wetland: The case study hamun wetland. *Ecol. Eng.* **2016**, *95*, 594–603. [CrossRef]
32. Novaczek, E.; Devillers, R.; Edinger, E.; Mello, L. High resolution habitat mapping to describe coastal denning habitat of a Canadian species at risk, atlantic wolffish (*Anarhichas lupus*). *Can. J. Fish. Aquat. Sci.* **2017**. [CrossRef]
33. Park, J.-K.; Das, A.; Park, J.-H. A new approach to estimate the spatial distribution of solar radiation using topographic factor and sunshine duration in South Korea. *Energy Convers. Manag.* **2015**, *101*, 30–39. [CrossRef]
34. Woo, D.G. A Study on Ecological Characteristics and Conservation of Yellow-Throated Marten (*Martes flavigula*) in Temperate Forests of Korea. Ph.D. Thesis, Seoul National University, Seoul, Korea, 2014.
35. Rumelhart, D.E.; McClelland, J.L.; Group, P.R. *Parallel Distributed Processing*; MIT Press: Cambridge, MA, USA, 1987; Volume 1.
36. Hines, J.; Tsoukalas, L.H.; Uhrig, R.E. *Matlab Supplement to Fuzzy and Neural Approaches in Engineering*; John Wiley & Sons, Inc.: Hoboken, NJ, USA, 1997.

37. Althuwaynee, O.F.; Pradhan, B.; Lee, S. A novel integrated model for assessing landslide susceptibility mapping using chaid and ahp pair-wise comparison. *Int. J. Remote Sens.* **2016**, *37*, 1190–1209. [CrossRef]
38. Kim, J.-C.; Lee, S.; Jung, H.-S.; Lee, S. Landslide susceptibility mapping using random forest and boosted tree models in Pyeong-chang, Korea. *Geocarto Int.* **2017**, 1–16. [CrossRef]
39. Lee, S.; Kim, J.-C.; Jung, H.-S.; Lee, M.J.; Lee, S. Spatial prediction of flood susceptibility using random-forest and boosted-tree models in Seoul metropolitan city, Korea. *Geomat. Nat. Hazards Risk* **2017**, 1–19. [CrossRef]
40. Tahmassebipoor, N.; Rahmati, O.; Noormohamadi, F.; Lee, S. Spatial analysis of groundwater potential using weights-of-evidence and evidential belief function models and remote sensing. *Arab. J. Geosci.* **2016**, *9*, 79. [CrossRef]

applied
sciences

MDPI

Article

Road Safety Risk Evaluation Using GIS-Based Data Envelopment Analysis—Artificial Neural Networks Approach

Syyed Adnan Raheel Shah [1,2,*], Tom Brijs [1] , Naveed Ahmad [2], Ali Pirdavani [4] , Yongjun Shen [3] and Muhammad Aamir Basheer [1]

[1]   Transportation Research Institute (IMOB), Hasselt University, Diepenbeek 3590, Belgium;
       tom.brijs@uhasselt.be (T.B.); muhammadaamir.basheer@student.uhasselt.be (M.A.B.)
[2]   Taxila Institute of Transportation Engineering, Department of Civil Engineering,
       University of Engineering & Technology, Taxila 47050, Pakistan; n.ahmad@uettaxila.edu.pk
[3]   School of Transportation, Southeast University, Nanjing 210096, China; yongjunshen@outlook.com
[4]   Faculty of Engineering Technology, Hasselt University, Diepenbeek 3590, Belgium; ali.pirdavani@uhasselt.be
*   Correspondence: syyed.adnanraheelshah@uhasselt.be or shahjee.8@gmail; Tel.: +32-465-591-407

Received: 31 July 2017; Accepted: 22 August 2017; Published: 29 August 2017

Abstract: Identification of the most significant factors for evaluating road risk level is an important question in road safety research, predominantly for decision-making processes. However, model selection for this specific purpose is the most relevant focus in current research. In this paper, we proposed a new methodological approach for road safety risk evaluation, which is a two-stage framework consisting of data envelopment analysis (DEA) in combination with artificial neural networks (ANNs). In the first phase, the risk level of the road segments under study was calculated by applying DEA, and high-risk segments were identified. Then, the ANNs technique was adopted in the second phase, which appears to be a valuable analytical tool for risk prediction. The practical application of DEA-ANN approach within the Geographical Information System (GIS) environment will be an efficient approach for road safety risk analysis.

Keywords: road safety; risk evaluation; data envelopment analysis; artificial neural networks; crash data analysis

---

1. Introduction

Crash injury severity has always been a major concern in highway safety research. To model the relationship between crash occurrence along with severity outcomes, related traffic features, and contributing factors, a large number of advanced models have been proposed. Road safety research incorporates a broad exhibit of research territories, and the most successful of them is crash information investigation. There have been a lot of discussion about crash information-based safety analysis and other distinguishable activity attributes have been proposed, more regularly than crashes, as an option. In any case, investigation of crash information remains the most broadly received way to deal with the safety of a transportation system (e.g., expressways, arterials, crossing points, etc.). The traditional approach is to build up connections between crash recurrence, traffic flow attributes, and geometry of the roads [1]. On the one hand, the impact of the geometric design on the probability of a driver behavior has been very much archived in conventional safety studies. This course of research is useful in settling on choices in such things as installing cautioning signs on roadway areas, etc. On the other hand, Average Annual Daily Traffic (AADT) is a generally used indicator for measuring the traffic movement conditions, as it is recorded by most organizations around the nation/the world, is accessible to all roadway areas, and gives a measure of introduction to the specific roadway segment. Crash recurrence examination in view of AADT is a total or aggregate approach to take a glance at the crash information where the recurrence of crashes is computed, by amassing the crash information over particular eras (months or years) and areas (particular roadway segments) [2].

During road safety analysis of a road, a major target is to locate those segments which are dangerous, and then to identify the factors influencing its safety level. This study focuses on the concept that crashes can be decreased by better assessment of road hazard incremental elements, and by recognizable proof of hazardous segments at the initial stage, and after that, assessment of very dangerous sections with reference to the major contributing components is conducted in the second stage. In doing so, a combination of Data Envelopment Analysis (DEA) with Artificial Neural Networks (ANNs) is applied to evaluate the performance of roads with reference to safety conditions. The outcome is able to help decision makers/safety engineers to build a valuable system to analyze risk and significant attributes. Although it is new in the road safety research field, such an integrated mechanism has been popular in other sectors like banks, hospitals, schools, and corporations. Some researchers have used a combination of DEA-ANNs to evaluate performance (efficiency/risk) of rail transport, power suppliers, etc. [3–5]. DEA-ANNs was also used for efficiency classification by different researchers for banks and corporate companies [5–7]. For analysis regarding hospitals and large companies, screening of training data was also evaluated by using the DEA-ANNs technique. In addition, DEA-ANNs was also introduced for data processing [8–13], and will be more useful when applied within a Geographical Information System (GIS) environment. In this study, this integrated concept, which is popular in other sectors, is introduced to evaluate the safety performance (risk evaluation) of motorways. This evaluation of risk helps decision makers to decide on economical investment for risky segments, along with related factors, and consequently to reduce the cost of risk evaluation.

2. Literature Review

2.1. Risk and Road Safety Analysis

Usually, road safety performance is evaluated on the basis of 'Risk' which is associated with the number of crashes and casualties, known as the road safety outcome. In the field of road safety, the risk is defined as 'the road safety outcome to the amount of exposure' as shown in Equation (1):

$$Risk = \frac{Road\ Safety\ Outcome}{Exposure} \tag{1}$$

Exposure can be measured using different parameters; while comparing the performance of road segments, it can be measured as vehicle miles traveled, vehicle hours traveled, volume and number of trips, etc., however for countries it can be passenger kilometers travelled, population and number of registered vehicles, etc. [14,15]. Risk assessment is necessary for road safety performance analysis. Although risk can be analyzed on the basis of direct calculation using outcome by exposure, in the case of multiple outcomes and multiple input, it is difficult to deal with the calculation. Crashes are random events, and their outcome can also vary, as in one crash there may be zero fatalities, or fifty or more fatalities. Thus, a method that can deal with multiple outputs can be beneficial in calculating risk for road safety performance analysis of different units.

2.2. DEA for Road Safety Analysis

Road safety performance analysis of highways is an important task for the safety of travelers. To analyze the safety performance of certain attributes, a benchmarking mechanism has remained a basic procedure to be adopted by researchers [16–18]. With reference to the applied techniques for this purpose, DEA has been a popular technique with its theoretical basis in linear programming. Evolving the concept of DEA from research work in 1978, Charnes et al. [19] applied a linear program to estimate an empirical production technology frontier (bench marking) for the first time [19]. In the basic DEA model, the definition of the best practices relies on the assumption that inputs have to be minimized and outputs have to be maximized (such as in the economics field). However, to use DEA for road safety risk evaluation, the target becomes the output, i.e., the number of traffic crashes, to be as low as possible with respect to the level of exposure to risk. Therefore, the DEA frontier based Decision Making Units(DMUs) or the best-performing road segments are those with minimum output levels given the input levels, and other segments' risk is then measured relative to this frontier [20].

Mathematically, to use DEA for road safety evaluation, the model is shown as follows:

$$
\begin{aligned}
\min R_0 \quad &= \quad \sum_{r=1}^{s} u_r y_{r0} \\
s.t. \quad &\quad \sum_{i=1}^{m} v_i x_{i0} = 1, \\
&\quad \sum_{i=1}^{m} v_i x_{ij} - \sum_{r=1}^{s} u_r y_{rj} \leq 0, \quad j = 1, \cdots, n \\
&\quad u_r, v_i \geq 0, \quad r = 1, \cdots, s, \quad i = 1, \cdots, m
\end{aligned}
\tag{2}
$$

where y_{rj} and x_{ij} are the rth output and ith input respectively of the jth DMU, u_r is the weight given to output r, and v_i is the weight given to input i.

In view of the model applications for road safety analysis, road safety condition was compared to 21 European countries [16] and an ideal trauma management record score was also calculated by using DEA [21]. Furthermore, using population, passenger-kilometers, and passenger cars as inputs, and the number of fatalities as output, DEA was used for the evaluation of risk level of countries [22]. Monitoring of yearly progress in road safety was also conducted by utilizing the DEA technique [23]. Adding to road safety determination on a national level for 27 Brazilian states, two fundamental indicators accessible in Brazil: death rate (fatalities per capita) and casualty rate (fatalities per vehicle and fatalities per vehicle kilometer traveled) were focused upon [24]. From the literature review on DEA application in the field of road safety, it was confirmed that DEA is one of the established techniques to evaluate the risk level of road safety.

2.3. ANNs for Road Safety Analysis

Artificial Neural Networks is a model instrument of nonlinear statistical data that can be used to model a complex relationship between input and output to seek patterns. ANNs has been often implemented in many fields of science for prediction [25]. In road safety research, ANNs was applied to investigate crashes with reference to driver, vehicle, roadway, and condition attributes [26]. After application of ANNs, the impact of factors like seatbelt usage, light condition, and driver's liquor utilization on driver's safety was evaluated [27]. ANNs was also applied to determine the relationship between crash severity and the model parameters including years, highway sections, section length (km), AADT, the degree of horizontal curvature, the degree of vertical curvature, heavy vehicles (percentage), and season summer (percentage). The results shown that degree of vertical curvature has strong impact on number of crashes [28]. By modeling AADT, SL (Posted speed limit), Gradient (Average segment gradient), and Curvature (Average segment curvature) against road crashes, it was concluded that ANN was superior to multivariate Poisson-lognormal models [29]. From the literature review, we can summarize that ANNs was previously used as a crash data analysis model, which was a useful technique to study road-related features, geometry, and other contributing factors to road safety.

2.4. DEA-ANNs Approach

The combination of DEA and ANNs has not been applied in the road safety field, but it is popular in other fields like banks and corporate sectors. From the previous studies it was concluded that DEA is powerful for efficiency calculation, but for prediction purposes ANNs is ahead, so a discussion started after [30] on combining these two techniques to obtain the best possible outputs, i.e., efficiency calculation for ranking and prioritizing and then efficiency prediction for factor analysis purpose. To validate this combination, efficiency prediction was performed for 50 companies [31], 19 power plants [32], 49 Indian business schools [5], 102 bank branches [7], and 45 countries [33]. Efficiency classification was also tested by studying 142 bank branches [34] and 23 supplier companies [35]. Following the similar pattern, the DEA-ANNs approach is selected in this study for road safety risk evaluation and analysis of factors affecting risk.

2.5. GIS for Road Safety Analysis

"While geometrical concept can be enriched by culture-specific devices like maps, or the terms of a natural language, underneath this variability lies a shared set of geometrical concepts. These concepts allow adults and children with no formal education, and minimal spatial language, to categorize geometrical forms and to use geometrical relationships to represent the surrounding spatial layout." (Elizabeth S. Spelke-Harvard). GIS has gained a reputation that provides a better visualization of a large data set for understanding and decision making processes. GIS-provided maps which helped in identifying the crash concentration areas, located along the major road in the main urban areas [36]. During the road safety analysis of motorway (M-25), GIS provided relevant data on road accidents, traffic and road characteristics for 70 segments [37]. High risky sections on the basis of potential crash cost for expressways of Shanghai with the application of GIS has been clearly mapped [38]. Zonal crash frequency has also been expressed through GIS, showing association with several social-economic, demographic, and transportation system factors [39]. Through spatial analysis of high risk areas, pedestrian crashes have also been mapped in Tehran [40]. In Belgium, through the use of GIS and point pattern techniques, mapping road-accident black zones has been conducted within urban agglomerations [41]. GIS has also been used to explore the spatial variations in relationship between Number of Crashes and other explanatory variables of 2200 Traffic Analysis Zones (TAZs) in the study area, Flanders, Belgium [42,43]. GIS was used for modelling crash data at a small-scale level in Belgium, which permitted the identification of several areas with exceptionally high crash data. It endorsed more effective reallocation of resources and more efficient road safety management in Belgium [44].

2.6. ANN-GIS Approach

ANNs has been introduced as a mapping tool to GIS to perform a predictive capability for joint operations [45]. Although GIS in combination with ANNs was popular in the fields of geoscience, irrigation, meteorology and Agriculture, it has been tested in the field of road safety by applying deep learning models using a Recurrent Neural Network (RNN) to predict the injury severity of traffic crashes for the North-South Expressway-Malaysia [46]. Previously this technique had been applied for sediment prediction in Gothenburg harbor [45], landslide susceptibility using the landslide occurrence factors produced with the help of a ANNs model [47], detection of flood hazards in the Blue Nile, White Nile, Main Nile, and River Atbara [48], macrobenthos habitat potential mapping regarding *Macrophthalmus dilatatus*, *Cerithideopsilla cingulata*, and *Armandia lanceolate* [49], learning the patterns of development in the region [50], tunneling performance prediction required in routine tunnel design works and performance in terms of stability as well as impact on surrounding environment [51], and deforestation maps production to determine the relationship between deforestation and various spatial variables such as the vicinity to roads and to expenditures, forest disintegration, elevation, slope, and soil type [52].

2.7. Research Gap

DEA is popular as an optimization tool with its theoretical background in linear programming. DEA is popular with reference to benchmarking mechanism for efficiency and risk evaluation [3,6]. Previously, DEA was popular with its multi stage properties, but it has shortcomings with respect to its prediction capabilities, which reduces its application. A powerful technique, ANN, has been joined with DEA to fill that gap. Finally, with the predictive potential of ANNs and the optimization capacity of DEA performing complementary features, a prominent modeling option is envisioned [3,6]. The performance of the DEA-ANNs technique in the field of road safety for decision making mechanisms for road safety performance analysis was evaluated. This is the first study for an application of the DEA-ANN approach within a GIS environment for road safety performance analysis, using a case study on Motorways. This will lead traffic engineers and decision makers to better visualize the risky sections and key factors for road safety condition improvement.

3. Materials and Methods

3.1. Basic Framework of the Analysis

Road authorities have to prioritize the sites which require safety treatment, due to budget limitations. So in this study, a two phase framework was proposed for road safety risk evaluation, as shown in Figure 1. In the first phase, the number of crashes and fatalities was evaluated against exposure variables, with the help of DEA to calculate the risk level of road segments. In the second phase, that risk was predicted and evaluated with the help of ANNs.

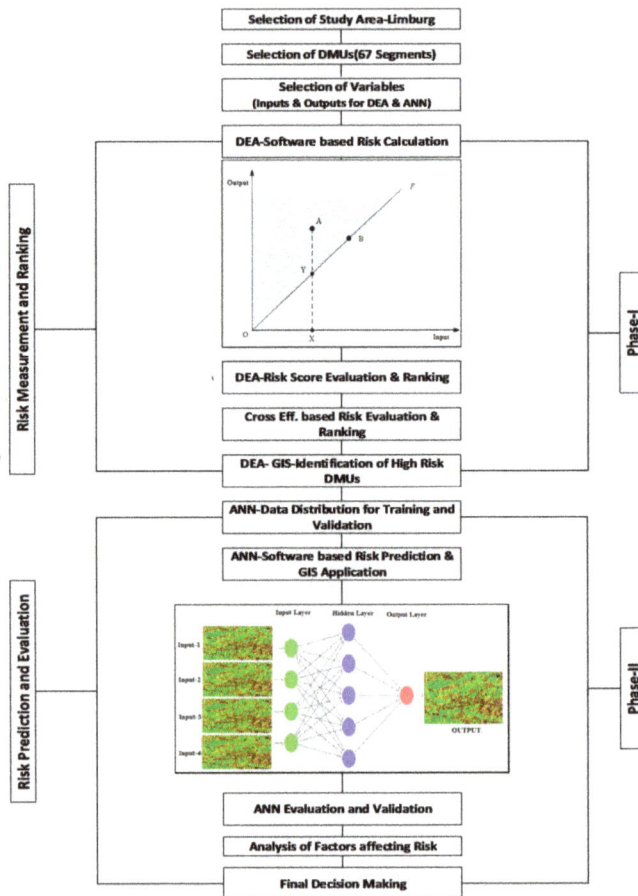

Figure 1. The Proposed Data Envelopment Analysis-Artificial Neural Networks (DEA-ANNs) Framework for Risk Evaluation.

3.2. Data Description and Selection of Variables

The study area selected for this study was two motorways in Belgium named E-313 and E-314 (Limburg Province Sections with a total length of 103 km). Each Motorway has segments, traffic-related characteristics, and road network segmentation derived from the FEATHERS model [53]. In this study, a segment with at least one crash was considered as a decision-making unit (DMU) to analyze the road safety condition. According to this criterion, 67 segments are selected for these two motorways. The crash data used in this study consisted of a geographically coded set of crash data that occurred between 2010 and 2012, which was provided by the Flemish Ministry of Mobility and Public Works, as shown in Table 1. The first and very critical step in conducting an analysis is the selection of inputs and outputs variables. For this purpose in the first stage (DEA), those variables which were the exposure variables and could not be directly affected by a traffic engineer/decision maker were selected to calculate risk, while in the second stage (ANN) those variables (i.e., Horz and Vert Curve design, speed, and flow) which could be altered or improved by directly changing certain parameters, were selected. So, the target while calculating risk was to reduce the number of crashes (NoC) and casualties (NoAP) with the increase of average volume to capacity on each segment (V/C), total daily vehicle miles travelled on each segment (VMT) and total daily vehicle hours travelled on each segment (VHT). A traffic engineer cannot directly change V/C, VMT, or VHT; however, the geometric design (Horz and Vert Curve), speed (speed limit) and flow (by controlling access) so practically, a selection of variables was targeted according to the feasibility of the problem's solution. To confirm the validity of the DEA model condition, an isotonicity test [54] was conducted. An isotonicity test comprises the intention of all inter-correlations between inputs and outputs for detecting whether increasing amounts of inputs lead to greater outputs. As positive correlations were established, the isotonicity test was accepted and the presence of the inputs and outputs was reasonable. However there are no diagnostic checks for improper model specification detection in DEA [55]. However, a general rule of thumb, the minimum number of DMUs is higher than three times the number of inputs plus outputs [56]. In our study with a total of three inputs and two outputs, so a set of 15 data points would be optimal; we have 67 data segments.

Table 1. Description Statistics of the Variables.

Stage	Variables	Description	Mean	SD	Min.	Max.
1st Stage DEA	NoC	No. of Crashes	9.58	13.12	1	74
	NoAP	No. of Affected Persons (Injured and Killed)	14.36	19.55	1	105
	V/C	Average Volume to Capacity on each segment	0.4405	0.1807	0.08	0.6435
	VMT	Total daily Vehicles Miles Travelled on each Segment	1828	1388	77	5186
	VHT	Total daily Vehicles hours Travelled on each Segment	1093	879	38	3616
2nd Stage ANN	Flow	Average annual daily traffic on each segment (vph)	968.1	449.6	31.5	1483.4
	Speed	Average Travel Speed for each segment (kph)	110.99	8.23	96.89	120
	Horz_Curve	0 = Tangent, 1 = Curve	–	–	0	1
	Vert_Curve	1 = Upward, 2 = Downward, 3 = Flat	–	–	1	3

3.3. Phase-I: Application of DEA for Risk Calculation and Ranking

As there were two major phases of modeling, we had decide on the variables for both phases. The initial target was to evaluate risk with reference to the variables that were basically exposure variables. In the basic DEA model, the definition of the best practice relied on the assumption that inputs had to be minimized, and outputs have to be maximized (such as in the economics field). However, to use DEA for road safety risk evaluation, the target became the output, i.e., the number of traffic crashes, to be as low as possible with respect to the level of exposure to risk.

There are two basic concepts in application of DEA, starting from efficiency as in Equation (3), and converting into calculation of risk as shown in Equation (4).

Efficiency: The basic concept of DEA-Efficiency calculation is as follows:

$$Efficiency = \frac{Weighted\ Sum\ of\ Output}{Weighted\ Sum\ of\ Input} = \frac{Maximize\ Output}{Minimize\ Input} \tag{3}$$

Risk: The basic concept of DEA-Risk calculation in connection between Equations (1) and (3):

$$Risk = \frac{Weighted\ Sum\ of\ Output}{Weighted\ Sum\ of\ Input} = \frac{Minimize\ Output}{Maximize\ Input} = \frac{Road\ Safety\ Outcome}{Exposure} \tag{4}$$

So the equation to calculate the Risk value through DEA is as follows:

$$Risk = \frac{U_2(NoC) + U_1(NoAP)}{V_1(V/C) + V_2(VMT) + V_3(VHT)} \tag{5}$$

where U_1 = weights for 1st output (NoC), U_2 = weights for 2nd output (NoAP); V_1 = Weights for 1st Input (V/C), V_2 = weights for 2nd Input (VMT), V_3 = weights for 3rd Input (VHT).

After calculation of Risk value, for ranking purposes, a cross-efficiency approach was one of the best methods to calculate A cross-risk value for ranking purposes. DEA has an attractive feature in that each DMU can have its own input and output weights, which leads to difficulty in making a comparison between DMUs. To compare DMUs, a Cross efficiency matrix (CEM) was developed as a DEA extension tool to assist in identifying the overall best or worst performer among all DMUs and rank them. Its basic idea is to apply DEA in a peer assessment instead of a self-assessment mode. Specifically, the CEM calculates the performance of a DMU with a concept by using not only its own optimal input and output weights, but also those of all other DMUs. Results can then be accumulated in a CEM as shown in Table 2. In the CEM, the element in the *ith* row and *jth* column signifies the risk scores of DMU *j* using the optimal weights of DMU *i*. The basic DEA risk is thus positioned in the principal diagonal. The average of each column of the CEM is calculated as a mean cross risk value for each DMU [20]. Since the same weighting process is applied for all the DMUs, their evaluations can then be made on a comparison basis, with a higher cross-risk score indicating a higher risky DMU.

Table 2. A Generalized Cross-Efficiency Matrix (CEM) [20].

Rating DMU	Rated DMU				
	1	2	3	... ...	n
1	E_{11}	E_{12}	E_{13}	... ...	E_{1n}
2	E_{21}	E_{22}	E_{23}	... ...	E_{2n}
.	.	.			.
n	E_{n1}	E_{n2}	E_{n3}	... ...	E_{nm}
Mean	$\overline{E_1}$	$\overline{E_2}$	$\overline{E_3}$	... ...	$\overline{E_n}$

For those DMUS, which have illogical weights in the basic DEA model, a relatively low or higher risk value will be calculated. Therefore, for ranking purpose, this method serves a type of sensitivity analysis by applying a method of a different set of weights to each DMU, with a back channel mechanism of self-generated weights rather than an externally imposed [20]. So the target value, which is a value of 1 to be considered for the best DMU, can now be changed, and after application of CEM it can vary, but the selection of best DMU (with the lowest Risk) will be easier.So after applying model (1) for calculating risk R_0 in road safety field, the lowest level has been considered as the frontier of safety. As explained above, for ranking purposes, a cross risk procedure [20] has been adopted to obtain the best ranking, as shown in Table 3.

The major advantage of using DEA here is that it can handle multiple inputs and multiple outputs. Moreover, DEA has some benefits as it does not require an assumption of a functional form relating inputs to outputs; DMUs considered in DEA are directly compared against a peer or combination of peers; Inputs and outputs used in DEA can have different measurement units. In this study, number of crashes (NoC) and number of affected persons-injured or killed (NoAP) are considered as two outputs, while exposure variables—average volume to capacity on each segment (V/C), total daily vehicle miles travelled on each segment (VMT), and total daily vehicle hours travelled on each segment (VHT)were considered as three inputs. Although the segment length also varied, it was not included here because it was already been involved in the backup calculation of VMT.

Table 3. DEA-Based Risk Evaluation and Ranking Segments.

DMUs	Input 1	Input 2	Input 3	Output 1	Output 2	CE-RISK VALUE	RANK
Road Seg.	V/C	VMT	VHT	NoC	NoAP		
1	0.368518	3039.221	1541.607	74	105	91.06902	1
29	0.139052	109.169	54.58458	6	8	71.72984	2
19	0.603085	183.7303	118.162	12	20	69.92395	3
2	0.384021	2494.327	1268.376	49	76	65.10151	4
34	0.07999	82.51051	41.25526	3	6	62.90294	5
5	0.277711	2190.904	1096.683	38	50	62.28254	6
25	0.139052	76.73981	38.36996	3	6	58.10739	7
26	0.236548	202.3093	101.2386	9	11	57.15026	8
3	0.360649	2683.937	1361.267	40	61	53.2604	9
21	0.53409	594.7792	336.9631	13	24	35.40002	10
-	-	-	-	-	-	-	-
53	0.631117	4275.086	3046.694	2	3	1.47492	64
67	0.592324	1093.968	734.8267	1	1	1.312419	65
49	0.498964	3214.268	1780.697	1	2	1.08319	66
66	0.574944	1714.219	1098.003	1	1	1.068806	67

Based on Model (1), the range of risk value began at 1 and proceeded to a higher value, so a segment with a value of 1 was considered safest, while the road segment with the highest value was considered the most dangerous. Moreover, the cross risk method [20] was used to make all the DMUs comparable. Table 3 presents the results. As the ranking was done on a priority basis to evaluate the safety condition of that segment, the risk value of 91.07 was the highest value in the table and was ranked first (i.e., the most risky segment). The top 10 riskiest segments are shown in Table 3 to explain an idea of risky segment selection for improvement.

Furthermore, risk value was normalized by applying natural log, and with the help of GIS, a complete spatial map of both motorways is shown in Figure 2. A straight line demonstration provides an insight in locating the most riskiest segment on a motorway or highway.

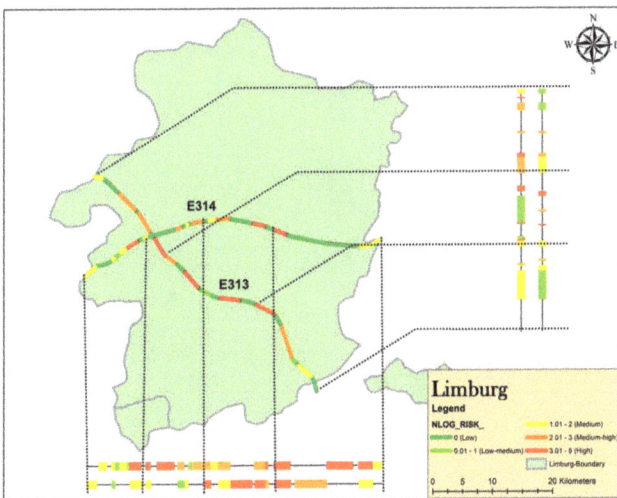

Figure 2. Risk based Straight Line Map for Motorways (E-313&314).

3.4. Phase-II: Application of ANNs Model for Risk Prediction and Evaluation

In the second phase, the dependent variable is the risk value generated by the DEA model, was transformed by applying natural log to have data normalized. For independent variables, speed could be controlled by controlling the speed limit; flow was directly related to the number of vehicles, and could be controlled by controlling access; horizontal curve could be removed or altered as per infrastructural changes, and the same was the case for vertical curve as a geometric design feature. So for the application of ANN, data was distributed on the basis of a K-Fold mechanism with five folds (i.e., distribution is as 53 segments-DMUs for Training and 14 segments-DMUs for validation).

ANNs, unlike other modeling platforms, requires some form of model validation to aid in the model-building process and to help prevent overfitting of the model. The basic idea behind validation (or cross-validation) is to hold a subset of the data out of the model-building process. This process forms two partitions of the data, a training set and a validation set (note that a third set, or test set, can also be used). The model is built using the training set, while the k-fold validation set is then used to assess how well the model performs, and to aid in model selection. The most mainstream decision for the quantity of concealed layers is used. A solitary concealed layer is typically adequate to catch even extremely complex connections between the indicators. The quantity of links in the shrouded layers likewise decides the level of multifaceted nature of the connection between the indicators that the system catches [57].

From one viewpoint, utilizing an excessive couple of links is not adequate to catch complex connections (e.g., review the unique instances of a straight relationship in direct and calculated relapse, in the extraordinary instance of zero links or no shrouded layer). Then again, an excessive number of links may prompt overfitting. A dependable guideline is, to begin with (number of indicators) links and reduce or increment gradually while checking for overfitting. Another approach is to start with the default neural model, with one layer and four nodes, and then run a much more complex model with two layers and several nodes, and different activation functions. If the fit statistics do not improve substantially with a more complex model, then a simpler model may suffice [57]. We applied a simpler model to check the performance of the ANNs model in our case of risk evaluation, as shown in Figure 3. After running the model as shown in Figure 3, we displayed the model structure. We saw input variables mapping to each of the activation functions in the hidden layer, and nodes in the hidden layer mapping to the output layer. The background mechanism in each of the nodes in the hidden layer designated that the Gaussian activation function was used. Model results for both the training and validation sets are shown in Table 4. The response variable (risk) for this model was continuous. Like other techniques, it was necessary to follow the validation mechanism. With the validation mechanism and separation of the data into two sets, unbiased results were provided.

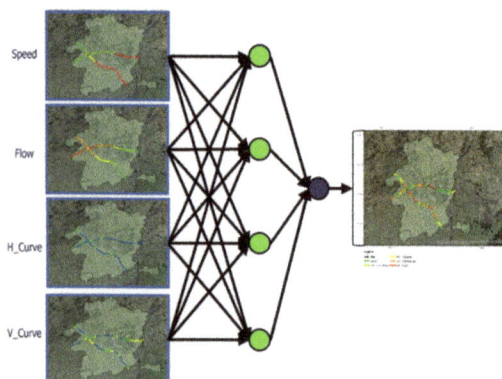

Figure 3. Risk Prediction GIS Map based on the ANNs Model.

In this study, as discussed earlier, original data were distributed into two parts. Out of 67 segments, 53 as the major data set were used for the training of the ANNs setup, while the remaining 14 were used for validation after model building. The training set was the part that estimated model parameters. The validation set was the part that assessed or validated the predictive ability of the model. In addition, the most critical validation was applied in this study. Specifically, the K-Fold technique was adopted, which divided the original data into K subsets. In turn, each of the K sets is used to validate the model fit on to the rest of the data, fitting a total of K models. The model giving the best validation statistic was chosen as the final model. This method was best for small data sets because it made efficient use of limited amounts of data [57,58].The ANN based predicted value of risk was mapped in the GIS environment as shown in Figure 4, showing a red line as the riskiest segments, while dark green segments are the safest as there zero crashes on these segments.

Table 4. Parametric estimates of the ANNs Model.

Parameters	Estimates-Hidden Layer				
	Code	H1_1	H1_2	H1_3	H1_4
Flow		0.258908	−2.00717	0.868246	4.984629
Speed		−1.26756	3.435496	−1.83834	2.048267
Horz_Curve	0	2.150204	13.66468	−1.03056	1.968045
Vert_Curve	1	2.141838	−2.07175	1.313892	0.014534
	2	−3.21511	7.986301	−3.09312	−0.53461
Intercept		1.90514	−1.87443	0.481882	−4.76064
	Int	H1_1	H1_2	H1_3	H1_4
NLog_Risk	2.221	−2.34861	6.612427	−1.8978	2.041027
Cross Validation					
Sample Size	Training	53	Validation	14	
R² (Training)	0.788	R² (Validation)	0.775	RMSE	0.624

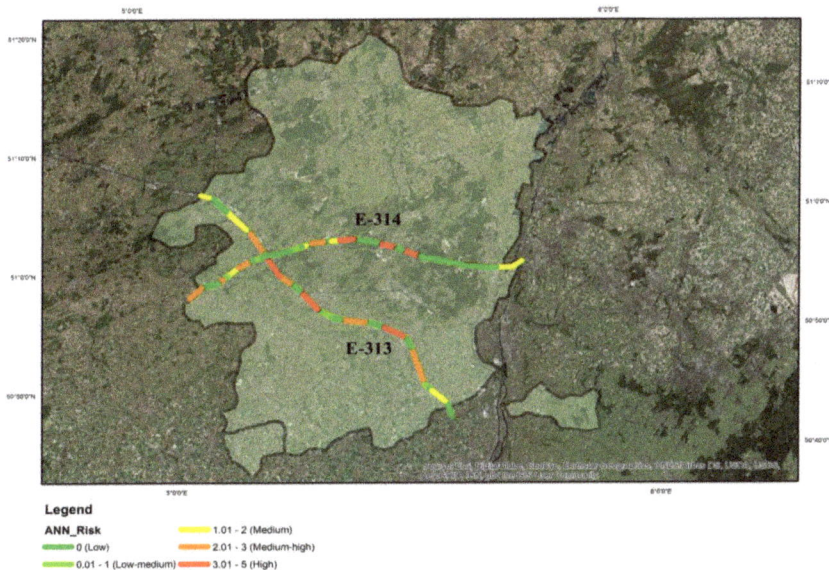

Legend

ANN_Risk
- 0 (Low)
- 0.01 - 1 (Low-medium)
- 1.01 - 2 (Medium)
- 2.01 - 3 (Medium-high)
- 3.01 - 5 (High)

Figure 4. Geographical information system (GIS)-based ANNs-predicted risk spatial map.

The values of R-square and Root Mean Square Error (RMSE) are the two basic validation indicators for testing the goodness-of-fit of the model. ANNs is a very flexible model and has a tendency to overfit data. When that happens, the model predicts the fitted data very well, but predicts future observations poorly. To mitigate overfitting, the neural platform applies a penalty on the model parameters and uses an independent dataset to assess the predictive power of the model. The applied technique to control overfitting is the squared method. This method is applied if it is considered that independent variables are contributing to the predictive ability of the model.

During the analysis, the graphical representation of data showed a better performance both in the case of training and validation data. The data distribution was adopted in five segments, having a distribution of 53 segments for training and 14 segments for validation. The plots showing the perfection of predictability were shown in Figure 5 for both training and validation data. The values of R-square were also almost similar for both major and training and validation data sets.

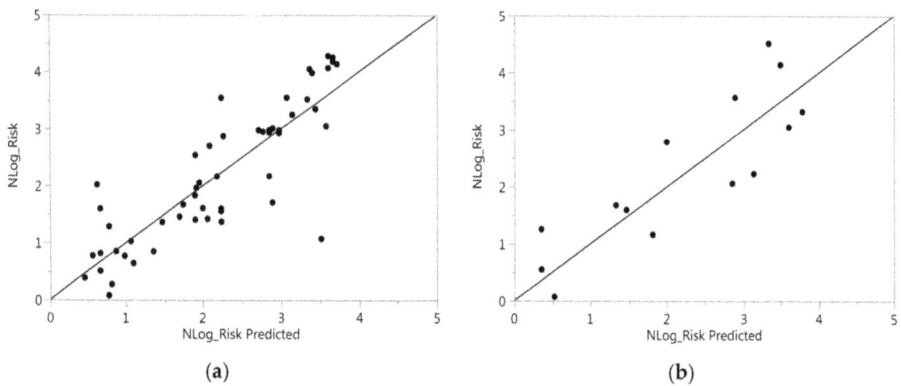

(a)   (b)

Figure 5. Actual By Predicted Plot (**a**) Training (**b**) Validation.

The contribution of factors associated with risk can be analyzed by the importance of the variables (i.e., Flow, Speed, Vertical and Horizontal Curve).The impact of variables is one of the necessary targets to analyze and improve the safety performance of the roads. Traffic safety engineers always search for the relationships between factors and safety performance indicators (i.e., Risk). The relationship can be observed in Table 5, which shows that speed and flow were two major factors which are having a high impact on risk.

Table 5. Factors Association with the Risk.

Factor	Main Effect	Total Effect	Comparison
Flow	0.224	0.908	
Speed	0.064	0.47	
Vert_Curve	0.072	0.426	
Horz_Curve	0.052	0.288	

A comprehensive analysis to overview the importance of factors provides traffic engineers to take a decision during road safety analysis and implementation procedure.

3.5. Model Selection Criteria

In order to assess the performance of the DEA-based Risk prediction models, a number of evaluation criteria were used to evaluate these models. These criteria were applied to measure how close the real values were to the values predicted using the developed models. They included Root

Mean Square Error (RMSE) and the correlation coefficient R or R^2. These are given in Equations (6) and (7) respectively [59].

$$RMSE = \sqrt{\frac{1}{n} \sum_{i=1}^{n} (y_i - \hat{y}_i)^2} \tag{6}$$

$$R = \frac{\sum_{i=1}^{n} (y_i - \bar{y})(\hat{y}_i - \bar{\hat{y}})}{\sqrt{\sum_{i=1}^{n} (y_i - \bar{y})^2} \sqrt{\sum_{i=1}^{n} (\hat{y}_i - \bar{\hat{y}})^2}} \tag{7}$$

where y is actual Risk values, $\hat{y}$ is the estimated Risk values using the proposed techniques, and n is the total number of observations of DMUs.

4. Results

In order to find the factors influencing the road safety risk, ANNs and multiple linear regression (MLR) were generated using the Road Traffic and Crash data obtained for European routes (E-313&E-314) of Limburg (Belgium). Although the basic target was to implement the ANNs model, regression analysis was also conducted to assess the performance of ANNs.

4.1. Performance of Model

The main objective of the methods (ANNs and MLR) was to fit an accurate model for risk prediction. The adequacy of such models are typically measured either by the coefficient of determination of the predictions against actual values (R^2) or by RMSE. Figure 6 shows the comparative diagram of prediction between ANNs and MLR.

The graph shown in Figure 6 suggests that ANNs is a better predictor than MLR. Moreover, if we considered the comparative assessment of model predicting capability, we see that R^2 from ANNs (0.788) is much higher than that from MLR (0.276). Another major tester of the capability of the model is the RMSE; a smaller value indicating better fit. It also indicated that ANNs (0.624) has performed better in comparison with MLR (1.0789), as shown in Table 6.

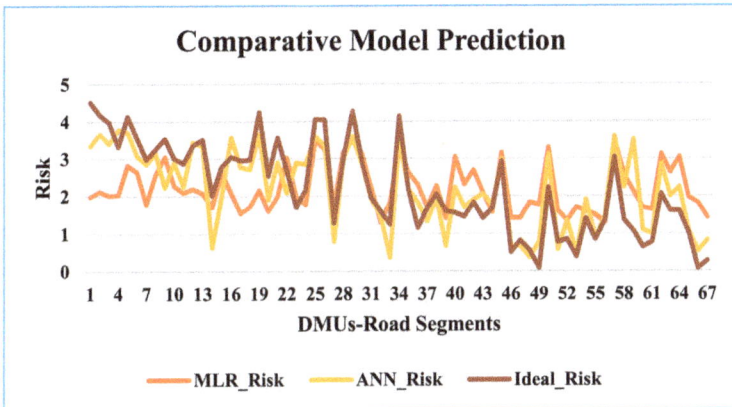

Figure 6. Comparative Analysis for Predicted Vs Actual Risk Values.

Table 6. Comparative Analysis of ANNs Vs Multiple Linear Regression (MLR).

Model	R^2 Predicted	R^2 (K-Fold) Validation	RMSE
Sample Size	53	14	
ANN	0.788	0.774	0.624109
MLR	0.276	0.147	1.0789985

Note: RMSE = Root Mean Square Error.

4.2. Analysis of Factors

As far as a solution to the problem is concerned, we can also analyze data on a graphical basis were the relationship can serve as a better understanding of our problem. After successfully applying the DEA-ANNs model for the road safety risk evaluation, we focused on the contributing factors used in the risk prediction. Decision makers/traffic safety engineer aim for low-cost treatments for problematic/risky segments. Thus from graphical analysis of the contributing factors, we saw that the majority of the crashes were on the curved portions of the motorways. Decision makers usually avoid infrastructural changes because redesigning and reconstruction is a costly procedure, so if they focus on the low-cost treatments, they can focus on speed and flow control. Figure 7 presents the relationship between the risk and the different contributing factors. The red lines represent mean speed levels, while the green lines represent mean flow level. We can see from Figure 7 that the risk level could be reduced by controlling just these two factors.

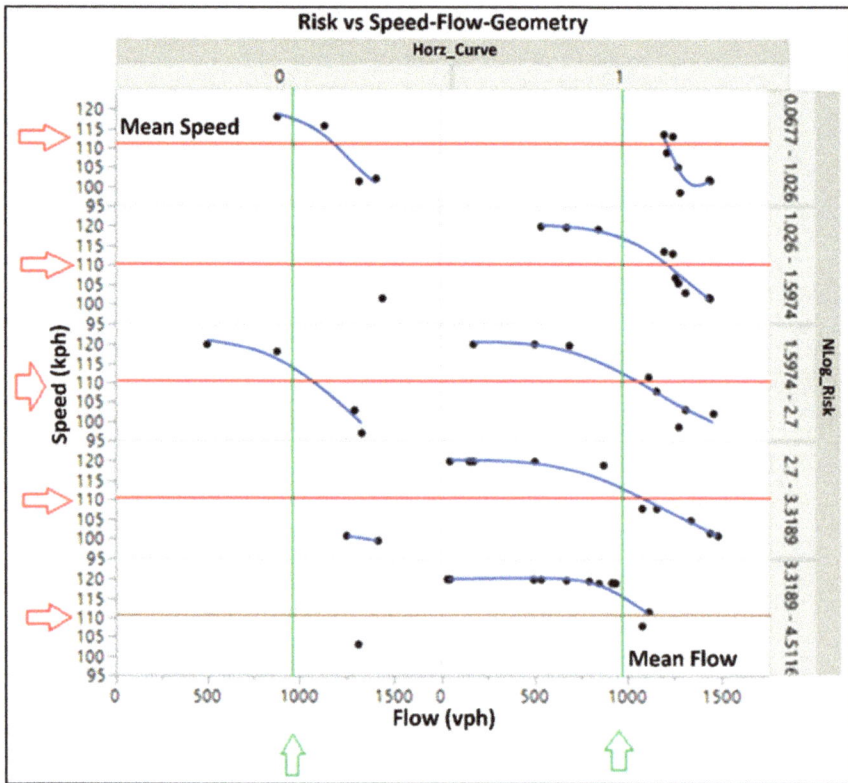

Figure 7. Contributing Factor based Risk Analysis.

4.2.1. Speed

In the case of motorways, a high speed limit is preferred to provide for free and easy maneuvering, but excessive speed is a very important factor having an impact on the number of crashes and injuries. In high-income countries, speed is one of the major factors (probably one third) of fatal and serious crashes [60–62]. We observed from the data that 35 out of 67 segments (52%) were above the mean speed limit 110 kph. So a reduction in speed limit could help in reducing risk level.

4.2.2. Flow

Flow is one of the major factors related to road safety, in parallel speed, the analysis showed that 39 out of 67 segments (58%) of the portion had above mean levels of traffic flow, i.e., 1000 vehicles. Traffic flow is one of the major contributing factor in road crashes [63,64]. "Based on the fluid mechanics theory of the traffic flow, the traffic flow parameters were specified, and the models of compressibility and viscosity of traffic flow were established respectively. Traffic control measures such as restricting the traffic flow at the upstream and downstream of the accident section should be carried out to control the crashes" [65]. So controlling the flow factor for the risky segments could assist in reducing the risk level of those segments.

4.2.3. Horizontal Curve

For the road safety analysis, the horizontal alignment designed cannot be ignored, especially the horizontal curve [66]. From previous research , the occurrence of accidents occurring on the curve is higher than the tangent (straight line), and it is necessary to design a horizontal curve [66]. In this analysis 80% of the risky level was along with the horizontal curves, so continuous marking of road marking signs for horizontal curves and straightening of curves can help in reducing crashes.

4.2.4. Vertical Curve

Research related to geometric characteristics showed that vertical curves had a significant effect on road crashes, and also while estimating speeds on highways [67]. Researchers also concluded that roads with vertical curves and higher speed limits tended to have more severe crashes [68]. Sometimes, a combination of horizontal and vertical curves is dangerous for road safety. The upward and downward gradient of the road contributed 76% in risk segment contribution, so a change in level could also help in reducing the risk level of road segments.

4.3. Safety Management and Financial Decision Making

Road safety management system and decision making is linked with econometrics i.e., funding and investments. Most countries need to enhance their understanding of spending on the significances of road safety, both by administration and organizations, and investment in road safety improvement. Road safety establishments need this knowledge to prepare financial and economic indication on the costs and usefulness of proposed solutions in order to win public and state support for funding road safety programs. There are prospects for targeted road safety funds that provide competitive revenues [69]. Road safety consultants and specialists develop business cases for this investment by applying such methods (i.e., the proposed DEA-ANNs method). A step change in funds invested in road safety management and in safer transport systems is compulsory to comprehend the success of motivated road safety targets in most of the world [69].

"Even though the implementation and maintenance costs of motorways vary significantly between the countries, in some cases also due to the different tendering systems, they are usually high, comparing to the implementation costs of other road safety road infrastructure related initiatives" [70]. During Cost-benefit analysis (CBA), the "cost-effectiveness of motorways also varies from case to case, especially due to the different implementation costs. In most cases, though, CBA results reveal relatively small ratios for new motorway development comparing to respective ratios regarding other road safety investments, mainly due to the very high implementation costs. However, even these ratios are considered as adequate to support the decisions for motorway development or the upgrade of existing rural network into motorways and apart from the strict financial criteria, the significant benefits for the road users can enhance the investment's effectiveness and should also be taken into account by the appropriate authorities" [70].

After safety analysis, we can target a different type of solutions: low-cost solutions, relatively costly solutions, and costly solutions. Speed limit change is considered as a low-cost solution because by changing sign boards for speed limit can help in the implementation of safety related alternatives. Consultants sometimes even recommend to installing permanent solution of electronic speed limit signs which help in controlling speed limit, some may be electronically related to the flow of the road and speed limit, to automatically change according to requirements. Flow limit is also another problem on the road, and can be solved by implementing the option of controlling access. The controlling flow option needs to have structures (e.g., toll installation) which lead to an investment higher than speed controlling signs. Flow can be controlled by applying tolls on those segments which are under high risk, which leads towards higher investments. Infrastructure change is one of the costly solutions during the safety solution process. Horizontal curve and vertical curve change can be backed by higher investments. Decision makers are always reluctant to change the structural pattern because a proper structural design change and construction is required to implement the decision. Sometimes, cost can increase by additional super elevation changes, in combination to horizontal and vertical changes.

4.4. Advantages and Limitations of Using the DEA-ANN Method

Since DEA offers some benefits to other approaches such "as "(1) DEA is able to handle multiple inputs and outputs (2) DEA does not require a functional form that relates inputs and outputs (3) DEA optimizes on each individual observation and compares them against the "best practice" observations. (3) DEA can handle inputs and outputs without knowing a price or knowing the weights and (4) DEA produces a single measure for every DMU that can be easily compared with other DMUs and also have some limitation as (1) DEA only calculates relative efficiency measures and (2) As a nonparametric technique statistical hypothesis test are quite difficult" [71,72]. "Neural networks offer a number of advantages, including requiring less formal statistical training, ability to implicitly detect complex nonlinear relationships between dependent and independent variables, ability to detect all possible interactions between predictor variables, and the availability of multiple training algorithms. Disadvantages include its "black box" nature, greater computational burden, proneness to overfitting, and the empirical nature of model development" [73]. However, overfitting can be controlled by the penalty method. Previously, DEA was popular with its multi-stage properties, but it has shortcomings with respect to its prediction capabilities, which reduces/limits its application. So, a powerful technique, ANNs, has been joined with DEA to fill that gap. Finally, the predictive potential of ANNs and the optimization capacity of DEA perform complementary features, thus envisioning a prominent modelling option [3,6].

5. Conclusions

This study focuses on road safety risk evaluation and connection between risk recurrence with respect to contributing factors. To enhance the estimation accuracy, a joint technique has been proposed and applied to achieve the risk evaluation, i.e., a benchmarking mechanism of DEA in combination with a prediction model of ANNs has been introduced to the road safety field. A crash dataset extracted from the Flemish Road safety department is stratified by two factors: the number of total crashes and number of affected persons, and is utilized to exhibit the proposed model formation of DEA and neural network performance. Notwithstanding the over-scattered crash information and the high relationship between the crash frequencies of the distinctive damage degrees, the outcomes demonstrate that comprehensive neural systems beat the multiple linear regression, shows in fitting and prescient execution. It demonstrates the neural system's prevalence over linear regression.

Risk has been calculated with the help of DEA for two motorways. Calculating risk has another advantage if segment length or volume of traffic is high or low; if we had just analyzed on the basis of number of crashes, it would not be a fair way to evaluated the most problematic segments out of a length of highway. Thus, using maximum information to evaluate an overall risk, DEA is a better option. In addition, we can rank them on the basis of risk value, and we can select our priorities on the basis that it could lead us to better decision making. So, for selecting problematic segments, it is a great achievement if we are able to indicate the most dangerous (risky) segments.

The predictability of risk values were checked with the assistance of ANNs: speed, flow, and horizontal and vertical curve. These are the most important factors which could be influenced by decision makers/transportation engineers. Selecting the contributing factors and changing the speed limit and flow limitation for risky segments could provide low-cost safety solutions. On the other hand, an infrastructural change like an amendment in the horizontal and vertical curve can cost much. However, this system can also help with designing better solutions as no one would prefer to change the structure of an entire highway (i.e., if a 100 km long highway), thus, selecting the most problematic section and solving the safety problem of only those segments would also provide low-cost decision making outcomes. Furthermore, combining ANN with GIS in a road safety analysis system can further encompass the functionality of the ANNs and, at the same time, increase the set of potential applications of GIS. The main advantage of using an ANNs system within a GIS environment for road safety and crash analysis includes the collection, manipulation, and analysis of the crash related data, which can be used effectively and resourcefully. The results of the overlay functions and spatial analysis performed by a GIS can be used as the input and training settings of a neural network, while the results of the neural network may be deployed by a GIS to produce a geospatial output. Each spatial input data and outcome of the neural network can be easily accumulated, normalized, rescaled, re-projected, and overlaid. It may accept different kinds of parameters (e.g., class, ordinal, continuous and categorical) as input or output values, and can handle deficient data [74]. The system is extremely flexible and self-adaptive, and capable of incorporating any improvement new data set. So, a joint approach of DEA-ANN within a GIS environment can provide an easy and an efficient output for decision makers for road safety data analysis and decision making for safety improvement.

Acknowledgments: This research is jointly supported by TITE and IMOB. and sponsored by IMOB for publication. Authors would like to thank HE-Boong Kwon (USA), one of pioneer of DEA-ANN method for his valuable guidance.

Author Contributions: Ali Pirdavani, Tom Brijs and Syyed Adnan Raheel Shah conceived and designed the concept; Syyed Adnan Raheel Shah and Naveed Ahmad performed literature review; Syyed Adnan Raheel Shah, Yongjun Shen and Tom Brijs applied DEA and ANN model; Muhammad Aamir Bashir and Ali Pirdavani contributed in data extraction-GIS application and contributed in analysis tools; Syyed Adnan Raheel Shah wrote the paper.

Conflicts of Interest: The authors declare no conflict of interest.

References

1. Songchitruksa, P.; Tarko, A.P. The extreme value theory approach to safety estimation. *Accid. Anal. Prev.* **2006**, *38*, 811–822. [CrossRef] [PubMed]
2. Golob, T.F.; Recker, W.W.; Alvarez, V.M. Tool to evaluate safety effects of changes in freeway traffic flow. *J. Transp. Eng.* **2004**, *130*, 222–230. [CrossRef]
3. Kwon, H.B. Exploring the predictive potential of artificial neural networks in conjunction with DEA in railroad performance modeling. *Int. J. Prod. Econ.* **2017**, *183*, 159–170. [CrossRef]
4. Hsiang, H.L.; Chen, T.Y.; Chiu, Y.H.; Kuo, F.H. A comparison of three-stage DEA and artificial neural network on the operational efficiency of semi-conductor firms in Taiwan. *Mod. Econ.* **2013**, *4*, 20.
5. Sreekumar, S.; Mahapatra, S. Performance modeling of Indian business schools: A DEA-neural network approach. *Benchmarking* **2011**, *18*, 221–239. [CrossRef]
6. Kwon, H.B. Performance modeling of mobile phone providers: A DEA-ANN combined approach. *Benchmarking* **2014**, *21*, 1120–1144. [CrossRef]

7. Azadeh, A.; Azadeh, A.; Saberi, M.; Moghaddam, R.T.; Javanmardi, L. An integrated data envelopment analysis–artificial neural network–rough set algorithm for assessment of personnel efficiency. *Expert Syst. Appl.* **2011**, *38*, 1364–1373. [CrossRef]
8. Mostafa, M.M. Modeling the efficiency of top Arab banks: A DEA–neural network approach. *Expert Syst. Appl.* **2009**, *36*, 309–320. [CrossRef]
9. Emrouznejad, A.; Anouze, A.L. Data envelopment analysis with classification and regression tree—A case of banking efficiency. *Expert Syst.* **2010**, *27*, 231–246. [CrossRef]
10. Samoilenko, S.; Osei-Bryson, K.M. Using Data Envelopment Analysis (DEA) for monitoring efficiency-based performance of productivity-driven organizations: Design and implementation of a decision support system. *Omega* **2013**, *41*, 131–142. [CrossRef]
11. Çelebi, D.; Bayraktar, D. An integrated neural network and data envelopment analysis for supplier evaluation under incomplete information. *Expert Syst. Appl.* **2008**, *35*, 1698–1710. [CrossRef]
12. Kuo, R.J.; Wang, Y.C.; Tien, F.C. Integration of artificial neural network and MADA methods for green supplier selection. *J. Clean. Prod.* **2010**, *18*, 1161–1170. [CrossRef]
13. Pendharkar, P.C. A hybrid radial basis function and data envelopment analysis neural network for classification. *Comput. Oper. Res.* **2011**, *38*, 256–266. [CrossRef]
14. Al Haji, G. *Towards a Road Safety Development Index (RSDI): Development of an International Index to Measure Road Safety Performance*; Linköping University Electronic Press: Linköping, Sweden, 2005; p. 113.
15. Yannis, G.; Papadimitriou, E.; Lejeune, P.; Treny, V.; Hemdorff, S.; Bergel, R.; Haddak, M.; Holló, P.; Cardoso, J.; Bijleveld, F.; et al. *State of the Art Report on Risk and Exposure Data. SafetyNet, Building the European Road Safety Observatory, Workp 2 Deliv D2*; European Road Safety Observatory: Brussels, Belgium, 2007; p. 120.
16. Elke, H.; Tom, B.; Geert, W.; Koen, V. Benchmarking road safety: Lessons to learn from a data envelopment analysis. *Accid. Anal. Prev.* **2009**, *41*, 174–182.
17. Wegman, F.; Oppe, S. Benchmarking road safety performances of countries. *Saf. Sci.* **2010**, *48*, 1203–1211. [CrossRef]
18. Shen, Y.; Hermans, E.; Bao, Q.; Brijs, T.; Wets, G. Serious injuries: An additional indicator to fatalities for road safety benchmarking. *Traffic Inj. Prev.* **2015**, *16*, 246–253. [CrossRef] [PubMed]
19. Charnes, A.; Cooper, W.W.; Rhodes, E. Measuring the efficiency of decision making units. *Eur. J. Oper. Res.* **1978**, *2*, 429–444. [CrossRef]
20. Shen, Y.; Hermans, E.; Brijs, T.; Wets, G.; Vanhoof, K. Road safety risk evaluation and target setting using data envelopment analysis and its extensions. *Accid. Anal. Prev.* **2012**, *48*, 430–441. [CrossRef] [PubMed]
21. Shen, Y.; Hermans, E.; Ruan, D.; Wets, G.; Brijs, T.; Vanhoof, K. Evaluating trauma management performance in Europe: A multiple-layer data envelopment analysis model. *Transp. Res. Rec.* **2010**, *2148*, 69–75. [CrossRef]
22. Shen, Y.; Hermans, E.; Bao, Q.; Brijs, T.; Wets, G. Road safety development in Europe: A decade of changes (2001–2010). *Accid. Anal. Prev.* **2013**, *60*, 85–94. [CrossRef] [PubMed]
23. Shen, Y.; Shen, Y.; Hermans, E.; Bao, Q.; Brijs, T.; Wets, G.; Wang, W. Inter-national benchmarking of road safety: State of the art. *Transp. Res. Part C* **2015**, *50*, 37–50. [CrossRef]
24. Bastos, J.T.; Shen, Y.; Hermans, E.; Brijs, T.; Wets, G.; Ferraz, A.C.P. Traffic fatality indicators in Brazil: State diagnosis based on data envelopment analysis research. *Accid. Anal. Prev.* **2015**, *81*, 61–73. [CrossRef] [PubMed]
25. Williams, J.; Li, Y. A case study using neural networks algorithms: Horse racing predictions in Jamaica. In Proceedings of the International Conference on Artificial Intelligence (ICAI 2008), Las Vegas, NV, USA, 14–17 July 2008.
26. Abdelwahab, H.; Abdel Aty, M. Development of artificial neural network models to predict driver injury severity in traffic accidents at signalized intersections. *Transp. Res. Rec.* **2001**, *1746*, 6–13. [CrossRef]
27. Chong, M.M.; Abraham, A.; Paprzycki, M. Traffic accident analysis using decision trees and neural networks. *arXiv* **2004**, arXiv:cs/0405050.
28. Yasin Çodur, M.; Tortum, A. An Artificial Neural Network Model for Highway Accident Prediction: A Case Study of Erzurum, *Turkey. Promet-Traffic Transp.* **2015**, *27*, 217–225.
29. Zeng, Q.; Huang, H.; Pei, X.; Wong, S.C. Modeling nonlinear relationship between crash frequency by severity and contributing factors by neural networks. *Anal. Sci Accid. Res.* **2016**, *10*, 12–25. [CrossRef]
30. Athanassopoulos, A.D.; Curram, S.P. A comparison of data envelopment analysis and artificial neural networks as tools for assessing the efficiency of decision making units. *J. Oper. Res. Soc.* **1996**, 1000–1016. [CrossRef]

31. Vaninsky, A. Combining data envelopment analysis with neural networks: Application to analysis of stock prices. *J. Inf. Optim. Sci.* **2004**, *25*, 589–611. [CrossRef]
32. Azadeh, A.; Javanmardi, L.; Saberi, M. The impact of decision-making units features on efficiency by integration of data envelopment analysis, artificial neural network, fuzzy C-means and analysis of variance. *Int. J. Oper. Res.* **2010**, *7*, 387–411. [CrossRef]
33. Ülengin, F.; Kabak, Ö.; Önsel, S.; Aktas, E.; Parker, B.R. The competitiveness of nations and implications for human development. *Socio-Econ. Plan. Sci.* **2011**, *45*, 16–27. [CrossRef]
34. Wu, D.D.; Yang, Z.; Liang, L. Using DEA-neural network approach to evaluate branch efficiency of a large Canadian bank. *Expert Syst. Appl.* **2006**, *31*, 108–115. [CrossRef]
35. Wu, D. Supplier selection: A hybrid model using DEA, decision tree and neural network. *Expert Syst. Appl.* **2009**, *36*, 9105–9112. [CrossRef]
36. Ciobanu, S.M.; Benedek, J. Spatial characteristics and public health consequences of road traffic injuries in Romania. *Environ. Eng. Manag.* **2015**, *14*, 2689–2702.
37. Wang, C.; Quddus, M.A.; Ison, S.G. Impact of traffic congestion on road accidents: A spatial analysis of the M25 motorway in England. *Accid. Anal. Prev.* **2009**, *41*, 798–808. [CrossRef] [PubMed]
38. Chen, C.; Li, T.; Sun, J.; Chen, F. Hotspot Identification for Shanghai Expressways Using the Quantitative Risk Assessment Method. *Int. J. Environ. Res. Public Health* **2016**, *14*, 20. [CrossRef] [PubMed]
39. Zhang, C.; Yan, X.; Ma, L.; An, M. Crash prediction and risk evaluation based on traffic analysis zones. *Math. Probl. Eng.* **2014**, *2014*, 9. [CrossRef]
40. Moradi, A.; Soori, H.; Kavousi, A.; Eshghabadi, F.; Jamshidi, E.; Zeini, S. Spatial analysis to identify high risk areas for traffic crashes resulting in death of pedestrians in Tehran. *Med. J. Islam. Repub. Iran* **2016**, *30*, 450. [PubMed]
41. Steenberghen, T.; Dufays, T.; Thomas, I.; Flahaut, B. Intra-urban location and clustering of road accidents using GIS: a Belgian example. *Int. J. Geogr. Inf. Sci.* **2004**, *18*, 169–181. [CrossRef]
42. Pirdavani, A.; Bellemans, T.; Brijs, T.; Wets, G. Application of geographically weighted regression technique in spatial analysis of fatal and injury crashes. *J. Transp. Eng.* **2014**, *140*, 04014032. [CrossRef]
43. Pirdavani, A.; Bellemans, T.; Brijs, T.; Kochan, B.; Wets, G. Assessing the road safety impacts of a teleworking policy by means of geographically weighted regression method. *J. Saf. Res.* **2014**, *39*, 96–110. [CrossRef]
44. Eksler, V.; Lassarre, S. Evolution of road risk disparities at small-scale level: Example of Belgium. *J. Pet. Sci. Eng.* **2008**, *39*, 417–427. [CrossRef] [PubMed]
45. Yang, Y.; Rosenbaum, M. Artificial neural networks linked to GIS for determining sedimentology in harbours. *J. Pet. Sci. Eng.* **2001**, *29*, 213–220. [CrossRef]
46. Sameen, M.I.; Pradhan, B. Severity Prediction of Traffic Accidents with Recurrent Neural Networks. *Appl. Sci.* **2017**, *7*, 476. [CrossRef]
47. Pradhan, B.; Lee, S.; Buchroithner, M.F. A GIS-based back-propagation neural network model and its cross-application and validation for landslide susceptibility analyses. *Comput. Environ. Urban Syst.* **2010**, *34*, 216–235. [CrossRef]
48. Elsafi, S.H. Artificial neural networks (ANNs) for flood forecasting at Dongola Station in the River Nile, Sudan. *Alex. Eng. J.* **2014**, *53*, 655–662. [CrossRef]
49. Lee, S.; Park, I.; Koo, B.J.; Ryu, J.H.; Choi, J.K.; Woo, H.J. Macrobenthos habitat potential mapping using GIS-based artificial neural network models. *Mar. Pollut. Bull.* **2013**, *67*, 177–186. [CrossRef] [PubMed]
50. Pijanowski, B.C.; Brown, D.G.; Shellito, B.A.; Manik, G.A. Using neural networks and GIS to forecast land use changes: A land transformation model. *Comput. Environ. Urban Syst.* **2002**, *26*, 553–575. [CrossRef]
51. Yoo, C.; Kim, J.M. Tunneling performance prediction using an integrated GIS and neural network. *Comput. Geotech.* **2007**, *34*, 19–30. [CrossRef]
52. Mas, J.F.; Puig, H.; Palacio, J.L.; Sosa López, A. Modelling deforestation using GIS and artificial neural networks. *Environ. Model. Soft* **2004**, *19*, 461–471. [CrossRef]
53. Janssens, D.; Wets, G.; Timmermans, H.J.; Arentze, T.A. Modelling short-term dynamics in activity-travel patterns: Conceptual framework of the Feathers model. In Proceedings of the 11th World Conference on Transport Research, Berkeley, CA, USA, 24–28 June 2007.
54. Avkiran, N.K. An application reference for data envelopment analysis in branch banking: Helping the novice researcher. *Int. J. Bank Mark* **1999**, *17*, 206–220. [CrossRef]

55. Galagedera, D.; Silvapulle, P. Experimental evidence on robustness of data envelopment analysis. *J. Oper. Res. Soc.* **2003**, *54*, 654–660. [CrossRef]

56. Raab, R.L.; Lichty, R.W. Identifying subareas that comprise a greater metropolitan area: The criterion of county relative efficiency. *J. Reg. Sci.* **2002**, *42*, 579–594. [CrossRef]

57. Shmueli, G.; Patel, N.R.; Bruce, P.C. *Data Mining for Business Analytics: Concepts, Techniques and Applications*; John Wiley & Sons: Hoboken, NJ, USA, 2016.

58. Jmp, A.; Proust, M. *Specialized Models*; AS Institute Inc.: Cary, NC, USA, 2013.

59. Tso, G.K.; Yau, K.K. Predicting electricity energy consumption: A comparison of regression analysis, decision tree and neural networks. *Energy* **2007**, *32*, 1761–1768. [CrossRef]

60. Elvik, R. Speed and road safety: Synthesis of evidence from evaluation studies. *Transp. Res. Rec.* **2005**, *1908*, 59–69. [CrossRef]

61. Kweon, Y.J.; Kockelman, K. Safety effects of speed limit changes: Use of panel models, including speed, use, and design variables. *Transp. Res. Rec.* **2005**, *1908*, 148–158. [CrossRef]

62. WHO. *World Report on Road Traffic Injury Prevention*; World Health Organization: Geneva, Switzerland, 2004.

63. Garber, N.; Ehrhart, A. Effect of speed, flow, and geometric characteristics on crash frequency for two-lane highways. *Transp. Res. Rec.* **2000**, *1717*, 76–83. [CrossRef]

64. Golob, T.F.; Recker, W.; Pavlis, Y. Probabilistic models of freeway safety performance using traffic flow data as predictors. *Saf. Sci.* **2008**, *46*, 1306–1333. [CrossRef]

65. Xie, F.; Feng, Q. Research of effects of accident on traffic flow characteristics. In Proceedings of the International Conference on Mechatronic Sciences, Electric Engineering and Computer (MEC), Shengyang, China, 20–22 December 2013.

66. Zhang, Y. Analysis of the Relation between Highway Horizontal Curve and Traffic Safety. In Proceedings of the International Conference on Measuring Technology and Mechatronics Automation (ICMTMA), Zhangjiajie, China, 11–12 April 2009.

67. Vayalamkuzhi, P.; Amirthalingam, V. Influence of geometric design characteristics on safety under heterogeneous traffic flow. *Transp. Res. Rec.* **2016**, *3*, 559–570. [CrossRef]

68. Ma, M.; Yan, X.; Abdel Aty, M.; Huang, H.; Wang, X. Safety analysis of urban arterials under mixed-traffic patterns in Beijing. Transportation Research Record. *Transp. Res. Rec.* **2010**, *2193*, 105–115. [CrossRef]

69. Zero, T. *Towards Zero: Achieving Ambitious Road Safety Targets through a Safe System Approach*; OECD: Paris, France, 2008.

70. Yannis, G.; Evgenikos, P.; Papadimitriou, E. *Best Practice for Cost-Effective Road Safety Infrastructure Investments*; Conference of European Directors of Road (CEDR): Paris, France, 2008.

71. Blumenberg, S. Benchmarking Financial Processes with Data Envelopment Analysis. 2005. Available online: www.is-frankfurt.de/publikationenNeu/BenchmarkingFinancialProcesses1208.pdf (accessed on 20 June 2017).

72. Charnes, A.; Cooper, W.W.; Lewin, A.Y.; Seiford, L.M. *Data Envelopment Analysis: Theory, Methodology, and Applications*; Springer Science & Business Media: New York, NY, USA, 2013.

73. Tu, J.V. Advantages and disadvantages of using artificial neural networks versus logistic regression for predicting medical outcomes. *J. Clin. Epidemiol.* **1996**, *49*, 1225–1231. [CrossRef]

74. Tsangaratos, P.; Benardos, A. Applying artificial neural networks in slope stability related phenomena. In Proceedings of the 13th International Congress-Bulletin of the Geological Society of Greece (BGSG), Chania, Greece, 5–8 September 2013; pp. 1901–1911.

applied
sciences

MDPI

Article

Optimized Neural Architecture for Automatic Landslide Detection from High-Resolution Airborne Laser Scanning Data

**Mustafa Ridha Mezaal [1], Biswajeet Pradhan [1,2,*] , Maher Ibrahim Sameen [1],
Helmi Zulhaidi Mohd Shafri [1] and Zainuddin Md Yusoff [1]**

[1]   Department of Civil Engineering, Faculty of Engineering, Universiti Putra Malaysia,
       Serdang 43400, Malaysia; gismustafa87@gmail.com (M.R.M.); maherrsgis@gmail.com (M.I.S.);
       helmi@eng.upm.edu.my (H.Z.M.S.); zmy@upm.edu.my (Z.M.Y.)
[2]   School of Systems, Management and Leadership, Faculty of Engineering and Information Technology,
       University of Technology Sydney, Building 11, Level 06, 81 Broadway,
       Ultimo NSW 2007 (P.O. Box 123), Australia
*   Correspondence: Biswajeet24@gmail.com or biswajeet@lycos.com

Academic Editor: Saro Lee
Received: 29 June 2017; Accepted: 13 July 2017; Published: 16 July 2017

Abstract: An accurate inventory map is a prerequisite for the analysis of landslide susceptibility, hazard, and risk. Field survey, optical remote sensing, and synthetic aperture radar techniques are traditional techniques for landslide detection in tropical regions. However, such techniques are time consuming and costly. In addition, the dense vegetation of tropical forests complicates the generation of an accurate landslide inventory map for these regions. Given its ability to penetrate vegetation cover, high-resolution airborne light detection and ranging (LiDAR) has been used to generate accurate landslide maps. This study proposes the use of recurrent neural networks (RNN) and multi-layer perceptron neural networks (MLP-NN) in landscape detection. These efficient neural architectures require little or no prior knowledge compared with traditional classification methods. The proposed methods were tested in the Cameron Highlands, Malaysia. Segmentation parameters and feature selection were respectively optimized using a supervised approach and correlation-based feature selection. The hyper-parameters of network architecture were defined based on a systematic grid search. The accuracies of the RNN and MLP-NN models in the analysis area were 83.33% and 78.38%, respectively. The accuracies of the RNN and MLP-NN models in the test area were 81.11%, and 74.56%, respectively. These results indicated that the proposed models with optimized hyper-parameters produced the most accurate classification results. LiDAR-derived data, orthophotos, and textural features significantly affected the classification results. Therefore, the results indicated that the proposed methods have the potential to produce accurate and appropriate landslide inventory in tropical regions such as Malaysia.

Keywords: landslide detection; LiDAR; recurrent neural networks (RNN); multi-layer perceptron neural networks (MLP-NN); GIS; remote sensing

1. Introduction

Landslides are dangerous geological disasters with catastrophic effects on human lives and properties. Landslides occur with high frequency in mountainous and hilly areas, such as the Cameron Highlands in Malaysia. Landslide incidence is related to a cluster of triggering factors, such as intense rainfall, volcanic eruptions, rapid snowmelt, elevated water levels, and earthquakes. Landslide inventory maps are crucial for measuring the magnitude and analyzing the susceptibility, hazard, and risk of earthquakes [1,2], as well as for examining distribution patterns and predicting the landscapes affected by landslide [3]. Mapping a landslide inventory in tropical areas is challenging because the dense vegetation cover in these regions obscures underlying landforms [4]. Moreover, the majority of available conventional landslide detection techniques are not rapid and accurate enough for inventory mapping given the rapid vegetation growth in tropical regions. Therefore, inventory mapping requires the use of more rapid and accurate techniques, such as light detection and ranging (LiDAR) [5], which uses active laser transmitters and receivers to acquire elevation data. In addition, LiDAR has the unique capability to penetrate densely vegetated areas [5] and provide detailed information on terrains with high point density. Moreover, it depicts ground surface features and provides useful information on topographical features in areas where landslide locations are obscured by vegetation cover [6,7].

Numerous studies have applied a multiresolution segmentation algorithm for the remote sensing of land features [8]. This algorithm requires the identification of three parameters (i.e., scale, shape, and compactness); the values of these parameters can be determined using the traditional trial-and-error method, which is very time consuming and laborious [5]. Moreover, using the algorithm to delineate the boundary of an object at different scales remains challenging [9]. Thus, optimal parameters for segmentation should be identified via semiautomatic and automatic approaches [10–12]. The automatic selection of segmentation parameters requires the use of the advanced supervised approach presented in [13].

Processing a large number of irrelevant features causes overfitting [14]. By contrast, the best classification results are obtained by selecting the most relevant feature [15]. Landslide identifcation in a particular area can be improved by selecting the most significant feature [15,16]. As shown in [2], selecting the most significant feature facilitates the differentiation of landslides from non-landslides. Accuracy can be improved by decreasing the number of features, as recommended in [17]. The efficiency of feature selection techniques for landslide detection has been proven in [18–20].

The neural network (NN) is effective in remote sensing applications [21], particularly in solving different image classification problems [22] specified by nonlinear mathematical fitting for function approximation. NN architectures are classified into the recurrent neural network (RNN), back-propagation neural network, probability neural network, and multilayer perceptron neural network (MLP-NN). NN-based classifiers can adapt to different types of data and inputs, and can overcome the issue of mixed pixels by providing fuzzy output and fit with multiple images [23,24]. These classifiers include parallel computation, which is superior to statistical classification approaches because it is non-parametric and does not require the prior knowledge of a distribution model for input data [25]. Moreover, NN-based classifiers can evaluate non-linear relationships between the input data and desired outputs and are distinguished by their fast generalization capability [26]. NN-based classifiers have been successfully in function approximation, prediction, pattern recognition, landslide detection, image classification, automatic control, and landslide susceptibility [27–32]. Authors of [33] found that MLP-NN can be effectively applied in landslide detection using multi-source data. The RNN model can effectively predict landslide displacement [34]. The above neural architecture techniques have not been extensively used for landslide detection using only LiDAR data. This research gap urged us to apply the RNN and MLP-NN models in landslide detection based on very high-resolution LiDAR data. To achieve this objective, we optimized multiresolution segmentation parameters via a supervised approach. Using the correlation-based feature selection (CFS) algorithm, we selected the most significant feature from high-resolution airborne laser scanning data.

2. Study Area

This study was performed in a small section of the Cameron Highlands, which is notorious for its frequent occurrence of landslides. The study area covers an area of 26.7 km^2. It is located on northern peninsular Malaysia within the zone comprising latitudes 4°26′3″ to 4°26′18″ and longitudes 101°23′48″ to 101°24′4″ (Figure 1). The annual average rainfall and temperature in this region are approximately 2660 mm and 24/14 °C (daytime/nighttime temperatures), respectively. Approximately 80% of its area is forested with a flat (0°) to hilly (80°) land form.

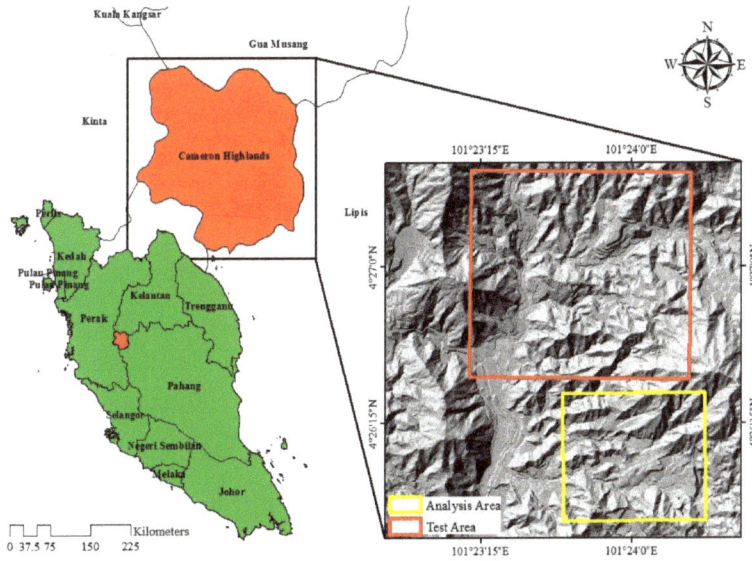

Figure 1. Location of the study area. The red boundary represents the analysis area and the yellow boundary represents the test area.

Two sites were selected to implement and test the proposed models (Figure 1). All the prerequisite considerations were taken in to account during test site selection to avoid missing any land cover classes. To obtain an accurate map of the analysis and test sites, the training sample size was measured via the stratified random sample method.

3. Methodology

3.1. Overall Methodology and Pre-Processing

LiDAR data and landslide inventories were first pre-processed to eliminate noise and outliers. A high-resolution digital elevation model (DEM) at 0.5 m was then derived from LiDAR point clouds to generate other LiDAR-derived products (i.e., slope, aspect, height or (normalized digital surface model (nDSM)), and intensity. LiDAR-derived products and orthophotos were then composited by rectifying their geometric distortions to generate one coordinate system and were finally prepared in geographic information system (GIS) for feature extraction. Suitable parameters (scale, shape, and compactness) at various levels of segmentation were obtained via a supervised approach, i.e., a fuzzy-based segmentation parameter optimizer (FbSP optimizer) [13]. The stratified random method was used to evaluate the training dataset in accordance with the procedure in [35]. The correlation-based selection algorithm (CFS) [36] was used to rank features from the most to least important. RNN and MLR-NN models were applied to detect landslide locations. The results of the models were validated using a 10-fold cross validation method. In addition, the models were evaluated in another part of the study area (i.e., the test site). Slope and aspect layers were overlaid with the results to identify other landslide characteristics (i.e., direction and run off). The study flow is illustrated in Figure 2.

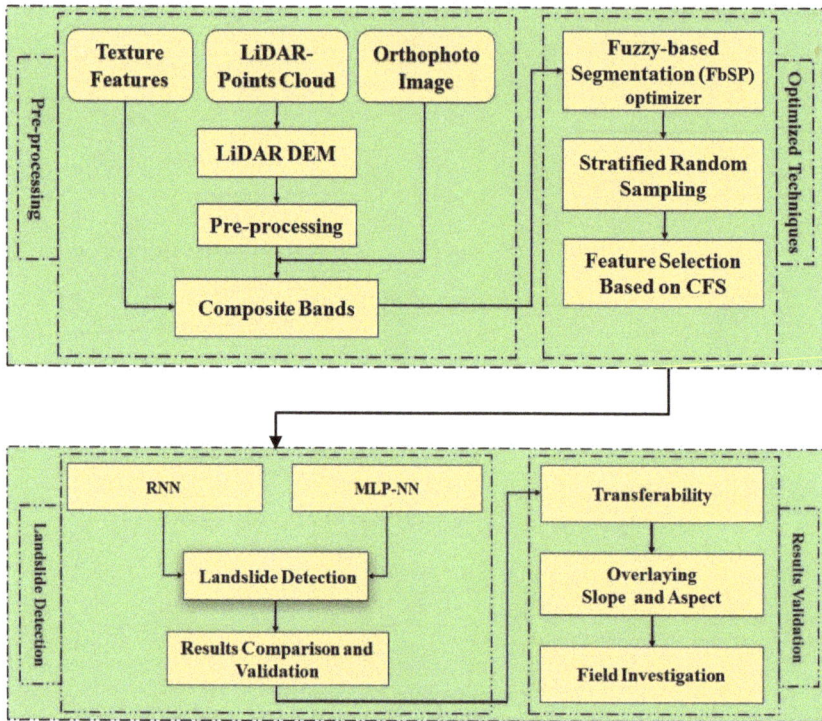

Figure 2. Overview of the proposed method. LiDAR: light detection and ranging; RNN: recurrent neural networks; MLP-NN: multi-layer perceptron neural networks; CFS: correlation-based feature selection; DEM: digital elevation model.

3.2. Landslide Inventory

The landslide inventory; produced previously by Pradhan and Lee, [39] was used to develop the proposed detection method and the total number of landslides is 21 in the study area covering 3781 m^2 (Figure 3).

3.3. Data

LiDAR point-cloud data were collected on 15 January 2015 at a point density of 8 points/m^2 and frequency pulse rate of 25,000 Hz. The absolute accuracy of the data (root-mean square errors) was restricted to 0.15 m and 0.3 m in the vertical and horizontal axes, respectively. Orthophotos were obtained using the same acquisition system that relied on the abovementioned cloud data. A DEM was derived from LiDAR point clouds with a spatial resolution of 0.5 m after non-ground points were removed using inverse distance weighting with a spatial reference of GDM2000/Peninsula RSO. Subsequently, LiDAR-based DEM was used to generate derived layers to facilitate the identification and characterization of landslide locations [37].

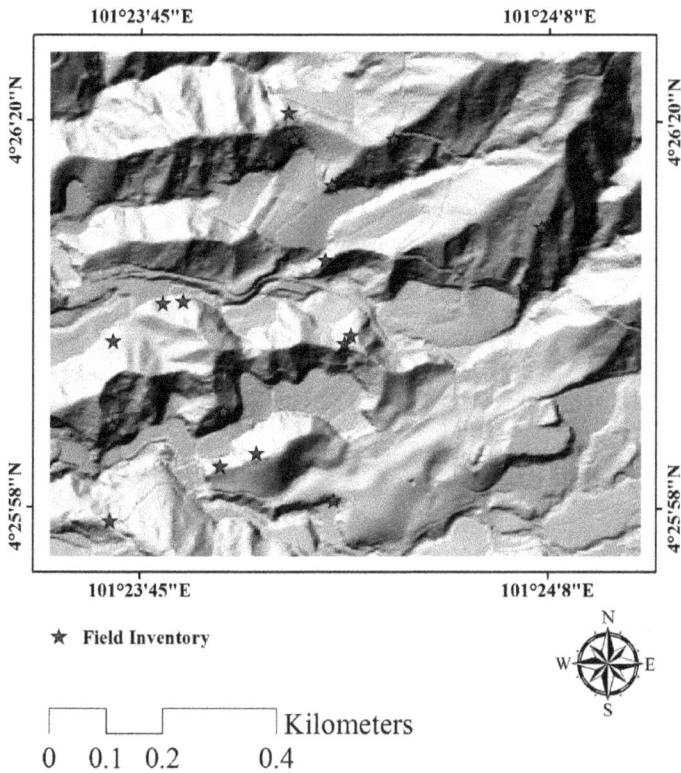

Figure 3. Shows the locations of landslide in the study area.

According to the authors of [38], slope directly and highly affects landslide phenomenology. The authors of [39] also inferred that slope is the principal factor that affects landslide occurrence. The author of [40] indicated that a hillshade map provides a good image of terrain movements, thus facilitating the development of landslide maps. Texture and geometric features are crucial for improving the classification accuracy of landslide mapping [14]. Landslide intensity and texture derived from LiDAR data are affected by the accuracy of landslide detection [9]. The accuracy and capacity of DEM to represent surface features are determined by terrain morphology, sampling density, and the interpolation algorithm [41]. In this study, hillshade, height (nDSM), slope, and aspect were generated from LiDAR-based DEM. As shown in Figure 4, landslide locations were detected using visible bands and texture features.

3.4. Image Segmentation

The sizes and shapes of image objects [42] are determined via image segmentation, the preliminary step in object-based classification. Optimal segmentation parameters depend on the environment under analysis, the selected application, and the underlying input data [8]. Previous studies have used the multiresolution segmentation algorithm with eCognition software for image segmentation [8,9]. Three parameters (scale, shape, and compactness) are defined in this algorithm. According to [5], these parameters can be obtained via the traditional trial-and-error method, which is time consuming and laborious. Therefore, the fuzzy logic supervised approach presented by [13] was adopted in this study.

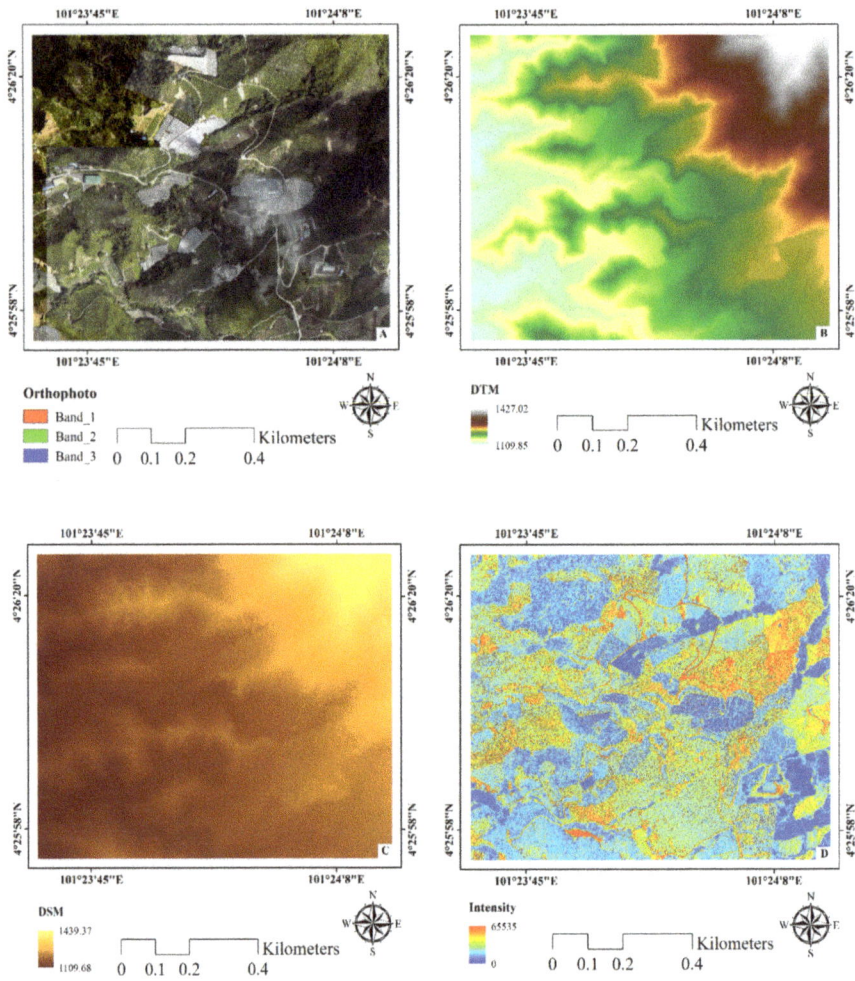

Figure 4. *Cont.*

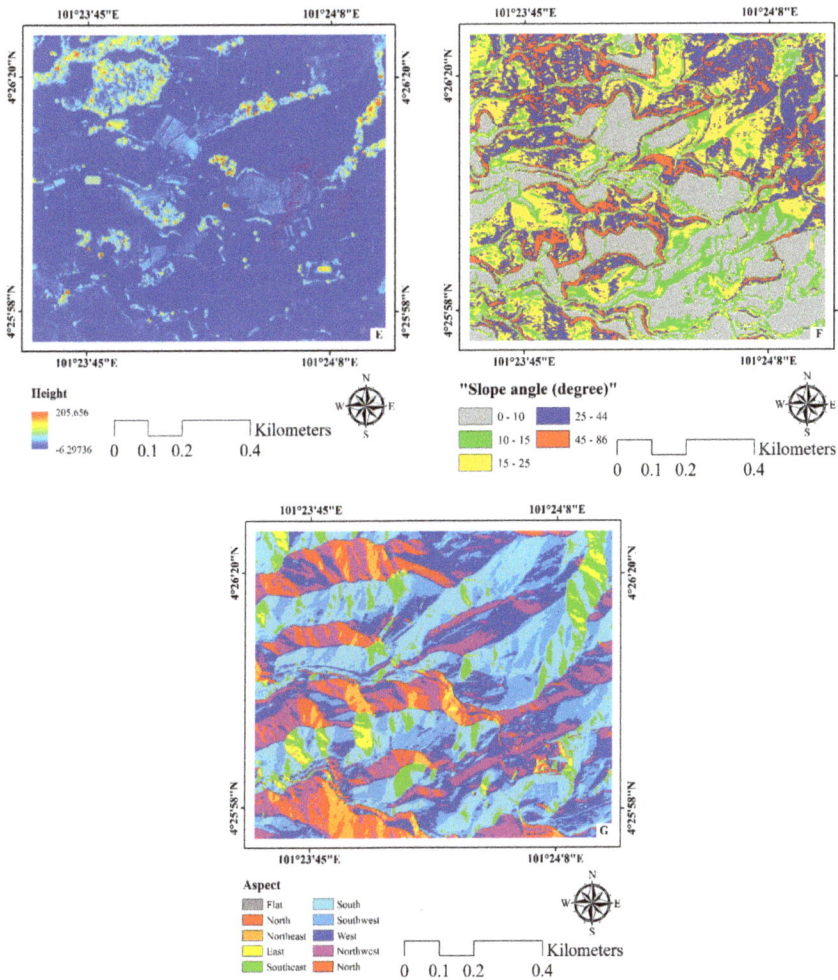

Figure 4. LiDAR-derived data; (**A**) Orthophotos; (**B**) digital terrain model (DTM); (**C**) digital surface model (DSM); (**D**) Intensity; (**E**) Height; (**F**) Slope; and (**G**) Aspect.

3.5. Training Sets

The authors of [35] suggested the use of stratified random sampling method to obtain an adequately sized training dataset for every class without any bias during sample selection. Accordingly, the present study adopted stratified random sampling to evaluate training samples and achieve high performance without strong bias. Four classes with different numbers of objects were set as shown in Table 1.

Stratified random sampling is a prerequisite to obtain prior knowledge of the two sites considered for landslide inventory. Hence, segmentation parameters were first optimized. Then, the landslide inventory was overlapped with the segmented layer for object labeling. ArcGIS 10.3 was used to construct sample sets automatically at each optimal scale. Subsequently, stratified random sampling was applied on the labeled objects. This process was performed 20 times at each optimal scale.

Table 1. Number of selected training objects in four classes.

Class Name	Number of the Object for Each Class
Landslide	52
Cut slope	67
Bare soil	80
Vegetation	150

3.6. Correlation-Based Feature Selection

The authors of [15] reported that the selection of only the most relevant features improves the quality of landslide identification and classification. Working with large numbers of features causes numerous problems. As reported in [43] and [14], some of these problems include the slow run time of algorithms due to the consideration of numerous resources, low accuracy when the number of features exceed the number of observation features, and overfitting when irrelevant features are used as inputs. Therefore, the most significant features should be selected to enhance the accuracy of feature extraction. In this study, relevant features were extracted using the CFS algorithm with Weka 3.7 software. Furthermore, the CFS algorithm was applied to all LiDAR-derived data, visible bands, and textural features, and was used to determine the feature subsets required to develop models for landslide identification. The CFS algorithm comprises two basic steps: the ranking of initial features and the elimination of the least important features through an iterative process.

3.7. MLP-NN

NNs are a family of biological learning models in machine learning. The NN model comprises interconnected neurons or nodes, which are structured into layers with random or full interconnections among successive layers [44]. The NN model comprises input, hidden, and output layers that are responsible for receiving, processing, and presenting results, respectively [44]. Each layer contains nodes connected by numeric weights and output signals. The weights are the functions of the sum of the inputs to the node modified by a simple activation function [45]. The possibility of learning is the most important feature that attracts researchers to use NNs.

Back-propagation, which was first proposed by Paul Werbos in 1974 and independently rediscovered by Rumelhart and Parker, is the most common learning algorithm used in NN. It aims to minimize the error function via the iterative approach as shown in Equation (1). NNs have been successfully used in remote sensing applications. However, this model has some limitations, specifically, high computational complexity and overlearning [46,47].

$$E = \frac{1}{2} \sum_{i=1}^{L} (d_i - o_i)^2 \qquad (1)$$

where d_i and o_i represent the desired output and the current response of node $"i"$ in the output layer, respectively. $"L"$ is the number of nodes in the output layer. Corrections to weight parameters were calculated and effected with the previous values in the iterative method, as demonstrated in Equation (2):

$$\begin{cases} \Delta w_{i,j} = -\mu \frac{\partial E}{\partial w_{i,j}} \\ \Delta w_{i,j}(t+1) = \Delta w_{i,j} + \alpha \Delta w_{i,j}(t) \end{cases} \qquad (2)$$

where delta rule $\Delta w_{i,j}$ is the weight parameter between nodes *i and j*; μ is a positive constant that controls the amount of adjustment and is referred to as learning rate; α is the momentum factor, which takes a value between 0 and 1; and *t* is the iteration number. α is referred to as the stabilizing factor because it smoothens quick changes between weights [48].

3.8. RNN

RNNs are designed to model sequences in NNs with feedback connections. They are very powerful in computational analysis and are biologically more reliable than other NN techniques given their lack of internal states. The memory of past activations in RNN is very effective with feedback connections, making them suitable for learning the temporal dynamics of sequential data. RNN is very powerful when used to map input and output sequences because it uses contextual information. However, traditional RNNs face the challenge of exploding or vanishing gradients. Hochreiter and Schmidhuber [49] proposed long short-term memory (LSTM) to tackle this issue.

Hidden units in LSTM are replaced with memory blocks that contain three multiplicative units (input, output, forget gates) and self-connected memory cells to allow for reading, writing, and resetting through a memory block and behavioral control. A single LSTM unit is shown in Figure 5. c_t is the sum of inputs at time step t and its previous time step activations. LSTM updates time step i given inputs x_t, h_{t-1}, and c_{t-1} as reported in [50].

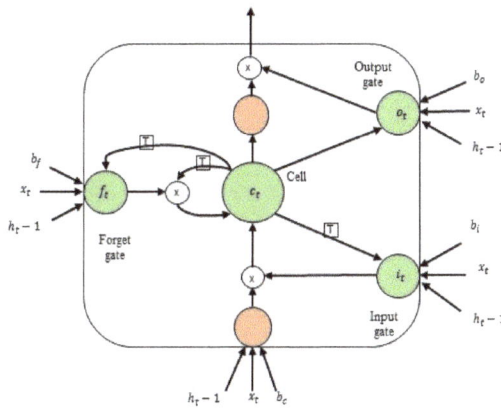

Figure 5. Structure of a memory cell in long short-term memory (LSTM)-RNN.

Input gates:
$$i_t = \sigma(W_{xi} \cdot x_t + W_{hi} \cdot h_{t-1} + W_{ci} \cdot c_{t-1} + b_i) \tag{3}$$

Forget gates:
$$f_t = \sigma\left(W_{xf} \cdot x_t + W_{hf} \cdot h_{t-1} + W_{xf} \cdot x_t + W_{cf} \cdot c_{t-1} + b_f\right) \tag{4}$$

Cell units:
$$c_t = i_t \cdot tanh\left(W_{xc} \cdot x_t + W_{hc} \cdot h_{t-1} + b_c + b_f \cdot c_{t-1}\right) \tag{5}$$

Output gates:
$$o_t = \sigma(W_{xo} \cdot x_t + W_{ho} \cdot h_{t-1} + W_{co} \cdot x_t + b_o) \tag{6}$$

The hidden activation (output of the cell) is also given by a product of the two terms:
$$c_t = o_t \cdot tanh(c_t) \tag{7}$$

where σ and tanh are an element-wise non-linearity, such as a sigmoid function and hyperbolic tangent function, respectively; W is the weight matrix; x_t refers to input at time step t; t, h_{t-1} represents the hidden state vector of the previous time step; and b_c denotes the input bias vector. The memory cell unit c_t is a sum of two terms: the previous memory cell unit c_{t-1}, which is modulated by f_t and c_t, a function of the current input, and previous hidden state, modulated by the input gate i_t due to i_t

and f_t being sigmoidal. Their values range within [0, 1], and i_t and f_t can be considered as knobs that the LSTM learns to selectively forget its previous memory or consider its current input, whilst o_t is an output gate that learns how much of the memory cell to transfer to the hidden layers.

3.9. Neural Network Models

3.9.1. MLP-NN

This study proposed the network architectures RNN and MLP-NN. Figure 6 depicts the MLP-NN model architecture, which has two hidden layers of 50 hidden units. Ten features were taken as inputs in the model to detect different types of objects, such as landslide, cut slope, bare soil, and vegetation. The MLP-NN model was trained through a back-propagation technique with the Adam optimizer and a batch size of 64. The hyper-parameters used in this NN were carefully selected through grid search and a 10-fold cross validation process.

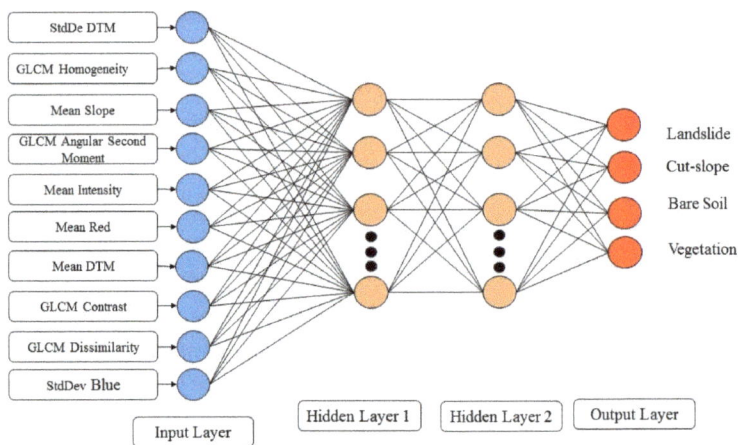

Figure 6. Architecture of the MLR-NN model; GLCM: gray level co-occurrence matrix, StdDev: standard deviation.

3.9.2. RNN

RNN is a sequence problem considered as the addition of loops to architecture. For example, in any layer under consideration, signals can be passed to each neuron and are subsequently forwarded to the next layer. The network output can be input to the network in the next input feature, and so on, as shown in Figure 7. In this study, RNN received 10 features as inputs to differentiate landslides from other objects (cut slope, bare soil, and vegetation). RNN consisted of an LSTM layer with 50 hidden units, two fully connected layers, a dropout layer, and a softmax layer. The back-propagation technique was used in trained the RNN model with Adam optimizer and a batch size of 128.

To avoid overfitting, a dropout layer was used in the RNN model and the NN learned weights from the training dataset. However, overfitting may occur when new data are inputted. The dropout layer randomly set some selected activations to zero, thus alleviating overfitting. The selected activations were used only during training and not during testing. The parameter was controlled by the number of activations that the dropout layer referred to as keep probability.

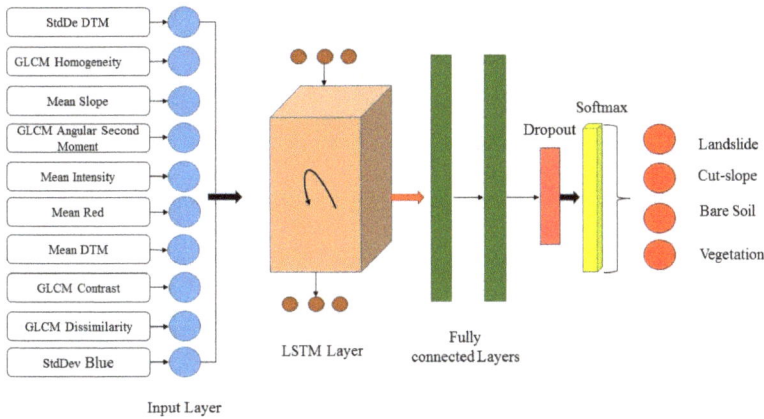

Figure 7. Architecture of the RNN model.

3.9.3. Optimization of Model Hyper-Parameters

The hyper-parameters of the RNN and MLP-NN models were optimized via a systematic grid search in scikit-learn [51] for 100 epochs. Despite its high computational cost, the systematic grid search provides better results because it systematically tunes the hyper-parameter values. Parameter combinations were selected for the models. The models were evaluated using a 10-fold cross-validation method. Among the evaluated parameters, the model with the highest validation accuracy was selected. Table 2 presents the most optimized parameters obtained for the models.

Table 2. Optimized model hyper-parameters; RNN: recurrent neural networks, MLP-NN: multi-layer perceptron neural networks.

Optimized Parameter	Suitable Value	Description
Minibatch size	126 (RNN) 64 (MLP-NN)	Number of training cases over which the Adam update is computed.
Loss function	categorical cross-entropy	The objective function or optimization score function is also called as multiclass legless, which is appropriate for categorical targets.
Optimizer	Adam	Adaptive moment estimation
dropout rates	0.6	Dropping out units (hidden and visible)

4. Results and Discussion

4.1. Supervised Approach for Optimizing Segmentation

The supervised approach was employed to optimize the parameters (i.e., scale, shape, and compactness) of the multiresolution segmentation algorithm for landslide identification and for differentiation from non-landslides (bare soil, cut slope, and vegetation). The optimized parameters rapidly increased the accuracy of classification to the optimum level by delineating the segmentation boundaries of the landslide. The application of optimized segmentation parameters allowed for the spatial and textural identification of features (landslide and non-slides). In our proposed method, accurate segmentation results should be first obtained prior to performing subsequent steps.

The optimal parameters of the multiresolution segmentation algorithm were obtained. The selected values for the three parameters are shown in Table 3. The initial segmentation parameters set in the supervised approach were 50, 0.1, and 0.1 for scale, shape, and compactness, respectively. After 100 iterations with these initial values, the optimal values obtained for scale, shape, and compactness were 75.52, 0.4, and 0.5, respectively in the analysis area. Meanwhile, the test area values were 100, 0.45 and 0.74, respectively. Figure 8a,b show the initial and optimal segmentation processes. The results of optimized segmentation accurately delineated landslide objects in the analysis and test areas.

Table 3. Multi-resolution segmentation parameters.

	Initial Parameters			Optimal Parameters		
Number	Scale	Shape	Compactness	Scale	Shape	Compactness
1	50	0.1	0.1	75.52	0.4	0.5
2	80	0.1	0.1	100	0.45	0.74

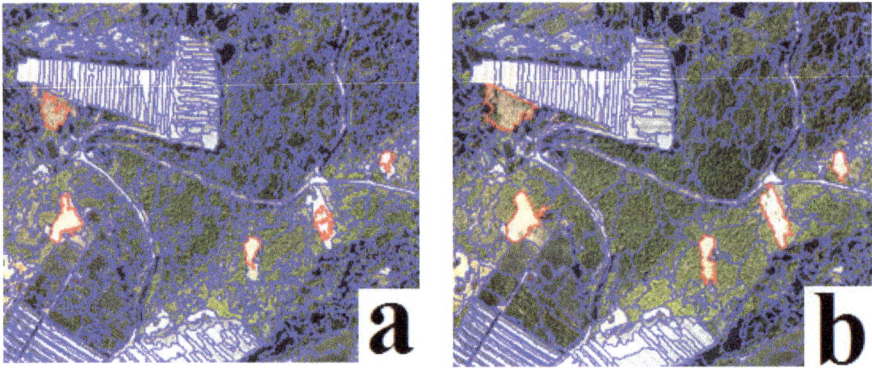

(a) Initial Segmentation   (b) Optimized Segmentation

Figure 8. Parameter optimization of the multiresolution segmentation algorithm: (a) initial segmentation and (b) optimized segmentation.

4.2. Relevant Feature Subset Based on a CFS Algorithm

In this study, the feature input consisted of 39 items of LiDAR-derived data (i.e., slope, height, and intensity), texture features (i.e., GLCM StdDev and GLCM homogeneity), and visible band. The optimal combination of features was selected via ten experiments using a CFS algorithm. Selection began from (1, 2, 4, 6, 8, 10, 12, 14, 16, 18, 20, 22, 24, 26, 28, 30, 32, 34, 36, 39) of the features. The most relevant feature subsets were obtained after 100 iterations in every experiment; this result is in line with the procedure proposed by Sameen et al. [52]. High classification accuracy was achieved when 10 of the features were applied, indicating that LiDAR-derived data, visible bands, and textural features were more effective in detecting the landslide location. Table 3 shows the most significant results of feature selection based on the CFS algorithm.

4.3. Results of Landslide Detection

Classification techniques affect the quality of the classification maps. Many classification algorithms have been established for each category, and each has its merits and demerits. In the present work, the RNN and MLP-NN models with optimized parameters were used for landslide detection with good accuracy. Figure 9 shows the classification results of the RNN and MLP-NN models in the analysis area. The qualitative assessment of the RNN model yielded high-quality results, as shown in Figure 9A. Well-defined landslide boundaries were detected and correctly differentiated from other objects (cut-slope, bare soil and vegetation). On the other hand, the qualitative assessment of MLP-NN produced low-quality results, as shown in Figure 9B.

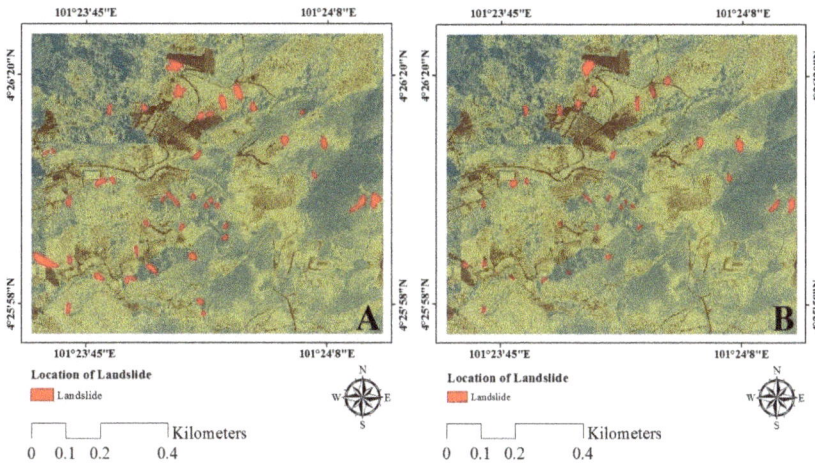

Figure 9. Results of the qualitative assessment of (**A**) RNN and (**B**) MLP-NN for the analysis area.

The proposed models were evaluated using another LiDAR dataset (test site) from the Cameron Highlands. All features (all existing objects) of the test area were carefully considered. Segmentation parameters were optimized using the FbSP optimizer. A 10-fold cross-validation approach introduced by Bartels et al. [53] was used to resolve this issue with high accuracy. Environmental conditions and differences in landslide characteristics resulted in misclassification [9]. Differences in the sensors used, illumination conditions, and the spatial resolutions of images are some of the challenges faced by the proposed NN models [54]. The results of qualitative assessment indicated that the proposed NNs with optimized techniques correctly detected landslide locations in the test site, as shown in Figure 10. The qualitative assessment of the RNN model yielded high-quality results, as shown in Figure 10A,B. On the other hand, the qualitative assessment of the MLP-NN model produced low-quality results, as shown in Figure 10.

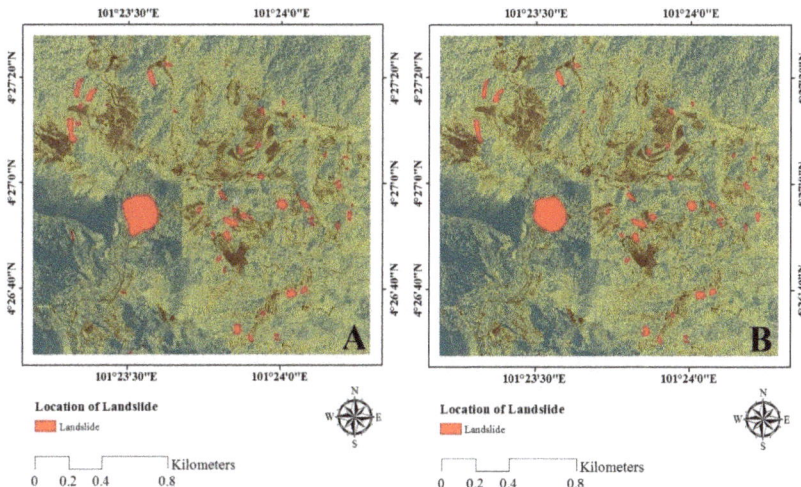

Figure 10. Results of the qualitative assessment of (**A**) RNN and (**B**) MLP-NN for the test area.

It is crucial to take the required measures to avoid the issue of the landslide separation from the bare land. The morphology characteristics of the landslide map is different from other types of land cover. For example, the shape, slope and other characteristics (i.e., dip direction, width and length) of the surface terrain may be changed after landslide occurs. Therefore, by using relevant features derived from very high resolution LiDAR, data such as texture and geometric features can be used to separate between landslides and bare land. In addition, applying different optimization techniques helped us to improve the classification accuracy in landslide detection over other landcover classes, such as bare land, man-made, etc., as described previously by Pradhan and Mezaal [9]. Their results demonstrated that using optimized techniques with very high resolution LiDAR data (0.5) enabled them to separate landslide and other types of land cover. In addition, the most relevant features in Table 4 were optimized during this study. Furthermore, authors of [16] suggested that using the object feature from LiDAR data is a suitable solution for landslide identification.

Table 4. Correlation-based feature selection (CFS) results for the most relevant feature subset at a scale of 75.52; StdDe: Standard deviation, DTM: Digital terrain model, GLCM: Gray level co-occurrence matrix.

Feature	Iteration	Rank
StdDe DTM	20	1
GLCM homogeneity	18	2
Mean slope	20	3
GLCM angular second moment	20	4
Mean intensity	17	5
Mean red	20	6
Mean DTM	20	7
GLCM contrast	18	8
GLCM dissimilarity	15	9
StdDev blue	20	10

The landslide detection results showed that the proposed model is robust. Optimizing the segmentation parameters, namely, scale, shape, and compactness, using the fuzzy logic supervised approach resulted in the effective differentiation of landslide from non-landslide (bare soil, cut slope and vegetation) objects. Creating accurate objects through the optimized segmentation process allowed the use of spatial, orthophoto, and textural features for feature detection. Landslides should be differentiated from non-landslides based on the accurate segmentation of spatial and textural features. The selection of relevant features in landslide detection relies on the experience of the analysts. Thus, a feature selection method is crucial for accurate and reliable landslide detection. The optimal features selected via the CFS method simplified landslide detection by the NN model. Computation time and reliance on the expert knowledge of the analyst were reduced. Moreover, the optimized parameters of the NN models improved the performance of the models, reduced the complexity of the models, and decreased overfitting in the training sample.

4.4. Performance of the MLP-NN and RNN

The models were implemented in Python using the open source TensorFlow deep learning framework developed by Google [26]. Meanwhile, the accuracy of the proposed NN models was tested using a 10-fold cross-validation method. The results are presented in Table 5. The best accuracy of 83.33% in the analysis area was achieved by the RNN model. The MLR-NN model achieved an accuracy of 78.38% in the same area. Furthermore, the RNN model outperformed the MLR-NN model in terms of stability of accuracy across different folds of the tested dataset. In the test area, accuracies of 81.11% and 74.56% were achieved with the RNN and MLP-NN models, respectively. These results indicated that the RNN model has better accuracy than the MLR-NN model in the analysis and test areas and indicated the high stability of the RNN model in detecting the spatial distribution of landslides.

Table 5. Cross-validation accuracy results of the proposed models.

Neural Network Model	Analysis Area	Test Area
RNN model	83.33%	81.11%
MLP-NN model	78.38%	74.56%

However, producing neural network models such as LSTM and convolution layers with fully connected networks is a crucial task. Complex networks with more hidden units and many modules often tend to have a better overfit due to the detection ability with respect to any possible interaction so the model becomes too specific to the training dataset. Thus, optimizing the network structures is very crucial for avoiding over-fitting. This study indicated that the hyperparameters in both models have a significant effect on their results. For example, the effect of learning rate varied from 0.1, 0.01, 0.05 and 0.001 in landslide detection. The highest accuracy was obtained when a learning rate reached 0.001. In contrast, increasing the learning rate to 0.1 significantly reduced the accuracy in both models. The batch size parameter in both models had significant effects on the result accuracy. The results of MLR-NN and RNN models showed high accuracies with batch sizes of 64 and 128, respectively. This indicates that RNN model achieved high accuracy with the increase of the batch size, whereas the accuracy of MLB-NN model was decreased.

Furthermore, it was revealed that the dropout rate had a substantial influence on the results of the RNN model. The RNN model showed higher accuracy when the dropout rate reached 0.6. The results of the RNN model indicated that the accuracy increased when the dropout rate parameter was increased.

The results of two models (Table 5), show that the accuracies of the RNN model outperformed the MLP-NN model in both study areas. This is due to several reasons, for example the fact that the MLP-NN model uses only local contexts and therefore it does not capture the temporal and spatial correlation in the dataset. Meanwhile, the hidden units of the RNN model contain historical information from the previous step. This indicates RNN model has more information about the data structure and accurate as compared to the MLP-NN model.

4.5. Sensitivity Analysis

The optimization of network architecture is necessary and should be considered over the use of standard parameters [28] because network architecture models are principally influenced by the analytical task and data type. Data could differ in size, relationships between independent and dependent variables, and complexity. Therefore, the neural architecture of the RNN and MLP-NN networks was enhanced using a grid search implemented in SciPy-python. The combinations of 10 parameters that can best identify landslide locations in densely vegetated areas were optimized.

The Adam optimizer is the most suitable algorithm for the optimization of the two NN models. Using the Adam optimizer with default parameters (learning rate $\mu = 0.001$, beta $\beta_1 = 0.9$, epsilon $\epsilon = 1e\text{-}08$ and weight decay $= 0.0$) yielded an accuracy of 0.77 and 0.825 for MLP-NN and RNN models, respectively, as shown in Figure 11. Rmsprop and Nadam optimizers also achieved excellent results for the two models. Overall, the Adam algorithm is more suitable for analyzing landslide data. However, better accuracy was obtained when Adadelta was used with the RNN model. Meanwhile, adding the weight decay in the neural network did not affect the results.

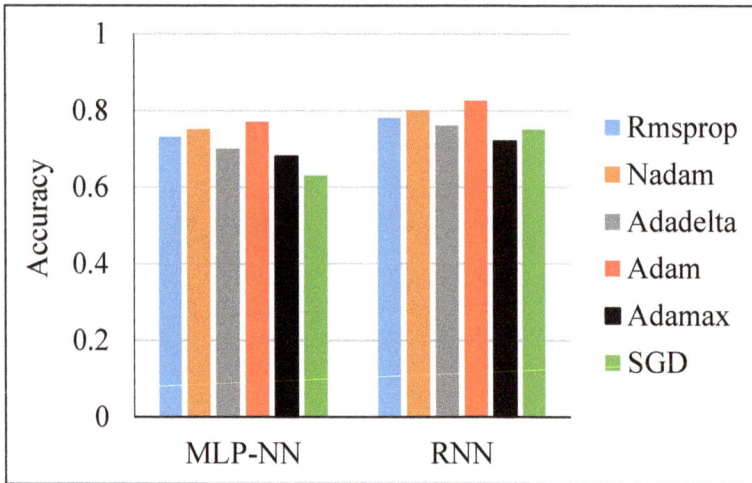

Figure 11. Impact of the optimization algorithm on the performance of MLP-NN and RNN models; SGD: Stochastic Gradient Descent.

Batch size, which refers to the number of training examples computed during optimization, has substantial effects on model accuracy. The results of the RNN and MLR-NN models are shown in Figure 12 and depict how increasing batch size from 2 to 128 (by ×2) affected model accuracy. The MLR-NN and RNN models exhibited the best accuracies with batch sizes of 64 and 128, respectively.

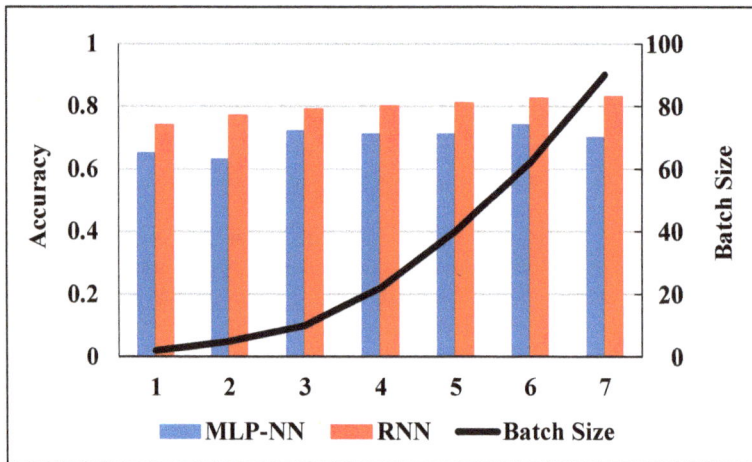

Figure 12. Impact of batch size on the performance of the MLP-NN and RNN models.

Overfitting can be avoided when dropouts are controlled through the number of parameters in the RNN model. Figure 13 illustrates the sensitivity analysis of the effects of dropout rate with various keep probability parameters on the RNN model. The results showed that the appropriate dropout rate is 0.6 for the RNN model. The selected dropout rate considerably affects the performance of NN models. The keep probability was selected in each dataset and analysis was conducted via a systematic grid search.

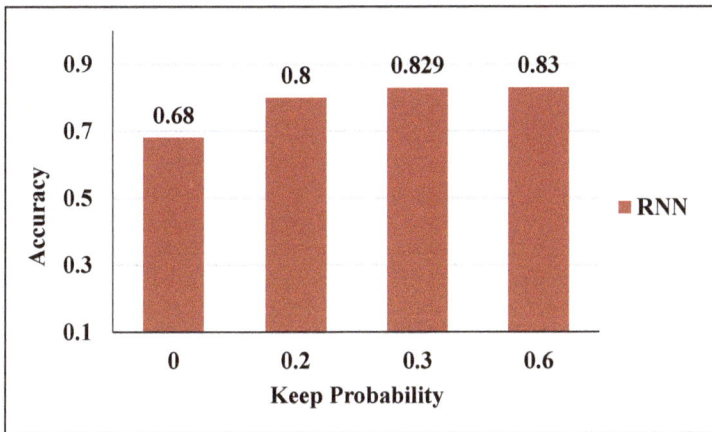

Figure 13. Influence of dropout rate on the performance of the RNN model.

4.6. Field Investigation

The reliability of the proposed methods was validated via field investigation using a handheld Global Position System (GPS) device (GeoExplorer 6000) to locate landslides (Figure 14) and to produce a precise and reliable inventory map of the Cameron Highlands. The more detailed information (landslide extent, source area, deposition, and volume) was obtained from in situ measurements which ultimately demonstrate the reliability of the produced inventory map in the field through use of a GeoExplorer 6000 handheld GPS. The results illustrated that the neural network techniques were able to detect true landslide locations which occurred in past years. Therefore, the results of this study verified that the proposed models can detect landslide locations and generate a reliable landslide inventory map.

Figure 14. Field photographs showing landslide locations during field investigation in (**A**) Tanah Rata and (**B**) Tanah Runtuh.

5. Conclusions

The Cameron Highlands, Malaysia form an ideal site for testing the feasibility of RNN and MLP-NN models for landslide detection based on high-resolution LiDAR data. The optimization of segmentation parameters is crucial for improving model performance and computational efficiency with different spatial subsets in the Cameron Highlands. Furthermore, optimization is essential for feature selection to improve the classification accuracy and the computational efficiency of the proposed methodology. The optimization of NN model parameters helped improve the performance of the model by reducing model complexity and preventing overfitting in the training sample. The RNN model exhibited better accuracy in the analysis and test areas than the MLR-NN model. This investigation showed that network architectures based on optimized techniques, very high resolution (VHR) airborne LiDAR-derived data, and spatial features could be used to effectively identify landslide locations in tropical regions. Therefore, this proposed automatic landslide detection method is a potential geospatial solution for managing landslide hazards and conducting landslide risk assessments.

Given that the proposed RNN model is more efficient than the MLP-NN model and has the potential to process the most relevant features, further studies should be conducted to fully optimize network structures for higher flexibility and eligibility for landslide detection. More theoretical tasks are recommended to enhance the representation of variables and data structure by the RNN model and the storage capacity of the data. Faster and more accurate NN techniques for landslide detection should be developed to overcome all the limitations related to accuracy and time. In addition, the RNN model can be integrated with other NN techniques to help improve other landslide applications.

Author Contributions: Mustafa Ridha Mezaal and Biswajeet Pradhan conceived and designed the experiments; Mustafa Ridha Mezaal, Biswajeet Pradhan and Maher Ibrahim Sameen performed the experiment and analyzed the data; Mustafa Ridha Mezaal and Biswajeet Pradhan contributed reagents/materials/analysis tools; Mustafa Ridha Mezaal and Biswajeet Pradhan wrote the paper. Biswajeet Pradhan is the main supervisor of PhD candidate Mustafa Ridha Mezaal; Helmi Zulhaidi Mohd Shafri and Zainuddin Md Yusoff supported as part of the co-supervisory committee of the first author.

Conflicts of Interest: The authors declare no conflict of interest.

References

1. Guzzetti, F.; Mondini, A.C.; Cardinali, M.; Fiorucci, F.; Santangelo, M.; Chang, K.-T. Landslide inventory maps: New tools for an old problem. *Earth-Sci. Rev.* **2012**, *112*, 42–66. [CrossRef]
2. Van Westen, C.J.; Castellanos, E.; Kuriakose, S.L. Spatial data for landslide susceptibility, hazard, and vulnerability assessment: An overview. *Eng. Geol.* **2008**, *102*, 112–131. [CrossRef]
3. Parker, R.N.; Densmore, A.L.; Rosser, N.J.; De Michele, M.; Li, Y.; Huang, R.; Petley, D.N. Mass wasting triggered by the 2008 Wenchuan earthquake is greater than orogenic growth. *Nat. Geosci.* **2011**, *4*, 449–452. [CrossRef]
4. Chen, R.-F.; Lin, C.-W.; Chen, Y.-H.; He, T.-C.; Fei, L.-Y. Detecting and Characterizing Active Thrust Fault and Deep-Seated Landslides in Dense Forest Areas of Southern Taiwan Using Airborne LiDAR DEM. *Remote Sens.* **2015**, *7*, 15443–15466. [CrossRef]
5. Pradhan, B.; Jebur, M.N.; Shafri, H.Z.M.; Tehrany, M.S. Data Fusion Technique Using Wavelet Transform and Taguchi Methods for Automatic Landslide Detection From Airborne Laser Scanning Data and QuickBird Satellite Imagery. *IEEE Trans. Geosci. Remote Sens.* **2016**, *54*, 1610–1622. [CrossRef]
6. McKean, J.; Roering, J. Objective landslide detection and surface morphology mapping using high-resolution airborne laser altimetry. *Geomorphology.* **2004**, *57*, 331–351. [CrossRef]
7. Whitworth, M.; Giles, D.; Murphy, W. Airborne remote sensing for landslide hazard assessment: A case study on the Jurassic escarpment slopes of Worcestershire, UK. *Q. J. Eng. Geol. Hydrogeol.* **2005**, *38*, 285–300. [CrossRef]
8. Blaschke, T. Object based image analysis for remote sensing. *ISPRS J. Photogramm. Remote Sens.* **2010**, *65*, 2–16. [CrossRef]
9. Pradhan, B.; Mezaal, M.R. Optimized Rule Sets for Automatic Landslide Characteristic Detection in a Highly Vegetated Forests. In *Laser Scanning Applications in Landslide Assessment*; Springer: New York, NY, USA, 2017; pp. 51–68.
10. Anders, N.S.; Seijmonsbergen, A.C.; Bouten, W. Segmentation optimization and stratified object-based analysis for semi-automated geomorphological mapping. *Remote Sens. Environ.* **2011**, *115*, 2976–2985. [CrossRef]
11. Belgiu, M.; Drăguţ, L. Comparing supervised and unsupervised multiresolution segmentation approaches for extracting buildings from very high resolution imagery. *ISPRS J. Photogramm. Remote Sens.* **2014**, *96*, 67–75. [CrossRef] [PubMed]

12. Drăguţ, L.; Tiede, D.; Levick, S.R. ESP: A tool to estimate scale parameter for multiresolution image segmentation of remotely sensed data. *Int. J. Geogr. Inf. Sci.* **2010**, *24*, 859–871. [CrossRef]

13. Zhang, Y.; Maxwell, T.; Tong, H.; Dey, V. Development of a supervised software tool for automated determination of optimal segmentation parameters for ecognition. In Proceedings of the ISPRS TC VII Symposium–100 Years, ISPRS, Vienna, Austria, 5–7 July 2010.

14. Chen, W.; Li, X.; Wang, Y.; Chen, G.; Liu, S. Forested landslide detection using LiDAR data and the random forest algorithm: A case study of the Three Gorges, China. *Remote Sens. Environ.* **2014**, *152*, 291–301. [CrossRef]

15. Kursa, M.B.; Rudnicki, W.R. Feature selection with the Boruta package. *J. Stat. Softw.* **2010**, *36*, 1–13. [CrossRef]

16. Li, X.; Cheng, X.; Chen, W.; Chen, G.; Liu, S. Identification of forested landslides using LiDar data, object-based image analysis, and machine learning algorithms. *Remote Sens.* **2015**, *7*, 9705–9726. [CrossRef]

17. Stumpf, A.; Kerle, N. Object-oriented mapping of landslides using Random Forests. *Remote Sens. Environ.* **2011**, *115*, 2564–2577. [CrossRef]

18. Borghuis, A.; Chang, K.; Lee, H. Comparison between automated and manual mapping of typhoon-triggered landslides from SPOT-5 imagery. *Int. J. Remote Sens.* **2007**, *28*, 1843–1856. [CrossRef]

19. Danneels, G.; Pirard, E.; Havenith, H.B. Automatic landslide detection from remote sensing images using supervised classification methods. In *Geoscience and Remote Sensing Symposium*; IEEE: Hoboken, NJ, USA, July 2007; pp. 3014–3017.

20. Moine, M.; Puissant, A.; Malet, J.-P. Detection of Landslides From Aerial and Satellite Images With a Semi-Automatic Method. Application to the Barcelonnette Basin (Alpes-de-Hautes-Provence, France). February 2009, pp. 63–68. Available online: https://halshs.archives-ouvertes.fr/halshs-00467545/document (accessed on 16 July 2017).

21. Pratola, C.; Del Frate, F.; Schiavon, G.; Solimini, D.; Licciardi, G. Characterizing land cover from X-band COSMO-SkyMed images by neural networks. In Proceedings of the Urban Remote Sensing Event (JURSE), Munich, Germany, 11–13 April 2011; pp. 49–52.

22. Singh, A.; Singh, K.K. Satellite image classification using Genetic Algorithm trained radial basis function neural network, application to the detection of flooded areas. *J. Vis. Commun. Image Represent.* **2017**, *42*, 173–182. [CrossRef]

23. Singh, K.K.; Singh, A. Detection of 2011 Sikkim earthquake-induced landslides using neuro-fuzzy classifier and digital elevation model. *Nat. Hazards* **2016**, *83*, 1027–1044. [CrossRef]

24. Mehrotra, A.; Singh, K.K.; Nigam, M.J.; Pal, K. Detection of tsunami-induced changes using generalized improved fuzzy radial basis function neural network. *Nat. Hazards* **2015**, *77*, 367–381. [CrossRef]

25. Benediktsson, J.A.; Sveinsson, J.R. Feature extraction for multisource data classification with artificial neural networks. *Int J. Remote Sens.* **1997**, *18*, 727–740. [CrossRef]

26. Yuan, H.; Van Der Wiele, C.F.; Khorram, S. An automated artificial neural network system for land use/land cover classification from Landsat TM imagery. *Remote Sens.* **2009**, *1*, 243–265. [CrossRef]

27. Fu, G.; Liu, C.; Zhou, R.; Sun, T.; Zhang, Q. Classification for High Resolution Remote Sensing Imagery Using a Fully Convolutional Network. *Remote Sens.* **2017**, *9*, 498. [CrossRef]

28. Sameen, M.I.; Pradhan, B. Severity Prediction of Traffic Accidents with Recurrent Neural Networks. *Appl. Sci.* **2017**, *7*, 476. [CrossRef]

29. Gorsevski, P.V.; Brown, M.K.; Panter, K.; Onasch, C.M.; Simic, A.; Snyder, J. Landslide detection and susceptibility mapping using LiDAR and an artificial neural network approach: A case study in the Cuyahoga Valley National Park, Ohio. *Landslides* **2016**, *13*, 467–484. [CrossRef]

30. Chang, K.T.; Liu, J.K.; Chang, Y.M.; Kao, C.S. An Accuracy Comparison for the Landslide Inventory with the BPNN and SVM Methods. *Gi4DM 2010*; Turino, Italy, 2010. Available online: https://www.researchgate.net/profile/Jin_King_Liu/publication/267709454_An_Accuracy_Comparison_for_the_Landslide_Inventory_with_the_BPNN_and_SVM_Methods/links/5511773f0cf29a3bb71de12c.pdf (accessed on 20 March 2017).

31. Robert, H.N. *Neurocomputing*; Addison-Wesley Pub. Co.: Boston, CA, USA, 1990; pp. 21–42.

32. Zurada, J.M. *Introduction to Artificial Neural Systems*; West Pub. Co.: Eagan, MN, USA, 1992; pp. 163–248.

33. Chang, K.T.; Liu, J.K. Landslide features interpreted by neural network method using a high-resolution satellite image and digital topographic data. In Proceedings of the XXth ISPRS Congress, Istanbul, Turkey, 12–23 July 2004.

34. Chen, H.; Zeng, Z.; Tang, H. Landslide deformation prediction based on recurrent neural network. *Neural Process. Lett.* **2015**, *41*, 169–178. [CrossRef]

35. Ma, H.-R.; Cheng, X.; Chen, L.; Zhang, H.; Xiong, H. Automatic identification of shallow landslides based on Worldview2 remote sensing images. *J. Appl. Remote Sens.* **2016**, *10*, 016008. [CrossRef]

36. Hall, M.A. Correlation-based Feature Selection for Machine Learning. Doctoral dissertation, The University of Waikato, Hamilton, New Zealand, April 1999.

37. Miner, A.; Flentje, P.; Mazengarb, C.; Windle, D. Landslide Recognition Using LiDAR Derived Digital Elevation Classifiers-Lessons Learnt from Selected Australian Examples. Available online: http://ro.uow. edu.au/cgi/viewcontent.cgi?article=1590&context=engpapers (accessed on 10 June 2017).

38. Martha, T.R.; Kerle, N.; van Westen, C.J.; Jetten, V.; Kumar, K.V. Segment optimization and data-driven thresholding for knowledge-based landslide detection by object-based image analysis. *IEEE Trans. Geosci. Remote Sens.* **2011**, *49*, 4928–4943. [CrossRef]

39. Pradhan, B.; Lee, S. Regional landslide susceptibility analysis using back-propagation neural network classifier at Cameron Highland, Malaysia. *Landslides* **2010**, *7*, 13–30. [CrossRef]

40. Olaya, V. Basic land-surface parameters. *Dev. Soil Sci.* **2009**, *33*, 141–169.

41. Barbarella, M.; Fiani, M.; Lugli, A. Application of LiDAR-derived DEM for detection of mass movements on a landslide. *Int. Arch. Photogramm. Remote Sens. Spat. Inf. Sci.* **2013**, *1*, 89–98. [CrossRef]

42. Duro, D.C.; Franklin, S.E.; Dubé, M.G. Multi-scale object-based image analysis and feature selection of multi-sensor earth observation imagery using random forests. *Int. J. Remote Sens.* **2012**, *33*, 4502–4526. [CrossRef]

43. Kohavi, R.; John, G.H. Wrappers for feature subset selection. *Artif. Intell.* **1997**, *97*, 273–324. [CrossRef]

44. Mokhtarzade, M.; Zoej, M.V. Road detection from high-resolution satellite images using artificial neural networks. *Int. J. Appl. Earth Obs. Geoinf.* **2007**, *9*, 32–40. [CrossRef]

45. Gardner, M.W.; Dorling, S.R. Artificial neural networks (the multilayer perceptron)—A review of applications in the atmospheric sciences. *Atmos. Environ.* **1998**, *32*, 2627–2636. [CrossRef]

46. Baczyński, D.; Parol, M. Influence of artificial neural network structure on quality of short-term electric energy consumption forecast. *IEE Proc.-Gener., Transm. Distrib.* **2004**, *151*, 241–245. [CrossRef]

47. Mia, M.M.A.; Biswas, S.K.; Urmi, M.C.; Siddique, A. An algorithm for training multilayer perceptron (MLP) for Image reconstruction using neural network without overfitting. *Int. J. Sci. Techno. Res.* **2015**, *4*, 271–275.

48. Yang, G.Y.C. *Geological Mapping from Multi-Source Data Using Neural Networks*; Geomatics Engineering, University of Calgary: Calgary, AB, Canada, 1995.

49. Hochreiter, S.; Schmidhuber, J. Long short-term memory. *Neural Comput.* **1997**, *9*, 1735–1780. [CrossRef] [PubMed]

50. Donahue, J.; Anne Hendricks, L.; Guadarrama, S.; Rohrbach, M.; Venugopalan, S.; Saenko, K.; Darrell, T. Long-term recurrent convolutional networks for visual recognition and description. In Proceedings of the IEEE Conference on Computer Vision and Pattern Recognition, Boston, MA, USA, 7–12 June 2015; pp. 2625–2634.

51. Pedregosa, F.; Varoquaux, G.; Gramfort, A.; Michel, V.; Thirion, B.; Grisel, O.; Vanderplas, J. Scikit-learn: Machine learning in Python. *J. Mach. Learn. Res.* **2011**, *12*, 2825–2830.

52. Sameen, M.I.; Pradhan, B.; Shafri, H.Z.; Mezaal, M.R.; bin Hamid, H. Integration of Ant Colony Optimization and Object-Based Analysis for LiDAR Data Classification. *IEEE J. Sel. Top. Appl. Earth Obs. Remote Sens.* **2017**, *10*, 2055–2066. [CrossRef]

53. Bartels, M.; Wei, H. Threshold-free object and ground point separation in LIDAR data. *Pattern Recognit. Lett.* **2010**, *31*, 1089–1099. [CrossRef]

54. Rau, J.Y.; Jhan, J.P.; Rau, R.J. Semiautomatic object-oriented landslide recognition scheme from multisensor optical imagery and DEM. *IEEE Trans. Geosci. Remote Sens.* **2011**, *52*, 1336–1349. [CrossRef]

applied
sciences

MDPI

Article

Application of Deep Networks to Oil Spill Detection Using Polarimetric Synthetic Aperture Radar Images

Guandong Chen [1], Yu Li [1,*], Guangmin Sun [1] and Yuanzhi Zhang [2,3,*]

[1] Faculty of Information Technology, Beijing University of Technology, Beijing 100124, China; cgd@emails.bjut.edu.cn (G.C.); gmsun@bjut.edu.cn (G.S.)
[2] National Astronomical Observatories, Chinese Academy of Sciences, Beijing 100012, China
[3] Key Laboratory of Lunar Science and Deep-space Exploration, Chinese Academy of Sciences, Beijing 100012, China
* Correspondence: yuli@bjut.edu.cn (Y.L.); yuanzhizhang@hotmail.com (Y.Z.); Tel.: +86-10-6480-7833 (Y.Z.)

Received: 29 July 2017; Accepted: 15 September 2017; Published: 21 September 2017

Featured Application: Using polarimetric synthetic aperture radar (SAR) remote sensing to detect and classify sea surface oil spills, for the early warning and monitoring of marine oil spill pollution.

Abstract: Polarimetric synthetic aperture radar (SAR) remote sensing provides an outstanding tool in oil spill detection and classification, for its advantages in distinguishing mineral oil and biogenic lookalikes. Various features can be extracted from polarimetric SAR data. The large number and correlated nature of polarimetric SAR features make the selection and optimization of these features impact on the performance of oil spill classification algorithms. In this paper, deep learning algorithms such as the stacked autoencoder (SAE) and deep belief network (DBN) are applied to optimize the polarimetric feature sets and reduce the feature dimension through layer-wise unsupervised pre-training. An experiment was conducted on RADARSAT-2 quad-polarimetric SAR image acquired during the Norwegian oil-on-water exercise of 2011, in which verified mineral, emulsions, and biogenic slicks were analyzed. The results show that oil spill classification achieved by deep networks outperformed both support vector machine (SVM) and traditional artificial neural networks (ANN) with similar parameter settings, especially when the number of training data samples is limited.

Keywords: oil spill; polarimetric synthetic aperture radar (SAR); deep belief network; autoencoder; remote sensing

1. Introduction

As one of the most significant sources of marine pollution, oil spills have caused serious environmental and economic impacts to the ocean and coastal zone [1]. Oil spills near the coast can be caused by ship accidents, explosion of oil rig platforms, broken pipelines, and deliberate discharge of tank-cleaning wastewater from ships. The NEREIDs program, sponsored by the European Commission, was the first robust attempt to use shipping, geological and metocean data to characterize oil spills in one of the major oil exploration areas of the world, prior to any major oil spill accident. Based on this data, oil spill models were established to simulate the development and trajectories of oil spills and investigate the susceptibility of coastal zone and find suitable measures to alleviate its impacts to the environment [2–5].

Early warning and near-real-time monitoring of oil slicks plays a very important role in cleaning up operation of oil spill to alleviate its impact to coastal environment [2,3]. Synthetic aperture radar (SAR) is one of most promising remote sensing systems for oil spill monitoring, for it can provide valuable information about the position and size of the oil spill [1]. Moreover, the wide coverage and all-day, all-weather capabilities make SAR very suitable for large scale oil spill monitoring and early warning [6–8].

In their early stages, studies of oil spill detection are mainly based on single polarimetric SAR images [9–12]. The theoretical rationale of SAR oil spill detection is that the presence of oil slicks on the sea surface dampens short-gravity and capillary waves, so the Bragg scattering from the sea surface is largely weakened. The ideal sea surface wind speed for oil spills detection is 3–14 m/s [13]. As a result, oil spills can be detected as "dark" areas in SAR images. However, some other manmade or natural phenomena can result in very similar low scattering areas on the sea surface, e.g., biogenic slicks, waves, currents and low-wind areas, etc. Conventional oil spill detection procedures use intensity, morphological texture, and auxiliary information to distinguish mineral oil and its lookalikes, with its processing chain divided into three main steps [13]: (1) dark spot detection; (2) features extraction; and (3) classification between mineral and its lookalikes.

Single polarimetric SAR-based oil spill detection algorithms need auxiliary information and large number of data samples to classify mineral oil and its lookalikes. Sometimes the shape and texture of oil slicks may vary, affecting the robustness of intensity-based oil spill classification algorithms. Polarimetric observation capabilities provided by advanced SAR sensors have much stronger capabilities for oil spills detection [14]. For instance, biogenic slicks and mineral oil are difficult to distinguish by single polarimetric SAR images. Yet, their polarimetric scattering mechanisms are largely different: for oil-covered areas, Bragg scattering is largely suppressed, and high polarimetric entropy can be documented. In the case of a biogenic slick, Bragg scattering is still dominant, but with a low intensity. Thus, similar polarimetric behaviors as those of oil-free areas should be expected in the presence of biogenic films. Hence, polarimetric features can largely help the image classification between mineral and biogenic lookalikes [14].

Various polarimetric features have been proposed to classify oil spills. The standard deviation of copolarized phase difference (phase difference between Vertical transmit and Vertical receive-VV and Horizonal transmit and Horizonal receive-HH channel) has shown a strong oil classification capability on C-, X-, and L-band data [15]. Nunziata et al. (2011) proposed pedestal height to describe the different polarization signature between mineral oil and biogenic lookalikes [16]. Minchew et al. (2012) took the advantage of copolarization ratio to study the mixing status of crude oil and sea water [17]. Zhang et al. (2011) used the conformity coefficient as a binary classifier [18]. Other polarimetric features such as degree of polarization, entropy, alpha angle, and Bragg likelihood angle were also used to classify oil spills [19–21].

Some previous studies conducted automatic oil-spill classification algorithms. Marghany (2001) developed models to discriminate textures between oil and water by using co-occurrence textures [22]. Gambardella et al. (2008) proposed one-class classification with an optimized feature selection algorithm and obtained a promising oil spill classification [23]. Frate et al. (2000) proposed a semiautomatic detection of oil spills by neural network [24]. Garcia-Pineda et al. (2008) developed the Textural Classifier Neural Network Algorithm (TCNNA) to map an oil spill in the Gulf of Mexico Deepwater Horizon accident [11]. Marghany (2013) used a genetic algorithm (GA) for automatic detection of an oil spill from ENVISAT ASAR (Advanced Synthetic Aperture Radar) data [25]. Li et al. (2013) used a Support Vector Machine (SVM) to detect oil spills based on morphological features on very limited data samples [26].

Polarimetric SAR features contain massive complementary and redundancy information. The extraction and optimization of them are closely related to the performance of oil spill classification [27]. Deep learning algorithms have very strong capabilities of exploring complex correlation between features and achieve very promising fitting result on complicated problems. It has been a very popular technique for image processing, computer vision, and natural language processing. According to the authors, deep learning has not been used in features optimization for oil spills detection based on polarimetric SAR data, and it should be a very promising research topic.

Deep neural network with multilayer neuron has powerful capabilities in describing complex functions compared with shallow networks [28]. However, the traditional gradient descent technique works poorly on a deep neural network when the weights are initialized randomly. The reason is that when the derivative is calculated using the back propagation method, the magnitude of the gradient (from the output layer to the initial layer of the network) decreases dramatically as the network depth increases. As the result, the gradient of the overall loss function, with respect to the weights of the first few layers, is very small. Thus, when the gradient descent method is used, the weights of the first layers change very slowly, so that they cannot learn effectively from the samples. This problem is often referred to as "gradient dispersion". In 2006, Hinton et al. proposed the deep belief network (DBN), which is a belief network composed of Restricted Boltzmann Machine (RBM) one layer at a time, to take the advantage of complementary priors of the data. Inspired by DBN, Beigio et al. (2006) used a stacked autoencoder, which is a deep multilayer neural network that initialized its weights by a greedy layer-wise unsupervised training strategy [29].

Moreover, feature dimension reduction can be seen as an early fusion step. Fusion at different stages of classification procedures is a booming research field that has shown capabilities for improvement of classification results. For instance, Vergara et al. fused the output of nonindependent detectors to derive the optimum classification result [30]. Late fusion of scores of several classifiers could be adapted to the proposed problem as a future research work.

The aims of this paper are exploring the capabilities of deep learning algorithms on polarimetric SAR-based marine oil spill detection. In Section 2, research methods including the representation of polarimetric SAR data, feature extraction methods and deep learning algorithms including DBN and SAE will be introduced. In Section 3, experiments were conducted on RADARSAT-2 data containing verified oil spills and biogenic lookalikes. The performance of different algorithms on various sample sizes for oil spill classification will be compared. Finally, conclusions are drawn in Section 4, and the significance and future work of the study will be briefly presented.

2. Methods

2.1. Foudamentals of Polarimetric SAR

The scattering characteristics of the observed target can be described by matrix, S; which links the scattered and incident electromagnet field, in the backscattered coordinate system:

$$\mathbf{E}^S = \frac{e^{-jkr}}{r}\mathbf{S}\mathbf{E}^i \tag{1}$$

where k is the wavenumber of the EM wave, r is the distance.

Fully polarimetric SAR observations can be achieved by quad-polarimetric mode, in which both horizontal and vertical polarized signals are transmitted alternatively and received coherently. The 2×2 scattering matrix is used to represent the single look complex quad-pol SAR data:

$$\mathbf{S} = \begin{pmatrix} S_{hh} & S_{hv} \\ S_{vh} & S_{vv} \end{pmatrix} \tag{2}$$

where S_{ij} describes the transmitted and received polarization, respectively, with h denoting the horizontal direction and v denoting the vertical direction.

To take advantage of statistical properties and reduce the effect of speckle noise of SAR data, covariance matrix is often derived from the scattering matrix by multilook its second order products:

$$C = \begin{pmatrix} \langle S_{hh}^2 \rangle & \langle \sqrt{2}S_{hh}S_{hv}^* \rangle & \langle S_{hh}S_{vv}^* \rangle \\ \langle \sqrt{2}S_{hv}S_{hh}^* \rangle & \langle 2S_{hv}^2 \rangle & \langle \sqrt{2}S_{hv}S_{vv}^* \rangle \\ \langle S_{vv}S_{hh}^* \rangle & \langle \sqrt{2}S_{vv}S_{hv}^* \rangle & \langle S_{vv}^2 \rangle \end{pmatrix} \tag{3}$$

where "*" is the symbol of conjugate, and "< >" stands for multilook by using an averaging window. Multilook is applied as a standard procedure to obtain the second order statistics (covariance matrix, coherence matrix) of the SAR data, an average window of 5×5 is normally used for balancing the multilook result and maintaining the spatial resolution.

2.2. Features Extraction for Oil Spills Detection

Previous studies proved experimentally that various SAR features could assist oil spill detection and classifications [31]. In this study, ten features including single VV channel intensity, entropy, alpha angle, degree of polarization, ellipticity, pedestal height, copolarized phase difference (CPD), conformity coefficient, correlation coefficient and coherence coefficient are extracted from the covariance matrix (or coherence matrix and Stokes vector deriving from the covariance matrix) [32] of polarimetric SAR data. The ten features investigated in this study, and their behavior on clean sea surface and sea surface covered by different materials, are given in Table 1. Detailed definitions and their behavior on different targets are provided explicitly in [27].

Table 1. Features investigated in this study.

Feature	Definition	For Mineral Oil	For Biogenic Slicks	For Clean Sea Surface
VV intensity	S_{VV}^2	Lower [1]	low	High
Entropy (H)	$P_i = \dfrac{\lambda_i}{\sum\limits_{j=1}^{3} \lambda_j}$	High	Low	Lower
Alpha (α)	$\alpha = P_1\alpha_1 + P_2\alpha_2 + P_3\alpha_3$	High	Low	Lower
Degree of Polarization (DoP)	$P = \dfrac{\sqrt{g_{i1}^2 + g_{i2}^2 + g_{i3}^2}}{g_{i0}}$	Low	High	High
Ellipticity (χ)	$\sin(2\chi) = -\dfrac{s_3}{m s_0}$	Positive	Negative	Negative
Pedestal Height (PH)	$NPH = \dfrac{\min(\lambda_1,\lambda_2,\lambda_3)}{\max(\lambda_1,\lambda_2,\lambda_3)}$	High	Low	Lower
Standard Deviation of CPD	CPD: $\varphi_c = \arg(\langle S_{HH}S_{VV}^* \rangle)$	High	Low	Lower
Conformity Coefficient (Conf. Co.)	$\mu \cong \dfrac{2(\mathrm{Re}(S_{HH}S_{VV}^*) - |S_{HV}|^2)}{|S_{HH}|^2 + 2|S_{HV}|^2 + |S_{VV}|^2}$	Negative	Positive	Positive
Correlation Coefficient (Corr. Co.)	$\rho_{HH/VV} = \left\| \dfrac{\langle S_{HH}S_{VV}^* \rangle}{\langle S_{HH}^2 \rangle \langle S_{VV}^2 \rangle} \right\|$	Low	High	Higher
Coherence Coefficient (Conf. Co.)	$Coh = \dfrac{|\langle T_{12} \rangle|}{\sqrt{\langle T_{11} \rangle \langle T_{22} \rangle}}$	Low	High	Higher

[1] Note: "lower" and "higher" mean that the property of the feature on a certain type of surface is close to the other surface that has the property of "low" or "high", but slightly lower or higher. "Std. copolarized phase difference (CPD)" stands for the standard deviation of CPD.

2.3. Deep Belief Network (DBN)

2.3.1. Restricted Boltzmann Machine

RBM is a neural perceptron consisting of visible and hidden layers, and the neurons between the visible layer (v_i, $i = 1, \ldots, N_v$) and the hidden layer (h_j, $j = 1, \ldots, N_h$) are bidirectional and fully connected. The basic structure of RBM is shown in Figure 1:

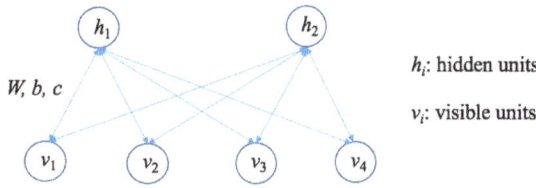

Figure 1. The illustration of a Restricted Boltzmann Machine (RBM) with two hidden units and four visible units.

In RBM, W represents the weight between any two connected neurons, in which each neuron has a bias coefficient b (of neurons) and c (of hidden neurons).

The energy of the RBM can be represented by:

$$E(v,h) = -\sum_{i=1}^{Nv} b_i v_i - \sum_{j=1}^{Nh} c_i j_i - \sum_{i,j=1}^{Nv,Nh} W_{i,j} v_i h_j \qquad (4)$$

And the probability of the activation of the hidden layer neuron h_j is:

$$P(h_j|v) = \sigma(b_j + \sum_i W_{i,j} x_i) \qquad (5)$$

Similarly, the neurons in the visible layer can also be activated by the bidirectional connected hidden neurons:

$$P(v_i|h) = \sigma(c_i + \sum_j W_{i,j} h_j) \qquad (6)$$

where σ is the activation function, e.g., sigmoid function:

$$\sigma(x) = \frac{1}{1+e^{-x}} \qquad (7)$$

Since for RBM the neurons of the same layer is not connected, they are independent:

$$P(h|v) = \prod_{j=1}^{Nh} P(h_j|v) \qquad (8)$$

$$P(v|h) = \prod_{i=1}^{Nv} P(v_i|h) \qquad (9)$$

Based on the input data vector x, the possibility of the activation of each hidden layer neuron can be calculated. Similarly, based on the activation state of hidden layer neurons, the activation state of visible layers can be calculated. Through a contrastive divergence algorithm [28], the parameters of the RBM: (b, c, W) can be set based on the input data vector x iteratively by a Gibbs sampling technique. An RBM can be seen as a feature detector, which is often used for dimensional reduction of the data. The training process of RBM is to find a probability distribution that can best produce training samples.

2.3.2. The Structure of DBN

DBN is a generative model which establishes a joint distribution between a label and the data sample. It not only considers P (label/observation), but also P (observation/label). In a DBN, several RBMs are connected. The hidden layer of the previous RBM is the next RBM's visible layer, and the output of the previous RBM is the input of the next RBM. During the pre-training process, the upper layer of RBM is trained before the training of the current layer. Usually when the top RBM is trained, the label information is also considered as the visible units.

2.3.3. The Fine-Tuning of DBN

Contrastive Wake-Sleep algorithms are usually used to fine-tune the pre-trained DBN. In the wake stage, the status of nodes of each layer is generated by external features and cognitive weights (upward), and the generated weights (downward) are modified using gradient descent algorithm. In the sleep stage, the state of the bottom neurons is generated through the top-level representation (the states learned by waking) and the weights generated in previous stage, then the cognitive weights of each layer are modified.

2.4. Stacked Autoencoder

2.4.1. Autoencoder

As shown in Figure 2, to build an autoencoder, three layers, namely, an input layer, a hidden layer and an output layer have to be established. The explanations of symbols used in Figure 2 are listed below:

n: the size of the input and output layer.
m: the size of the hidden layer.
$x \in \mathbb{R}^n$, $h \in \mathbb{R}^m$, $y \in \mathbb{R}^n$ stand for the data vector of the input, hidden and output layers, respectively.
$b \in \mathbb{R}^m$, $c \in \mathbb{R}^n$ stand for the bias vector of the hidden and output layers, respectively.
$W \in \mathbb{R}^{m \times n}$ stands for weights matrix between the input and hidden layer.
$\widetilde{W} \in \mathbb{R}^{n \times m}$ stands for weights matrix between the hidden and input layer.

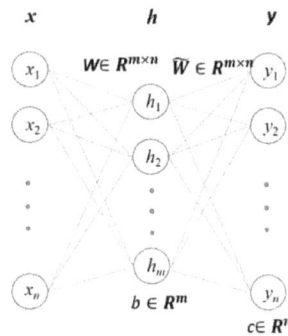

Figure 2. The structure of an autoencoder.

From the input layer to the output of hidden layer, the input signal is encoded. And from hidden layer to the output, the output of hidden layer is decoded by:

$$h = f(x) = s_f(Wx + b) \tag{10}$$

$$y = g(x) = s_g(\widetilde{W}x + c) \tag{11}$$

In Equations (10) and (11), $f()$ and $g()$ stand for the encoding and decoding functions, respectively. S_f and S_g are the corresponding activation functions of the encoder and decoder. sigmoid function can be chosen as the activation function and W^T can be taken as the weights $\widetilde{W}$ of the decoder.

Given input vectors, the autoencoder aims to minimize the difference between an input x and the output y. The reconstruction error can be described by the cross-entropy function:

$$L(x, y) = -\sum_{i=1}^{n} [x_i \log(y_i) + (1 - x_i) \log(1 - y_i)] \tag{12}$$

For the training set, S; the average reconstruction error can hence be established as:

$$\mathcal{L}(\theta) = \sum_{x \in S} L(x, g(f(x))) \tag{13}$$

By minimizing $\mathcal{L}(\theta)$, the parameter $\theta = \{W, b, c\}$ of the autoencoder can be fitted. The learning of an autoencoder does not need the label information, so it is an unsupervised procedure. The output of the hidden layer h can be seen as a representation of input x.

2.4.2. The Stacking of Autoencoders

In a SAE, autoencoders are stacked so that they take the output $h(k)$ of one hidden layer of the former autoencoder as the input for its successive autoencoder. Each layer is trained by a greedy unsupervised layer-wise training strategy, and the upper layers are the representations of relevant high-level abstractions (Figure 3). Stacked autoencoders can establish the deep neural network more efficiently by initializing its weights in a region near its local minimum.

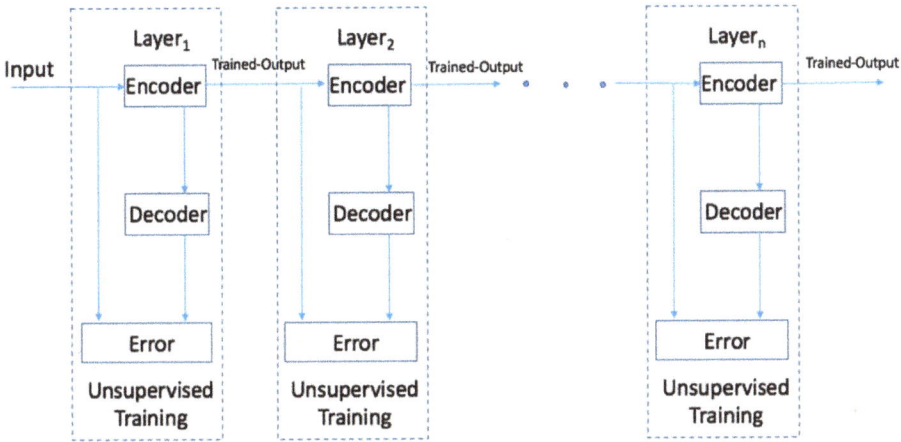

Figure 3. The demonstration of stacked autoencoder.

2.4.3. Fine-Tuning of the SAE

Normally the last layer of the SAE is connected to a classifier, it can be a neural network, Softmax classifier, SVM, etc. Finally, a fine-tuning process is also taken on either the whole network or only the classifier by taking the advantage of the label information through a supervised classification, using the back-propagation algorithm.

3. Experimental Results

3.1. The Experiment Data

In this study, RADARSAT-2 quad-pol SAR data acquired during the 2011 Norwegian oil-on-water experiment (59°59' N, 2°27' E) were used for analysis. The data was received at 17:27 of 8 June 2011 UTC in fine-quad polarimetric mode, with the spatial resolution of 4.7 × 4.8 m in range and azimuth directions. The incident angle of the image is 34.5–36.1° and the local wind speed is 1.6–3.3 m/s. For the convenience of processing and display, a data sample with 2000 × 2000 pixels was picked from the single look complex (SLC) data. The pseudo RGB image of the RADARSAT-2 data on the Pauli basis are provided in Figure 4. In the scene, three verified slicks were present; from left to right, they were: biogenic film, emulsions and mineral oil [33]. The biogenic film was simulated by Radiagreen plant oil. Emulsions were made of Oseberg blend crude oil mixed with 5% IFO380 (Intermediate Fuel Oil), released 5 h before the radar acquisition. Additionally, the Balder crude oil was released 9 h before the radar acquisition [34].

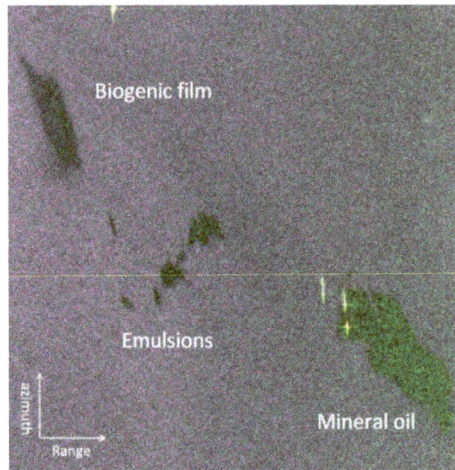

Figure 4. Pauli RGB image of RADARSAT-2 data. (RADARSAT-2 Data and Products © Macdonald, Dettwiler and Associates Ltd., Vancouver, BC, Canada, 2011—All Rights Reserved. RADARSAT is an official mark of the Canadian Space Agency).

3.2. The Experiment Procedure

The SLC quad-polarimetric SAR data was firstly multi-looked, and then the covariance matrix and coherency matrix of the data samples were generated. As mentioned before, 10 features are extracted and saved as a 10-dimension vector for each pixel.

As shown in Figure 5, the 24,000 data samples were picked up from the image, including 12,000 verified positive (mineral oil) and 12,000 negative (clean sea surface and biogenic slick) samples. The data samples were picked by squared boxes with the size of 20 × 20 for convenience and keeping the purity of the sample, and then their order was shuffled.

In order to test the performance of different algorithms and avoid over-fitting, a six-fold cross-validation was applied. We first divided the training set into six subsets equally. Then the five-sixths of the data samples were used as training set and the rest were taken as testing set. Sequentially, we repeat the classification and another one-sixth data sample were used as testing set. The experiment was conducted six times until each instance of the whole training set is predicted once. Finally, the cross-validation accuracy is the overall percentage of data which are correctly classified.

In order to test the performance of these algorithms on smaller sample sizes, the whole dataset was divided into smaller groups. All the 24,000 data samples were divided into 5 and 25 groups randomly. Then classifications were conducted on these groups, namely 4000 training, 800 testing and 800 training, and 160 testing samples respectively. In the experiment, the classification accuracy of smaller sample size was obtained by averaging the classification result on each group respectively.

In the experiment, two previously introduced deep learning algorithms (i.e., DBN and SAE) were tested on their performance of oil spill detection and classification. In addition, two traditional supervised classifiers including neural network (NN) and SVM were compared.

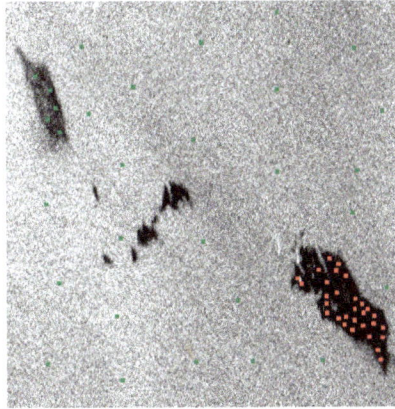

Figure 5. Demonstration of a selected area for analysis (taking VV^2 image as background); 24,000 pixels are picked as data samples.

The key parameters of these applied classifiers are shown in Tables 2–5:

Table 2. Parameter settings of the neural network.

Parameter	Value
Sizes of layers	[10, 8, 6, 2]
Activation function	Sigmoid
Learning rate	1
Number of epochs	100
Batch size	100

Table 3. Parameter settings of the support vector machine (SVM).

Parameter	Value
Type of SVM	C-SVC (n kind classification)
Type of kernel	Radial Basis Function (RBF): $\exp\left(-\gamma \times |u - v|^2\right)$
γ of the RBF	$1/k$ (k: number of features)
Cost	1
ε (termination criterion)	0.001
Weight w_i	1 (set the parameter C of class i to $w_i \times C$)
Shrinking h	1 (use the shrinking heuristics)

Table 4. Parameter settings of the stacked autoencoder (SAE).

Parameter	Value
Sizes of SAE layers	[8, 6]
Activation function SAE	Sigmoid
Learning rate of SAE	1
Input Zero Masked Fraction of SAE	0.5
Number of epochs when training SAE	10
Batch size when training SAE	100
Size of the whole network	[10, 8, 6, 2]
Activation function of the neural network	Sigmoid
Learning rate when making fine-tuning	3
Number of epochs when making fine-tuning	100
Batch size when making fine-tuning	100

Table 5. Parameter settings of the deep belief network (DBN).

Parameter	Value
Sizes of RBM layers	[8, 6]
Number of epochs when training RBM	10
Batch size when training RBM	100
Momentum for RBM	0
Learning rate alpha of RBM	1
Size of the whole network	[10, 8, 6, 2]
Activation function of the neural network	Sigmoid
Number of epochs when making fine-tuning	100
Batch size when making fine-tuning	100

LIBSVM-a library for Support Vector Machines [33] was used to implement the SVM algorithm. The parameters C and γ were derived by shrinking heuristics search technique. The neural network has ten input neurons, two hidden layers and two output neurons. The initialization of the former layers of deep learning algorithms SAE and DBN were carried out by unsupervised pretraining, and then the outputs were connected to a neural network with two output neurons.

3.3. Results and Discusion

To examine the feature dimension reduction capability of the deep neural networks, scatter plots of the main original features and the features derived by principal component analysis (PCA), DBN and SAE are shown in Figure 6. Two of the most discriminative features, conformity coefficient and degree of polarization (DoP) of HH and VV transition/receiving combinations, as two of the most effective feature in oil spill classification [31], are plotted in Figure 6a. Scatter plots of the first two components derived by PCA are shown in Figure 6b. In this paper, taking the advantage of DBN and SAE, the dimension of polarimetric features are reduced to six, then they are put into fully connected neural network. To show these features in a scatter plot, PCA was implemented on the six features, and then the first two components are shown in Figure 6c,d. It can be observed that deep neural network algorithms effectively extracted the information from high dimensional features and improved their separability to distinguish mineral oil and none mineral samples.

The classification results are shown in Table 6 statistically, some key findings and discussions are listed as follow:

- SAE achieved the highest classification accuracy (lowest testing error) among all the algorithms on different sample sizes. DBN achieved a close performance to SAE. SAE and DBN applied in the experiment had similar structures and both of them took the advantage of greedy unsupervised layer-wise pretraining, so very similar performances were achieved. The unsupervised pretraining worked as a feature optimizer, which can reveal the latent relationship and reduction of noise in features. It helps to improve the performance of the followed supervised classification procedure.
- On the small training data set, deep learning algorithms have much higher performance than neural networks. When the number of data set is reduced, the parameters of traditional NN cannot be sufficiently tuned. Based on unsupervised pretraining, deep learning algorithms such as SAE and DBN have much stronger capability to achieve the optimized solution of the learning problem.
- When the number of data sample size is reduced, the classification error will increase (i.e., the accuracy is reduced). When the number of data sets reduced, the characteristics of the studied object cannot be sufficiently expressed by the limited number of data samples, so the classification performance is reduced.
- On the large training data set, NN have a close performance to deep learning algorithms. With large number of training data, the parameters of NN can be sufficiently adjusted. In this experiment, the NN have a few hidden layers, the gradient of objective function could pass to the

layers in the front effectively. As the result, comparable classification performance was achieved by NN on large data set.

- SVM has better performance on small sample sizes than NN. SVM is based on structural risk minimization, which has superior performance on relative small data sets. It maximizes the classification margin, which is decided by a few support vectors and could successfully avoid the risk of the "curse of dimensionality". However, although the SVM has several advantages, it is equivalent to a NN with one hidden layer, so on learning complicated relationships its performance is no better than the other three more complex classifiers applied in the experiment.

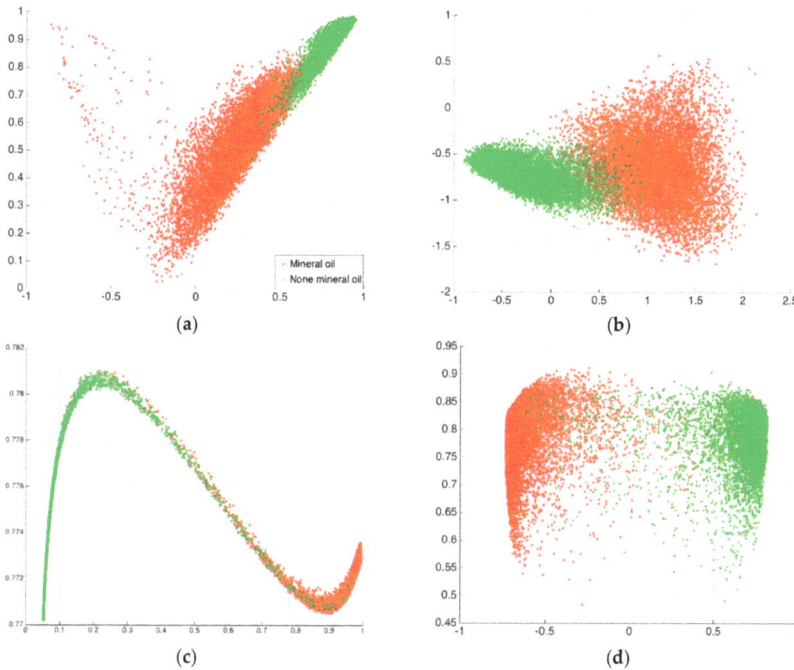

Figure 6. Scatter plots of the main original features (**a**) and the features derived by principal component analysis (PCA), deep believe network (DBN) and stacked autoencoder (SAE) (**b–d**).

Table 6. Testing error (1-accuracy) of classification achieved by different classifiers and on different sizes of data samples.

Classifier	Number of Data Samples (Training-Testing) Accuracy (Execution Time/Seconds)		
	a. 20,000–4000	b. 4000–800	c. 800–160
SVM (support vector Machine)	1.26% (1.16)	1.38% (0.07)	1.64% (0.01)
NN (nerual network)	1.16% (6.52)	1.68% (1.33)	2.13% (0.34)
PCA (principal component analysis)-SVM	1.20% (0.9)	1.33% (0.06)	1.58% (0.01)
PCA-NN	1.17% (6.2)	1.48% (1.2)	1.90% (0.35)
SAE (stacked autoencoder)	1.05% (6.33)	1.33% (1.24)	1.39% (0.32)
DBN (deep believe network)	1.16% (5.63)	1.42% (1.18)	1.53% (0.28)

The confusion matrix of the cross-validation testing result is shown in Tables 7–10. The best classification results were achieved by SAE on the largest data set: 20,000 training and 4000 testing samples. On the 24,000 testing set, 251 pixels were wrongly classified. 101 pixels of these 251 pixels were false positive (commission errors) and 150 pixels were false negative (omission errors). Similar false positive was achieved by DBN, with slightly higher false negative rate. In the confusion matrix achieved by NN and SAE, it can be discovered that compared with deep learning algorithms, they achieved lower false negative and higher false positive rates.

Table 7. Confusion Matrix (20,000 training, 4000 testing, six-fold) derived by SAE.

Confusion Matrix of SAE	Mineral Oil	Non Mineral Oil	Total
Mineral oil (truth)	11,844	150	12,000
Nonmineral oil (truth)	101	11,896	12,000
Total	11,948	12,052	24,000

Table 8. DBN Confusion Matrix (20,000 training, 4000 testing, six-fold) derived by DBN.

Confusion Matrix of DBN	Mineral Oil	Non Mineral Oil	Total
Mineral oil (truth)	11,819	181	12,000
Non mineral oil (truth)	101	11,899	12,000
Total	11,948	12,052	24,000

Table 9. SAE Confusion Matrix (20,000 training, 4000 testing, six-fold) derived by NN.

Confusion Matrix of NN	Mineral Oil	Non Mineral Oil	Total
Mineral oil (truth)	11,881	119	12,000
Non mineral oil (truth)	160	11,840	12,000
Total	11,948	12,052	24,000

Table 10. SAE Confusion Matrix (20,000 training, 4000 testing, six-fold) derived by SVM.

Confusion Matrix of SVM	Mineral Oil	Non Mineral Oil	Total
Mineral oil (truth)	11,893	107	12,000
Non mineral oil (truth)	201	11,799	12,000
Total	11,948	12,052	24,000

From the binary output that achieved by SAE (Figure 7), it can be observed that a few pixels in the area covered by the biogenic slick are classified as mineral oil. The possible reason of these "misclassifications" is the affection of signal noise on space-borne SAR data or the uniform distribution of the mineral oil and biogenic slicks. This misinterpretation can be further eliminated by a simple postprocessing step. Corrosion and swelling algorithms can be applied on the binary classification result to fix the small holes (missing alarm) in large oil-covered areas and isolated positive targets (false alarm) in the sea surface area.

The receiver operating characteristics (ROC) of these classifiers in oil spill detection are approached based on 20,000 training, 4000 testing (Figure 8a,b) and 4000 training, 800 testing samples (Figure 8c,d) respectively. All the ROC curves are very close to the upper-left corner of the ROC map (Figure 8a,c). In the zoomed-in map (Figure 8b,d), some minor differences can be observed. Compared with other classifiers, SVM achieved a lower true positive rate under low false positive rate requirements, and higher true positive rate under high false positive rate requirements. And for NN the situation is just opposite. Deep neural networks, SAE and DBN achieved a modest true positive rate in the whole false positive rate range.

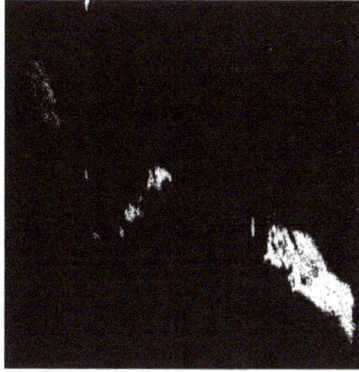

Figure 7. Classification result achieved by SAE, 0 stands for nonoil and 1 stands for mineral oil.

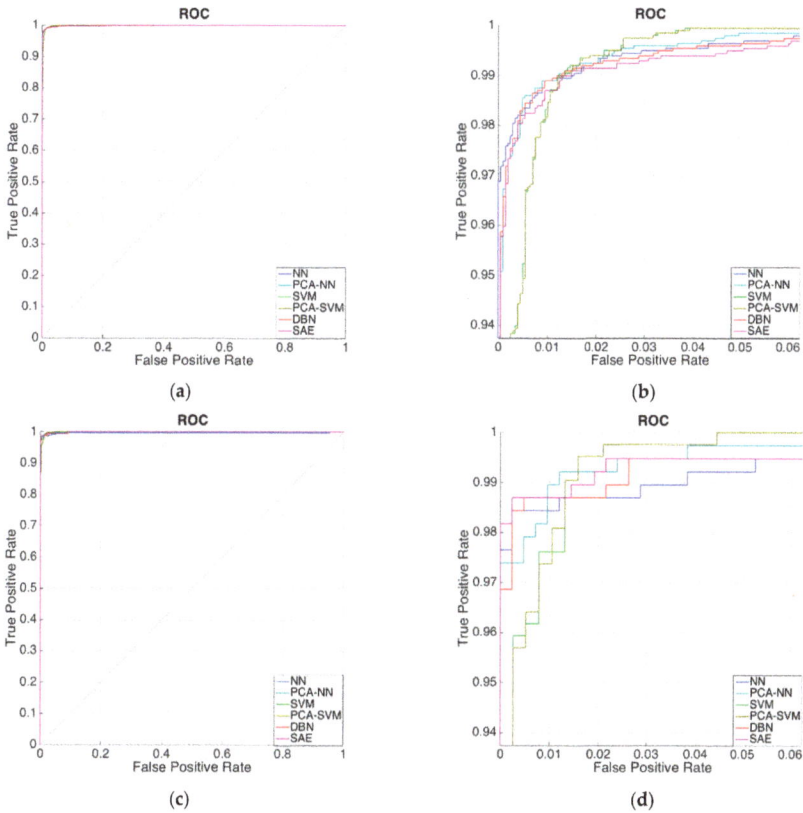

Figure 8. Receiver operating characteristics (ROC) curves of the classifiers. (**a**) ROC curve achieved based on 20,000 training, 4000 testing samples; (**b**) Zoomed in map of (**a**); (**c**) ROC curve achieved based on 4,000 training, 800 testing samples; (**d**) Zoomed in map of (**c**).

4. Conclusions

In this paper, the capability of polarimetric SAR to detect and classify marine oil spills was investigated. Potential features were extracted from a covariance matrix, a coherence matrix and a Stokes vector of the original SLC quad-pol SAR data. Deep learning algorithms together with classic classifiers were compared and analyzed. A key discovery of this paper is that given insufficient number of data samples, deep learning algorithms such as SAE and DBN can achieve better performance than traditional algorithms by initializing their parameters from a position closer to the optimum solution. Polarimetric SAR data confirmed its strong capacity in distinguishing mineral oil and its biogenic lookalikes. This can be achieved by a one-step operation, with no need to firstly segment and then classify data samples based on auxiliary information. The advantages demonstrated by polarimetric SAR can greatly boost the efficiency and accuracy of marine oil spill detection. Further studies will be conducted on features extracted from compact polarimetric SAR modes, with wider swath width, to achieve larger monitoring areas and shorter revisit times: two of the prime requirements for marine surveillance through large areas.

Acknowledgments: The RADARSAT-2 data provided by CSA and MDA is highly appreciated. This research is jointly supported by the National Key Research and Development Program of China (2016YFB0501501) and the Natural Scientific Foundation of China (41471353 and 41706201). The authors would like to thank the anonymous reviewers for their valuable comments and suggestions that helped improve the quality of this manuscript.

Author Contributions: Yu Li and Yuanzhi Zhang conceived and designed the experiments. Guandong Chen performed the experiments and analyzed the data. Yu Li, Guangmin Sun and Yuanzhi Zhang wrote the paper.

Conflicts of Interest: The authors declare no conflicts of interest.

References

1. Fingas, M. *The Basics of Oil Spill Cleanup*; Lewis Publisher: Boca Raton, FL, USA, 2001.
2. Alves, T.M.; Kokinou, E.; Zodiatis, G.; Lardner, R.; Panagiotakis, C.; Radhakrishnan, H. Modelling of oil spills in confined maritime basins: The case for early response in the Eastern Mediterranean Sea. *Environ. Pollut.* **2015**, *206*, 390–399. [CrossRef] [PubMed]
3. Lardner, R.; Zodiatis, G. Modelling oil plumes from subsurface spills. *Mar. Pollut. Bull.* **2017**. [CrossRef] [PubMed]
4. Soomere, T.; Döös, K.; Lehmann, A.; Meier, H.M.; Murawski, J.; Myrberg, K.; Stanev, E. The potential of current- and wind-driven transport for environmental management of the baltic sea. *Ambio* **2014**, *43*, 94–104. [CrossRef] [PubMed]
5. Chrastansky, A.; Callies, U. Model-based long-term reconstruction of weather-driven variations in chronic oil pollution along the German North Sea coast. *Mar. Pollut. Bull.* **2009**, *58*, 967–975. [CrossRef] [PubMed]
6. Alpers, W.H.; Espedal, A. Oils and Surfactants. In *Synthetic Aperture Radar Marine User's Manual*; Christopher, R.J., Apel, J.R., Eds.; US Department Commerce: Washington, DC, USA, 2004.
7. Gade, M.; Alpers, W. Using ERS-2 SAR for routine observation of marine pollution in European coastal waters. *Sci. Total Environ.* **1999**, *237*, 441–448. [CrossRef]
8. Ferraro, G.; Meyer-Roux, S.; Muellenhoff, O.; Pavliha, M.; Svetak, J.; Tarchi, D.; Topouzelis, K. Long term monitoring of oil spills in European seas. *Int. J. Remote Sens.* **2009**, *30*, 627–645. [CrossRef]
9. Migliaccio, M.; Tranfaglia, M.; Ermakov, S.A. A physical approach for the observation of oil spills in SAR images. *IEEE J. Ocean. Eng.* **2005**, *30*, 496–507. [CrossRef]
10. Topouzelis, K.; Karathanassi, V.; Pavlakis, P.; Rokos, D. Detection and discrimination between oil spills and look-alike phenomena through neural networks. *ISPRS J. Photogramm. Remote Sens.* **2007**, *62*, 264–270. [CrossRef]
11. Garcia-Pineda, O.; MacDonald, I.R.; Li, X.; Jackson, C.R.; Pichel, W.G. Oil spill mapping and measurement in the gulf of Mexico with Textural Classifier Neural Network Algorithm (TCNNA). *IEEE J. Sel. Top. Appl. Earth Obs. Remote Sens.* **2013**, *6*, 2517–2525. [CrossRef]
12. Pan, G.; Tang, D.; Zhang, Y. Satellite monitoring of phytoplankton in the East Mediterranean Sea after the 2006 Lebanon oil spill. *Int. J. Remote Sens.* **2012**, *33*, 7482–7490. [CrossRef]
13. Solberg, A.H.S. Remote sensing of ocean oil-spill pollution. *Proc. IEEE* **2012**, *100*, 2931–2945. [CrossRef]

14. Migliaccio, M.; Gambardella, A.; Tranfaglia, M. SAR polarimetry to observe oil spills. *IEEE Trans. Geosci. Remote Sens.* **2007**, *45*, 506–511. [CrossRef]

15. Migliaccio, M.; Nunziata, F.; Gambardella, A. On the co-polarized phase difference for oil spill observation. *Int. J. Remote Sens.* **2009**, *30*, 1587–1602. [CrossRef]

16. Nunziata, F.; Migliaccio, M.; Gambardella, A. Pedestal height for sea oil slick observation. *IET Radar Sonar Navig.* **2011**, *5*, 103–110. [CrossRef]

17. Minchew, B.; Jones, C.E.; Holt, B. polarimetric analysis of backscatter from the deepwater horizon oil spill using L-band synthetic aperture radar. *IEEE Trans. Geosci. Remote Sens.* **2012**, *50*, 3812–3830. [CrossRef]

18. Zhang, B.; Perrie, W.; Li, X.; Pichel, W. Mapping sea surface oil slicks using RADARSAT-2 quad-polarization SAR image. *Geophys. Res. Lett.* **2011**, *38*. [CrossRef]

19. Skrunes, S.; Brekke, C.; Eltoft, T. An experimental study on oil spill characterization by multi-polarization SAR. In Proceedings of the 9th European Conference on Synthetic Aperture Radar (2012 EUSAR), Nuremberg, Germany, 23–26 April 2012; pp. 139–142.

20. Shirvany, R.; Chabert, M.; Tourneret, J.-Y. Ship and oil-spill detection using the degree of polarization in linear and hybrid/compact dual-pol SAR. *IEEE J. Sel. Top. Appl. Earth Obs. Remote Sens.* **2012**, *5*, 885–892. [CrossRef]

21. Salberg, A.-B.; Rudjord, O.; Solberg, A.H.S. Model based oil spill detection using polarimetric SAR. In Proceedings of the IEEE International Geoscience and Remote Sensing Symposium (IGARSS), Munich, Germany, 22–27 July 2012; pp. 5884–5887.

22. Marghany, M. RADARSAT automatic algorithms for detecting coastal oil spill pollution. *Int. J. Appl. Earth Obs. Geoinf.* **2001**, *3*, 191–196. [CrossRef]

23. Gambardella, A.; Giacinto, G.; Migliaccio, M. On the Mathematical Formulation of the SAR Oil-Spill Observation Problem. In Proceedings of the IEEE International Geoscience and Remote Sensing Symposium (IGARSS), Boston, MA, USA, 7–11 July 2008; pp. III-1382–III-1385.

24. Frate, F.; Petrocchi, A.; Lichtenegger, J.; Calabresi, G. Neural networks for oil spill detection using ERS-SAR data. *IEEE Trans. Geosci. Remote Sens.* **2000**, *38*, 2282–2287. [CrossRef]

25. Marghany, M. Genetic Algorithm for Oil Spill Automatic Detection from Envisat Satellite Data. In Proceedings of the Computational Science and Its Applications—ICCSA, Ho Chi Minh City, Vietnam, 24–27 June 2013; Murgante, B., Misra, S., Carlini, M., Carmelo, M.T., Nguyen, H., Taniar, D., Apduhan, B.O., Gervasi, O., Eds.; Springer: Berlin/Heidelberg, Germany, 2013; pp. 587–598.

26. Li, Y.; Zhang, Y. Synthetic aperture radar oil spill detection based on morphological characteristics. *Geospat. Inf. Sci.* **2014**, *17*, 8–16. [CrossRef]

27. Zhang, Y.; Li, Y.; Liang, X.S.; Tsou, J. Comparison of oil spill classifications using fully and compact polarimetric SAR images. *Appl. Sci.* **2017**, *16*, 193. [CrossRef]

28. Hilton, G.; Salakhutdinov, R.R. Reducing the dimensionality of data with neural networks. *Science* **2006**, *313*, 504–507.

29. Bengio, Y.; Lamblin, P.; Popovici, D.; Larochelle, H. Greedy Layer-Wise Training of Deep Networks. In *Advances in Neural Information Processing Systems 19 (NIPS'06)*; MIT Press: Cambridge, MA, USA, 2006.

30. Vergara, L.; Soriano, A.; Safont, G.; Salazar, A. On the fusion of non-independent detectors. *Digit. Signal Process.* **2016**, *50*, 24–33. [CrossRef]

31. Li, Y.; LIN, H.; Chen, J.; Zhang, Y. Comparisons of circular transmit and linear receive compact polarimetric SAR features for oil slicks discrimination. *J. Sens.* **2015**, *2015*, 631561. [CrossRef]

32. Li, Y.; Zhang, Y.; Chen, J.; Zhang, H. Improved compact polarimetric SAR quad-pol reconstruction algorithm for oil spill detection. *IEEE Geosci. Remote Sens. Lett.* **2014**, *11*, 1139–1142. [CrossRef]

33. Chang, C.-C.; Lin, C.-J. LIBSVM: A library for support vector machines. *ACM Trans. Intell. Syst. Technol.* **2011**, *2*, 27. [CrossRef]

34. Skrunes, S.; Brekke, C.; Eltoft, T. Characterization of marine surface slicks by Radarsat-2 multipolarization features. *IEEE Trans. Geosci. Remote Sens.* **2014**, *52*, 5302–5319. [CrossRef]

applied
sciences

MDPI

Article

Application of Artificial Neural Networks to Ship Detection from X-Band Kompsat-5 Imagery

Jeong-In Hwang [1], Sung-Ho Chae [1,2], Daeseong Kim [1] and Hyung-Sup Jung [1,*]

[1] Department of Geoinformatics, University of Seoul, Seoul 02504, Korea; happy9680@uos.ac.kr (J.-I.H.); cshbe90@uos.ac.kr (S.-H.C.); kds2991@uos.ac.kr (D.K.)

[2] Center for Environmental Assessment Monitoring, Environmental Assessment Group, Korea Environment Institute (KEI), Sejong-si 30147, Korea

* Correspondence: hsjung@uos.ac.kr; Tel.: +82-2-6490-2892

Received: 31 July 2017; Accepted: 18 September 2017; Published: 20 September 2017

Abstract: For ship detection, X-band synthetic aperture radar (SAR) imagery provides very useful data, in that ship targets look much brighter than surrounding sea clutter due to the corner-reflection effect. However, there are many phenomena which bring out false detection in the SAR image, such as noise of background, ghost phenomena, side-lobe effects and so on. Therefore, when ship-detection algorithms are carried out, we should consider these effects and mitigate them to acquire a better result. In this paper, we propose an efficient method to detect ship targets from X-band Kompsat-5 SAR imagery using the artificial neural network (ANN). The method produces the ship-probability map using ANN, and then detects ships from the ship-probability map by using a threshold value. For the purpose of getting an improved ship detection, we strived to produce optimal input layers used for ANN. In order to reduce phenomena related to the false detections, the non-local (NL)-means filter and median filter were utilized. The NL-means filter effectively reduced noise on SAR imagery without smoothing edges of the objects, and the median filter was used to remove ship targets in SAR imagery. Through the filtering approaches, we generated two input layers from a Kompsat-5 SAR image, and created a ship-probability map via ANN from the two input layers. When the threshold value of 0.67 was imposed on the ship-probability map, the result of ship detection from the ship-probability map was a 93.9% recall, 98.7% precision and 6.1% false alarm rate. Therefore, the proposed method was successfully applied to the ship detection from the Kompsat-5 SAR image.

Keywords: synthetic aperture radar (SAR); ship detection; artificial neural network (ANN); Kompsat-5

---

1. Introduction

Ship-detection algorithms from SAR images have been proposed by many researchers [1–5]. Most methods exploit the fact that ships on SAR images are much brighter than the sea surface due to the corner-reflection effect [6,7]. That is, the ship is a target having a higher backscatter coefficient, and hence the ship is very bright in SAR imagery, while the sea surface is very dark in SAR imagery because it has a lower backscatter coefficient.

Constant false-alarm rate (CFAR) has been widely used for ship detection. The CFAR detects ships by finding a threshold value using the probability density function (PDF) of background clutter [8–11]. However, detection results from the CFAR algorithm can be biased by statistical variables such as mean and standard deviation, because the false-alarm rate must be fixed [12]. In addition, the CFAR algorithm has the disadvantages that it is very complex and time-consuming [13]. The adaptive threshold approach is very similar to CFAR in that it estimates the mean and standard deviation of an image using constant value c, but the approach has a drawback in that it does not consider the characteristics of background noise [14].

Appl. Sci. **2017**, *7*, 961

Recently, many studies have relied on image-based feature extraction (machine learning) methods, such as the artificial neural network (ANN) and support vector machine (SVM) methods [15,16]. The machine-learning analyzes the input data by calculating weighting factors from the relation between input data and training set. Thus, results from the machine learning can be varied according to input data and training set [17]. This means that the performance of the machine-learning algorithm is largely dependent on the selection of input data and training sets. In order to choose the training set for ship detection, we can use automatic identification system (AIS) data if it is available, whereas we can select the training set from a map which is manually created if AIS data is not available [18]. Thus, the selection of training set used for ship detection is not very difficult. However, the selection of input layers used for ship detection is not easy. The limitations of ship detection using the machine-learning algorithm are as follows:

Firstly, image characteristics have been ignored when the ANN approach is applied to the detection of a target. For a better detection of the detected target using machine learning, image characteristics of the detected target must be emphasized, while those of other targets should be not underlined. Consequently, the input layers of ANN must be created from the image after target-emphasis procedures are done. This means that we need to emphasize ships in an SAR image in order to detect ships from SAR images. However, in most previous studies, input layers have not been created from SAR images without ship emphasis processing [19–22]. The SAR images have speckle noise, the ghost phenomena, the side-lobe effect, and so on. The SAR image characteristics affect the precision of ship detection. To remarkably reduce the negative effects in the ANN approach, we need to minimize these effects. Moreover, the Kompsat-5 SAR image has severe ghost phenomena and side-lobe effects. Additional steps to mitigate the effects should be applied to the Kompsat-5 image.

Secondly, subset images have usually been used to validate the ship detection algorithm in the previous studies [17,23]. When ships are detected from SAR images, the most important thing is to minimize the negative effects from background noise, the side-lobe effect, wave effect, land area effect and so on. Thus, it is a better choice that the algorithm performance is validated by using the full scene. Therefore, for a better performance of the ship detection using the ANN approach, the input layers of ANN should be created by minimizing the negative effects of the Kompsat-5 SAR image, and the detection performance should be tested and validated by using the full scene rather than the sub-image.

In this paper, we propose an efficient method to detect ship targets from X-band Kompsat-5 SAR imagery using the ANN approach. The proposed method creates two input layers for ship detection through optimal image processing in order to consider Kompsat-5 SAR image characteristics. The method is tested and validated by using a full-scene Kompsat-5 SAR image. The detection procedure is as follows: (1) two input layers, an intensity differential map and a texture differential map, are generated from an SAR image using the azimuth low-pass filter, the non-local (NL)-means filter, the median filter, the sum-of-square (SS) operation, and so on; (2) a ship-probability map is created by the ANN approach; and (3) ships are detected from the map by using a threshold value. The NL-means filter and median filter were used to reduce phenomena related to the false detections. The NL-means filter effectively reduced noise on SAR imagery without smoothing edges of the objects, while the median filter was used to remove ship targets in SAR imagery. For the performance validation of the proposed method, the X-band Kompsat-5 SAR data with horizontal transmitting and horizontal receiving (HH) polarization was used in the study. A total number of 78 ship targets were extracted from the test full scene via a visual analysis and used to calculate the precision, recall and false detection rate.

2. Test Data

X-band Kompsat-5 SAR data with the HH single polarization of the standard mode was used in this study (Figure 1). The data was provided in the single look complex (SLC) Level 1A format, and the detailed information for the data is summarized in Table 1. As seen in Figure 1, ships on sea and artificial structures on the land are very bright due to the corner reflection effect in SAR imagery. Especially, the ghost phenomenon is very severe in the Kompsat-5 SAR image as seen Figure 1b, and the side-lobe effect is severe in the image as seen Figure 1a. The side-lobe effect is caused by the matched filtering when the SAR image compression is applied [24,25]. The compression of the raw radar signal is represented as a sinc function, and the side lobes of the sinc function are relatively high. The ghost phenomena are caused by aliasing of the Doppler phase history of each target [26]. If the ghost phenomena occur in a SAR image, after-images of land objects having high backscattering coefficients appear in the ocean, and hence, the after-images can be misunderstood as ships because they are much brighter than the sea surface. Therefore, the side-lobe effect and ghost phenomena should be mitigated before detecting ship targets, because they degrade the quality of an image and are an obstacle to ship detection.

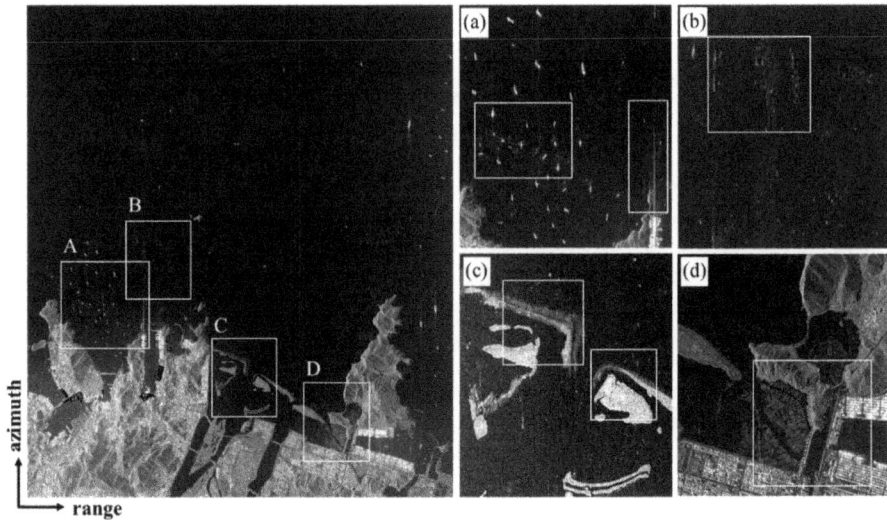

Figure 1. Kompsat-5 single look complex (SLC) image used for this study. (**a–d**) indicate the sub-images magnified from the A to D boxes in the Kompsat-5 SLC image.

Table 1. Kompsat-5 single look complex (SLC) synthetic aperture radar (SAR) data information. HH: horizontal transmitting and horizontal receiving.

Imaging Mode	Standard (Strip)
Polarization	HH
Incidence angle (deg.)	41.98
Azimuth pixel spacing (m)	1.75
Range pixel spacing (m)	1.29
Ground-range pixel spacing (m)	1.93
Azimuth processing bandwidth (Hz)	3100
Pulse repetition frequency (Hz)	4032
Orbit	Ascending

The shuttle radar topography mission digital elevation model (SRTM DEM) with a spatial resolution of 30 m was used for this study [27]. A synthetic SAR image was simulated from the SRTM DEM and used to mask land area in the SAR image, in order to remove false alarms from the land area. However, some areas, such as fish farms, offshore bars and so on, were still left in the SAR image. The structures over the areas could bring out false detection, as shown in

Figure 1c,d. Therefore, we need to mitigate the false detection from the structures for a better ship detection precision.

3. Methods

In this study, we proposed an efficient method to detect ships from Kompsat-5 SAR imagery through the ANN approach. A key of the proposed method is that the input layer of the ANN approach uses two ship-emphasized images through minimizing the negative effects of the SAR image characteristics. Figure 2 presents the detailed workflow of the proposed method. The proposed method is categorized into (1) input-layer generation from a SAR image via SAR image processing, (2) ship-probability map creation through the ANN approach and (3) ship identification from the probability map using threshold value.

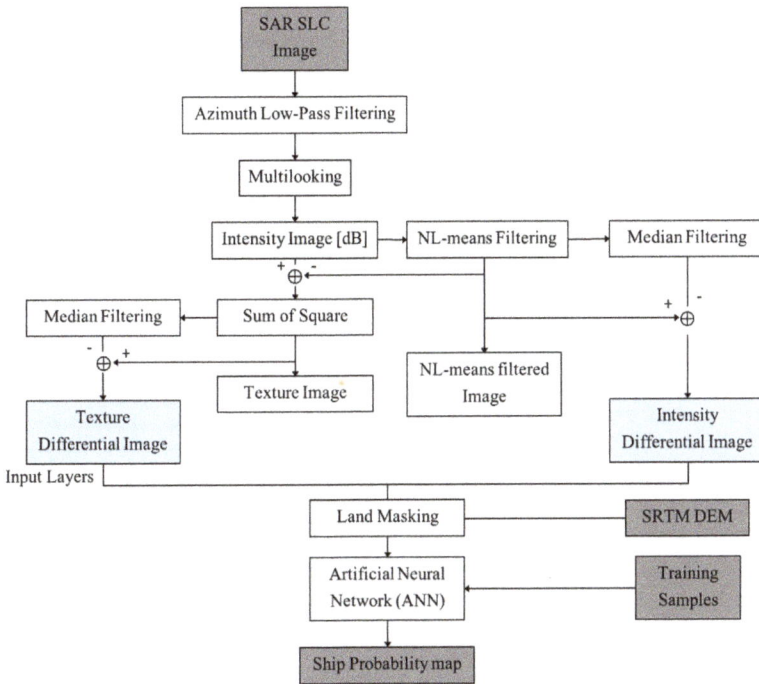

Figure 2. Detailed workflow of the proposed method in this study. SAR: synthetic aperture radar; SRTM DEM: shuttle radar topography mission digital elevation model; NL: non-local.

For more-optimal input-layer generation, we need to (1) enhance ships in the SAR image and (2) reduce the negative effects of the SAR image characteristics, such as the speckle noise, ghost phenomena, side-lobe effect, and so on. Especially, since the negative effects are very severe in the X-band Kompsat-5 image, the effects should be corrected. The input-layer generation procedure is as follows:

(1) azimuth low-pass filtering, which is designed to mitigate the ghost phenomena,
(2) multi-look processing, which is normally used for speckle noise reduction,
(3) sigma-naught conversion, which includes a unit conversion,
(4) intensity differential image creation,
(5) texture differential image creation.

The azimuth low-pass filtering is used to mitigate the ghost phenomenon. This filter is designed to reduce the azimuth processing bandwidth. If the azimuth processing bandwidth is large, the spatial resolution is high while the ghost effect is severe. Otherwise, low spatial resolution and reduced ghost effect can be seen in the SAR image. For the filter, the SAR SLC image is Fourier transformed in the azimuth direction, and then multiplied with the Hanning window having the bandwidth of $n \cdot$PRF, where n is a fraction factor and PRF is the pulse repetition frequency. The fraction factor n depends on the characteristics of SAR sensor. Finally, we obtain the ghost-mitigated SLC image after the inverse Fourier transform is applied. The multi-look processing is used for the speckle noise reduction. It is well-known that this step improves SAR image quality [28,29]. The multi-looked SAR image is converted into the sigma-naught image, which has a decibel unit.

In order to create intensity- and texture-differential images, the NL-means and median filters are used. The PDF of the sigma-naught image is a normal distribution, and hence the NL-means filter can be properly applied to the sigma-naught image. The NL-means filter smooths images using weighting factors that are estimated from a similarity between the center and adjacent pixels. For this, the filter uses two moving window kernels, unlike conventional filters. The smoothing factors of the filter can be determined from the standard deviation of noisy areas. Generally, conventional smoothing filters reduce the noise component of images while they smooth the edge of objects. However, the NL-means filter can reduce the noise without smoothing the object edges. Thus, when the NL-means filter is applied to the ship detection, the edge of ships can be preserved while the noise can be well-reduced. The median filter estimates the median value from the window kernel, and hence the filter does effectively remove outliers. The filter has been widely used for smoothing SAR images due to the speckle noise characteristics. The median filter is used to remove small bright objects, such as ship objects, from the SAR image. The details for the intensity- and texture-differential image generation will be followed in the subsections.

After the two input layers are created, the land areas in the input layers are masked out by using the SRTM DEM. Then, they are used for the ANN approach, and finally, the ship-probability map is created. If ground-truth data, such as the automatic information system (AIS), are available, they can be used for training and verification. However, since in-situ data is not available in this study, the training set for the ANN approach is determined by using a statistical threshold technique. Training samples of the ship detection are determined from the range calculated from the intensity-differential image. The ship training samples (P_S) are selected as follows:

$$P_S(i,j) = \begin{cases} 1, & if \ P(i,j) > MAX(P) - T_S \times Range(P) \\ 0, & otherwise \end{cases}, \tag{1}$$

where $P(i,j)$ is the pixel value at the range i and azimuth j, T_S is the threshold value for the ship samples, $MAX(P)$ is the maximum value of all of the pixels and $Range(P)$ is the range, which is defined as $MAX(P) - MIN(P)$. The non-ship training samples (P_{NS}) are selected by:

$$P_{NS}(i,j) = \begin{cases} 1, & if \ P(i,j) < MIN(P) + T_{NS} \times Range(P) \\ 0, & otherwise \end{cases}, \tag{2}$$

where T_{NS} is the threshold value for the non-ship samples. The threshold values of $T_S = 0.4$ and $T_{NS} = 0.1$ are used for this study, because the standard deviation of sea objects is a smaller value whereas that of ship objects is a higher value.

The final ship detection is performed by applying a threshold value to the ship-probability map. The threshold value is estimated from the histogram of the ship-probability map. This step is done by finding a point that the two PDFs from the ship and non-ship objects cross.

3.1. Intensity Differential Image

The intensity-differential image is created from the sigma-naught image by using the NL-means and median filter operations. Then, the image is used as the input layer of the ANN approach. The intensity-differential image is created by the following procedure:

(1) applying the NL-means filter to the sigma-naught image,
(2) applying the median filter to the NL-means-filtered image,
(3) subtracting the median-filtered image from the NL-means-filtered image.

In this process, it is important to determine the optimal parameters of NL-means and median filters. The NL-means filter is designed to minimize noise effects in the ANN processing. That is, the reason why the NL-means filter is used is to reduce noise without smoothing object edges. Thus, the optimal parameters used for the NL-means filter should be determined by considering the noise reduction as well as the edge preservation. The median filter is used to remove ship objects. Thus, the kernel size of the median filter should be larger than the size of ship objects, because the median value should not come from ship-object pixels but sea-object pixels. If the filter is successful, we will not be able to see any ships in the median-filtered image. The median filter does not remove objects larger than ships, including fish farms and offshore bars, but it smooths the objects. If the larger objects disappear in the image, the ship false-detection might increase. Finally, the intensity-differential image can be created by subtracting the median-filtered image from the NL-means-filtered image. The background is almost zero in the intensity-differential image, and the brightness values of fish farms and offshore bars are low in the image. The intensity-differential image enhances ship targets, and hence it must increase the ship-detection probability.

3.2. Texture Differential Image

The texture-differential image is created from the NL-means-filtered image by using the median filter and sum-of-square operation. The image is used as one of the input layers. The texture differential image is created as follows:

(1) applying the NL-means filter to the sigma-naught image,
(2) calculating the difference between the NL-means-filtered image and the sigma-naught image,
(3) creating the texture image by applying the sum-of-square operation to the difference image,
(4) applying the median filter to the texture image,
(5) subtracting the median-filtered image from the texture image.

The texture image is calculated by summing the squares of the pixel values of the difference image within a window kernel. The difference image is acquired by the difference between the sigma-naught image and NL-means-filtered image. The texture differential image is generated by subtracting the median-filtered image from the texture image. The kernel size of the NL-means filter in the texture-differential image generation may be same as, or smaller than, the intensity-differential image generation. The NL-means filter is utilized to enhance the object edges in the texture-image processing, unlike its use for noise reduction in the intensity-image processing. Thus, a little weak filtering can be imposed in the NL-means filtering. If the optimal processing is applied, most ship pixels in the texture image are still bright, except some pixels within large ships. Thus, the ship objects have higher pixel values through the sum-of-square calculation. The brightness values of background pixels are reduced by using the texture-differential image, which is created by subtracting the median-filtered image from the texture image. In this texture-differential image, the ship objects are highlighted, while the side-lobe and ghost effects, fish farms and offshore bars have low brightness values. Therefore, the texture-differential image can be efficiently used to reduce the misdetection rate in some areas.

3.3. Short Description of ANN

ANN is a computing system that is inspired by the biological neural network. It is composed of interconnecting artificial neurons. In general, a neural network function creates a linear output pattern, given a particular input layer [30]. A multilayer perceptron (MLP) is commonly used as several types of neural. The MLP is a feed-forward artificial neural-network model that maps input data onto an appropriate output. An MLP is comprised of multiple node layers. Each layer connects a network with the next layer. In many cases, the units of these networks use a sigmoid function as an activation function. For training the network, MLP utilizes back propagation [31]. The back-propagation algorithm consists of a propagation and weight update phases. In the propagation step, the calculated error, which is the difference between the target value and the output value, is propagated to each layer. By using this propagated error, the weight is modified in a weight-update phase. They are repeated until the network performance is good enough [32]. We implemented the ANN with MATLAB software. The used ANN method is a two-multilayer perceptron with sigmoid-function and four neurons in the hidden layer and one linear-function output neuron. It is also trained with the back-propagation algorithm.

4. Results and Discussion

The X-band Kompsat-5 SLC image listed in Figure 1 was used for the validation of the proposed method. Since the Kompsat-5 image has severe ghost phenomena, the azimuth band-pass filtering was applied first. The fraction factor $n = 0.5$ was used for the filter, and hence the azimuth processing bandwidth of about 3100 Hz was reduced to that of about 2419 Hz. The ghost phenomena were effectively mitigated in the SAR image, while the spatial resolution in the azimuth direction was degraded from about 2.29 to 2.94 m. In order to reduce the speckle noise of the image, the multi-look operation of 5×5 looks in the range and azimuth directions was applied to the image by considering the ratio of azimuth and ground-range pixel spacing. The sigma-naught image was created by considering the incidence angle, and converted into the decibel unit.

To generate the intensity-differential image from the sigma-naught image, the parameters of the NL-means filter was estimated from the standard deviation of some land areas where the intensity of noise components is relatively high. The kernel size of 5×5 and the filtering parameter of 5.0 were used for the NL-means filter. The NL-filtered image enhanced the edge of ship objects and reduced the image noise and the side-lobe effects, as shown in Figure 3b, Figure 3f or Figure 3j. The median filter was used to remove ships from the NL-means-filtered image. The kernel size of the median filter of 21×21 in both the azimuth and range directions was used for this study, because the maximum length of ships was about 20 pixels. As shown in the Figure 3c, all of the ships disappeared in the median-filtered image, while the fish farms and offshore bars were smoothed but still preserved in the median-filtered image in Figure 3g or Figure 3k. Consequently, the pixel values of the ships in the intensity-differential image were much higher than those of the sea (Figure 3d). Moreover, the side-lobe effect was remarkably reduced, as seen in Figure 3d, and most of the land area, fish farms and offshore bars were removed in Figure 3h or Figure 3l. The intensity-differential image can raise the possibility of ship detection.

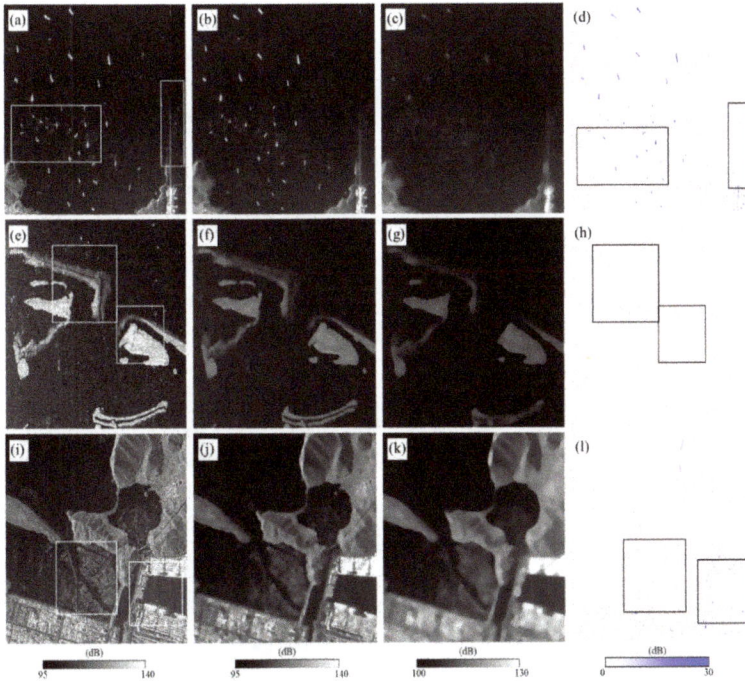

Figure 3. Procedure for the generation of a differential image, one of the input layers of artificial neural network (ANN): (**a,e,i**) the intensity images, (**b,f,j**) the NL-means filtered images, (**c,g,k**) the median filtered images and (**d,h,l**) differential images in the boxes A, C and D (see Figure 1).

One of the input layers, the texture image, was calculated by summing the squares of the difference between the sigma-naught image and the NL-means-filtered image. In order to enhance the ship objects in the texture image, the kernel size of 5×5 and the filtering parameter of 3.0 were used for the NL-means filter. After the NL-means filter was applied to the sigma-naught image, the difference map was generated by subtracting the NL-means-filtered image from the sigma-naught image. Figure 4a, Figure 4e or Figure 4i show the sub-images of the difference image in the boxes A, C and D of Figure 1. In the figures, the object edges were enhanced, and most bright pixels came from ships or buildings. Since most ships and buildings are small, their pixel values could be preserved because most of their pixels can be considered as edges. Moreover, the object edges were enhanced in the texture image by the sum-of-square calculation using the window kernel of 5×5, as shown in Figure 4b, Figure 4f or Figure 4j. As seen in Figure 4b, all of the ship pixels in the texture image were much brighter than other objects. This means that the ship objects in Figure 4b were emphasized through the sum-of-square calculation. The median filter, having the kernel size of 21×21, was used for the removal of the ships in the texture image. The ships effectively disappeared in Figure 4c. Thus, the ships were much enhanced in the texture-differential image that was created by the difference between the texture image and median-filtered image (Figure 4d), while the pixel values in the fish farms and offshore bars were remarkably reduced (Figure 4d, Figure 4h or Figure 4l). This image can increase the ship-detection rate as well as reduce the false-detection rate. However, the texture-differential image was a little noisy due to the sum-of-square calculation. An image-processing step can be further applied for the noise reduction. In this study, we did not consider the additional noise reduction step because the noise pixel values were very small.

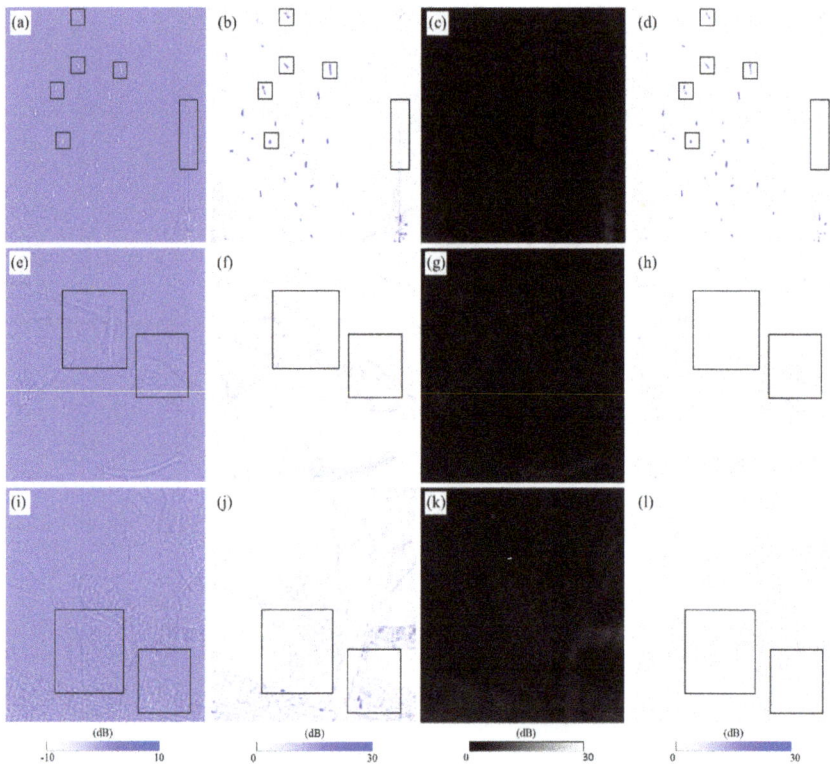

Figure 4. Procedure for the generation of a texture-differential image, which is another input layer used for ANN: (**a,e,i**) the difference between sigma-naught and NL-means-filtered images, (**b,f,j**) the texture images, (**c,g,k**) the median-filtered texture images and (**d,h,l**) the texture-differential images in the boxes A, C and D (see Figure 1).

The intensity- and texture-differential images were applied as the input layers to the ANN approach. The training set was obtained by the statistical-threshold approach of Equations (1) and (2). The threshold values of 15.16 and 2.53 for ship and non-ship training selections, respectively, were estimated from the intensity-differential image, and then 67 training ship samples were selected from the statistical-threshold approach. The ANN approach was applied to the ship detection using the two input layers and the training set. Figure 5 shows the ship-probability map estimated by using ANN. The pixels in the ship-probability map have values between 0 and 1. The pixel value of '0' means that the pixel is not a ship at all, while the pixel value of '1' denotes that the pixel is definitely a ship. The sea surface had values as low as 0.1, while the ship objects had values as high as 0.85. If accurate coastline maps or large-scale topographic maps are available, the bridges can be masked out. In Figure 6a,b, the ghost and side-lobe effects were as low as 0.3. As shown in Figure 6c,d, the probability value in the fish farms and offshore bars was lower than the ship objects.

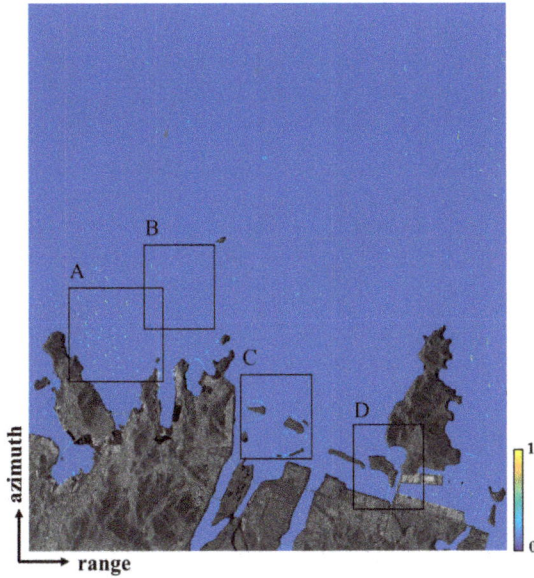

Figure 5. Ship-probability map obtained by using the ANN approach. The boxes A to D are used for the detailed analysis.

In order to estimate the ship-thresholding value from the ship-probability map, the histogram of the ship-probability map was created (Figure 7). From the histogram, the point that the PDFs of ship and non-ship objects cross was determined, as shown in Figure 7. The threshold value of 0.67 was applied by determining the ship objects from the probability map.

We calculated the recall and precision to validate the performance of the proposed method. The recall (sensitivity) measures how many objects are detected, and the precision denotes how many objects are correct among the detected objects. The recall and precision have a value between 0 and 1. The recall and precision can be defined as given by:

$$\begin{cases} Recall = N_r / N_t \\ Precision = N_r / N_g \end{cases} \quad (3)$$

where N_g is the number of ground truth, N_t is the number of ships detected by ANN and N_r is the number of correctly detected ships.

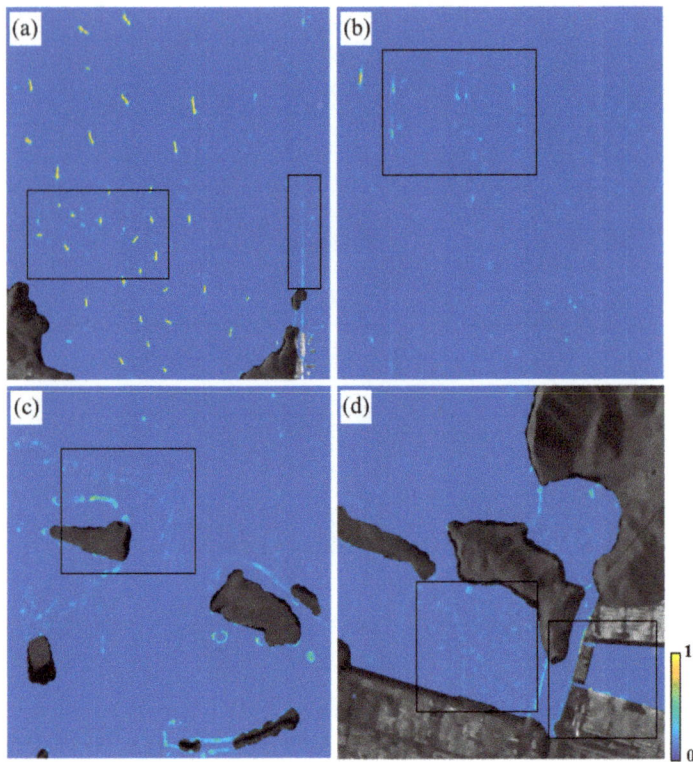

Figure 6. (**a–d**) The ship-probability maps magnified from the A to D boxes in Figure 5.

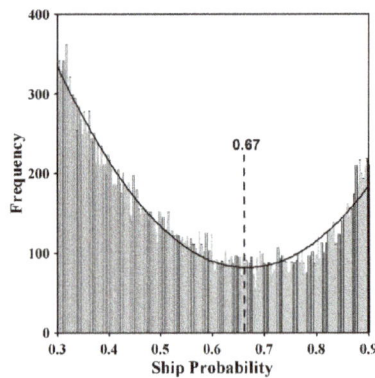

Figure 7. Ship-probability map obtained by using the ANN approach.

We found the total number of 78 ships from the SAR image through visual analysis and identified ships by using threshold values of 0.67. When the threshold value of 0.67 was used, 82 objects were identified as ship objects, and 77 objects among them were real ships (while five objects were from ghost phenomena). Thus, in this case, the recall and precision were about 93.9% and 98.7%, respectively. The false-detection rate was 6.1%. Table 2 summarizes the recall, precision and false detection rate from the detection result. The result means that about 98% of all ships can be detected, and the detected assignments are actual ships about 97% of the time. The main reason for false detection was the ghost phenomena. Even though the ghost effect was reduced by the azimuth low-pass filtering, some targets still had a higher brightness value. One of the false detections was due to the barge. The one non-detected ship had a lower value in the NL-means-filtered images, unlike the other ships, because the brightness value of the ship target had a lower value.

Table 2. Recall, precision and false detection rate from the detection result of the test Kompsat-5 image. N_g: number of ground truth, N_t: number of detected ships, N_r: number of correctly detected ships

Threshold	N_g	N_t	N_r	Recall (%)	Precision (%)	False Detection Rate (%)
0.67	78	82	77	93.9	98.7	6.1

Kompsat-5 SAR imagery has severe ghost phenomena and side-lobe effects. Due to these effects, it is very hard to identify ship targets. To improve the ship-detection rate, we applied the proposed method to the Kompsat-5 SAR image. Almost all the ships were detected by the proposed method, although some of the ghost objects were detected as ship objects. This means that the proposed method was successfully applied to ship detection from the Kompsat-5 SAR image.

5. Conclusions

We showed an efficient method to detect ship targets from the Kompsat-5 SAR image using the artificial neural network (ANN) approach. The method is composed of three main steps: (1) input-layer generation, (2) the ship-probability map generation using the ANN approach and (3) the ship-object identification. For the reduction of the side-lobe effect, ghost phenomena and speckle noises, we used the azimuth band-pass, NL-means filters and the median filter. The intensity- and texture-differential images were generated by image processing from the SAR image, and used as the input layers to enhance ship objects. The training set was derived by the statistical approach and applied to the ANN approach. The ship-probability map was generated and the ship objects were highlighted in the probability map. We have tried to estimate an optimal thresholding value from the probability map. For this, the histogram of the map was created. From the histogram, the point that the PDFs of ship and non-ship objects cross was determined as the optimal threshold value. The threshold value was 0.67.

The performance validation of our proposed method was carried out. The total number of 78 ships were found in the SAR image through visual analysis and identified ships. Through the process, 82 objects were identified as ship objects, and 77 objects among them were real ships, while five objects were from the ghost phenomenon. Thus, in this case, the recall and precision were about 93.9% and 98.7%, respectively. The false-detection rate was 6.1%. This means that about 99% of all ships can be detected, and the detected ships are true ships about 94% of the time.

It is very important to reduce the ghost and side-lobe effects and speckle noises when we find ship objects from the Kompsat-5 SAR imagery. Their reduction can enhance the ship objects in SAR imagery. The proposed method effectively reduced these noise effects, and consequently, the ship objects were well-enhanced as well as detectable. Therefore, the proposed method was successfully applied to ship detection from the Kompsat-5 SAR image. Moreover, the proposed method can be applied to other SAR satellite images such as COSMO-SkyMed and TerraSAR-X by adjusting the filtering parameters of the proposed method. It would be also expected that the proposed method can be applied not only to ANN but also to other deep learning techniques, such as convolutional neural networks (CNN).

Acknowledgments: This research (NRF-2015R1A2A2A01005018) was supported by Mid-career Researcher Program through National Research Foundation of Korea (NRF) grant funded by the Ministry of Education, Science and Technology (MEST).

Author Contributions: Hyung-Sup Jung conceived and designed the experiments. Jeong-In Hwang performed the experiments. Hyung-Sup Jung and Jeong-In Hwang analyzed the data. Sung-Ho Chae and Daeseong Kim helped the data processing for the SAR imagery. Jeong-In Hwang and Hyung-Sup Jung wrote the paper.

Conflicts of Interest: The authors declare no conflict of interest.

References

1. Hwang, J.; Kim, D.; Jung, H.-S. An efficient ship detection method for KOMPSAT-5 synthetic aperture radar imagery based on adaptive filtering approach. *Korean J. Remote Sens.* **2017**, *33*, 89–95. [CrossRef]
2. Wang, X.; Chen, C. Adaptive ship detection in SAR images using variance WIE-based method. *Signal Image Video Process* **2016**, *10*, 1219–1224. [CrossRef]
3. Eldhuset, K. An Automatic Ship and Ship Wake Detection System for Spaceborne SAR Images in Coastal Regions. *IEEE Trans. Geosci. Remote Sens.* **1996**, *34*, 1010–1019. [CrossRef]
4. Jiang, Q. Ship detection in RADARSAT SAR imagery using PNN-model. In Proceedings of the ADRO Symposium'98, Montreal, QC, Canada, 13–15 October 1998.
5. Kuo, J.M.; Chen, K.S. The Application of Wavelets Correlator for Ship Wake Detection in SAR Images. *IEEE Trans. Geosci. Remote Sens.* **2003**, *41*, 1506–1511. [CrossRef]
6. Wang, C.; Jiang, S.; Zhang, H.; Wu, F.; Zhang, B. Ship detection for high-resolution SAR images based on feature analysis member. *IEEE Geosci. Remote Sens. Lett.* **2014**, *11*, 119–123. [CrossRef]
7. Ai, J.; Qi, X.; Yu, W.; Deng, Y.; Liu, F.; Shi, L. A new CFAR ship detection algorithm based on 2-D joint log-normal distribution in SAR images. *IEEE Geosci. Remote Sens. Lett.* **2010**, *7*, 806–810. [CrossRef]
8. Crisp, D.J. *The State-of-the-Art in Ship Detection in Synthetic Aperture Radar Imagery*; Australian Government, Department of Defence: Edinburgh, Australia, 2004; p. 115.
9. Brusch, S.; Lehner, S.; Fritz, T.; Soccorsi, M.; Soloviev, A.; van Schie, B. Ship surveillance with TerraSAR-X. *IEEE Trans. Geosci. Remote Sens.* **2011**, *49*, 1092–1103. [CrossRef]
10. Fei, C.; Liu, T.; Lampropoulos, G.A. Markov chain CFAR detection for polarimetric data using data fusion. *IEEE Trans. Geosci. Remote Sens.* **2012**, *50*, 397–408. [CrossRef]
11. Novak, L.M.; Sechtin, M.B.; Cardullo, M.J. Studies of target detection algorithms that use polarimetric radar data. *IEEE Trans. Aerosp. Electon. Syst.* **1989**, *25*, 150–165. [CrossRef]
12. Tao, D.; Anfinsen, S.; Brekke, C. Robust CFAR detector based on truncated statistics in multiple-target situations. *IEEE Trans. Geosci. Remote Sens.* **2016**, *54*, 117–134. [CrossRef]
13. Xiangwei, X.; Kefeng, J.; Huanxin, Z.; Jixiang, S. A fast ship detection algorithm in SAR imagery for wide area ocean surveillance. In Proceedings of the 2012 IEEE Radar Conference (RADAR), Atlanta, GA, USA, 7–11 May 2012; pp. 570–574. [CrossRef]
14. Ouchi, K. A brief review on recent trend of synthetic aperture radar applications to maritime safety: Ship detection and classification. In Proceedings of the Application of Remote Sensing to Maritime Safety and Security, Jeju, Korea, 20–22 April 2016.
15. Kaplan, L. Improved SAR target detection via extended fractal features. *IEEE Trans. Aerosp. Electron. Syst.* **2001**, *37*, 436–451. [CrossRef]
16. Howard, D.; Roberts, S.; Brankin, R. Target detection in SAR imagery by genetic programming. *Adv. Eng. Softw.* **1999**, *30*, 303–311. [CrossRef]
17. Martín-de-Nicolás, J.; Mata-Moya, D.; Jarabo-Amores, M.P.; del-Rey-Maestre, N.; Bárcena-Humanes, J.L. Neural network based solutions for ship detection in SAR images. In Proceedings of the 18th International Conference on Digital Signal Processing, Fira, Greece, 1–3 July 2013.
18. Khesali, E.; Enayati, H.; Modiri, M.; Aref, M.M. Automatic ship detection in Single-Pol SAR Images using texture features in artificial neural networks. *ISPRS Arch.* **2015**, *40*, 395–399. [CrossRef]
19. Ødegaard, N.; Knapskog, A.O.; Cochin, C.; Louvigne, J.-C. Classification of ships using real and simulated data in a convolutional neural network. In Proceedings of the Radar Conference (Radar-Conference) 2016, Philadelphia, PA, USA, 2–6 May 2016.
20. Osman, H.; Pan, L.; Blostein, S.D.; Gagnon, L. Classification of ships in airborne SAR imagery using back propagation neural networks. In Proceedings of the SPIE 1997, San Diego, CA, USA, 27 July–1 August 1997; pp. 126–136.

21. Dao-Duc, C.; Xiahvi, H.; Mor'ere, O. Maritime vessel Images Classification using deep convolutional neural network. In Proceedings of the 6th International Symposium on Information and Communication Technology, Dubai, UAE, 3–4 December 2015; pp. 276–281.

22. Zou, Z.X.; Shi, Z.W. Ship detection in spaceborne optical image with SVD networks. *IEEE Trans. Geosci. Remote.* **2016**, *54*, 5832–5845. [CrossRef]

23. Bentes, C.; Frost, A.; Velotto, D.; Tings, B. Ship-iceberg discrimination with convolutional neural networks in high resolution SAR images. In Proceedings of the EUSAR 2016: 11th European Conference on Synthetic Aperture Radar, Hamburg, Germany, 6–9 June 2016.

24. Freeman, A. On ambiguities in SAR design. In Proceedings of the EUSAR 2006: 6th European Conference on Synthetic Aperture Radar, Dredsden, Germany, 16–18 May 2006.

25. Carsey, F.D. *Microwave Remote Sensing of Sea Ice*; American Geophysical Union: Washington, DC, USA, 1992; p. 462.

26. Marchand, M.A.J. SAR image quality assessment. *Revista de Teledeteccin* **1993**, *2*, 12–18.

27. Farr, T.; Rosen, P.; Caro, E. The shuttle radar topography mission. *Rev. Geophys.* **2000**, *45*, 37–55. [CrossRef]

28. Oliver, C.; Quegan, S. *Understanding Synthetic Aperture Radar Imagery*; Artech House: Norwood, MA, USA, 1998.

29. Mansourpour, M.; Rajabi, M.; Blais, J. Effects and performance of speckle noise reduction filters on active radar and SAR Images. In Proceedings of the ISPRS Ankara Workshop, Ankara, Turkey, 14–16 February 2006; pp. 14–16.

30. Majumdar, A.; Majumdar, P.K.; Sarkar, B. Application of linear regression, artificial neural network and neuro-fuzzy algorithms to predict the breaking elongation of rotor-spun yarns. *Indian J. Fiber Text. Res.* **2005**, *30*, 19–25.

31. Spencer, M.; Eickholt, J.; Cheng, J. A deep learning network approach to AB initio protein secondary structure prediction. *IEEE/ACM Trans. Comput. Biol. Bioinform.* **2015**, *12*, 103–112. [CrossRef] [PubMed]

32. Koskela, T.; Lehtokangas, M.; Saarinen, J.; Kaski, K. Time series prediction with multilayer perceptron, FIR and Elman neural networks. In Proceedings of the World Conference Neural Network, San Diego, CA, USA, 15–20 September 1996; INNS Press: San Diego, CA, USA, 1996; pp. 491–496.

applied sciences

MDPI

Article

A New Damage Assessment Method by Means of Neural Network and Multi-Sensor Satellite Data

Alessandro Piscini *, Vito Romaniello, Christian Bignami and Salvatore Stramondo

Istituto Nazionale di Geofisica e Vulcanologia, 00143 Roma, Italy; vito.romaniello@ingv.it (V.R.);
christian.bignami@ingv.it (C.B.); salvatore.stramondo@ingv.it (S.S.)
* Correspondence: alessandro.piscini@ingv.it; Tel.: +39-06-51-860-630

Received: 29 June 2017; Accepted: 26 July 2017; Published: 1 August 2017

Abstract: Artificial Neural Network (ANN) is a valuable and well-established inversion technique for the estimation of geophysical parameters from satellite images. After training, ANNs are able to generate very fast products for several types of applications. Satellite remote sensing is an efficient way to detect and map strong earthquake damage for contributing to post-disaster activities during emergency phases. This work aims at presenting an application of the ANN inversion technique addressed to the evaluation of building collapse ratio (CR), defined as the number of collapsed buildings with respect to the total number of buildings in a city block, by employing optical and SAR satellite data. This is done in order to directly relate changes in images with damage that has occurred during strong earthquakes. Furthermore, once they have been trained, neural networks can be used rapidly at application stage. The goal was to obtain a general tool suitable for re-use in different scenarios. An ANN has been implemented in order to emulate a regression model and to estimate the CR as a continuous function. The adopted ANN has been trained using some features obtained from optical and Synthetic Aperture Radar (SAR) images, as inputs, and the corresponding values of collapse ratio obtained from the survey of the 2010 M7 Haiti Earthquake, i.e., as target output. As regards the optical data, we selected three change parameters: the Normalized Difference Index (NDI), the Kullback–Leibler divergence (KLD), and Mutual Information (MI). Concerning the SAR images, the Intensity Correlation Difference (ICD) and the KLD parameters have been considered. Exploiting an object-oriented approach, a segmentation of the study area into several regions has been performed. In particular, damage maps have been generated by considering a set of polygons (in which satellite parameters have been calculated) extracted from the open source Open Street Map (OSM) geo-database. The trained ANN has been proposed for the M6.0 Amatrice earthquake that occurred on 24 August 2016, in central Italy, by using the features extracted from Sentinel-2 and COSMO-SkyMed images as input. The results show that the ANN is able to retrieve a building collapse ratio with good accuracy. In particular, the fusion approach modelled the collapse ratio characterized by high values of CR (more than 0.5) over the historical center that agrees with observed damages. Since the technique is independent from different typologies of input data (i.e., for radiometric or spatial resolution characteristics), the study demonstrated the strength of the proposed approach for estimating damaged areas and its importance in near real time monitoring activities, owing to its fast application.

Keywords: earthquake; damage assessment; neural networks; satellite data; SAR; Sentinel-2

1. Introduction

Artificial neural networks (ANN), computational modelling tools, have found wide acceptance in many disciplines due to their adaptability to complex real world problems.

ANNs have demonstrated their ability to model non-linear physics systems [1], involving complex physical behaviors, and were applied to the analysis of remotely sensed images with promising results. Some examples of applications are: retrieval of soil moisture and agricultural variables from microwave radiometry [2], estimation of snow water equivalent and snow water depth from microwave images [3], retrieval of leaf area index (LAI) and other biophysical variables from the MEdium Resolution Imaging Spectrometer (MERIS) and MODerate-resolution Imaging Spectroradiometer (MODIS) instruments [4], estimation of chlorophyll from MERIS [5], retrieval of volcanic ash and sulphur dioxide from hyperspectral data [6]. ANNs are also used, with good results, for rainfall prediction involving other geophysical data [7,8].

Since an earthquake usually acts in a nonlinear way, the neural network algorithm can be an appropriate method for damage estimation purposes since it demonstrated to be a good non-linear approximator [9]. Recently, ANNs have been applied to detect damaged buildings, following an earthquake, by using high spatial resolution optical images acquired after the seismic event [10]. A neural network based-approach is being implemented to assess the status of buildings after earthquake excitation, predicting the displacement at different floors considering the wave energy propagating only into the ground floor [11].

By using change features from satellite images, accurate and reliable damage mapping can be obtained, exploiting both optical and radar sensors [12–14]. In Romaniello et al., 2016, an unsupervised algorithm for damage classification purposes has been developed [14].

Currently, quantitative estimation of earthquake damage level as a continuous function, using an ANN, has not yet been exploited and the present study represents a first attempt at applying an ANN to both optical high resolution Sentinel-2 and Synthetic Aperture Radar (SAR) remote sensing data for collapse ratio modelling. To the best of our knowledge, the methodology based on ANN has not been utilized in modelling earthquake damage assessment.

In the present study, two different neural networks using different Earth Observation (EO) datasets have been realized in order to model, as a continuous function, building damage. The first neural network (NN) experiment used, as input, only features obtained from optical data, whilst the second one (in a data fusion approach) also the features obtained from SAR images. NDI, KLD and MI features from optical data, and ICD and KLD from SAR data were used. Regarding the SAR data, KLD and MI parameters are very suitable features that can contribute to damage estimation [15]; the ICD demonstrated itself a very good damage proxy [13,16]. Concerning the optical data, the most significant performances are related to the NDI, KLD [17], and MI indexes [18]. These features show very good sensitivity to the collapse ratio.

The case study is the Central Italy strong earthquake, which took place on August 2016. On 24 August 2016 at 1:36 UTC, a M6.0 earthquake occurred in the Apennines of Central (hereafter Amatrice earthquake) Italy at depth 8 km, over a NNW-SSE striking, WSW dipping normal fault [19], destroying the closest towns to the epicenter—Amatrice, Accumoli and Arquata del Tronto—and causing near 300 fatalities. This earthquake revealed the importance of a rapid earthquake damage assessment, right after a seismic event, which can address the civil protection interventions towards the most affected areas. This work allowed to quantitatively evaluate the performance of NNs in terms of CR retrieval accuracy and generalization capability.

2. Neural Network Approach and Employed Features

ANNs are based on the concept of the single artificial neuron, the 'Perceptron', introduced by [20] to solve problems in the area of character recognition [21]. Using supervised learning, with the Error-Correction Learning (ECL) rule for network weights adjustment, those networks can learn to map from one data space to another using examples [22]. One of the most common and reliable learning techniques is the back-propagation (BP) algorithm [23]. BP consists of two phases: in the feedforward pass, an input vector is presented to the network and propagated forward to the output; in the back-propagation phase, the network output is compared to a desired output; network weights are then adjusted in accordance with an ECL rule [23–25]. Cross validation can be used to detect when over-fitting starts during supervised training of a neural network; training is then stopped before convergence to avoid over-fitting (early stopping). Early stopping using cross validation was done by splitting the training data into a training set, a cross-validation set, and a test set, and then training the networks only using the training set and evaluating the per-example error on the test set on a sample basis after a defined number of epochs. Finally, training was stopped when the error—the difference between neural network output and target—on the cross validation set was higher than the previous error value [26].

The performance of a trained ANN is generally assessed by computing the root mean square error (RMSE) between expected values and activation values at the output nodes or, in the case of classification, a percentage of correctly classified examples of the validation set. At the time our study was carried out, the ground truth was unavailable for Amatrice earthquake, so a visual inspection using high resolution images has been adopted.

From studies of past literature and development activities performed in a European research project (APhoRISM, www.aphorism-project.eu), we identified and employed a set of features that demonstrated a good sensitivity to damage at object scale. Regarding optical data, we used the NDI, KLD and MI change indexes. As far as SAR data is concerned, we considered the KLD and ICD parameters. Note that all change indexes have been calculated at object scale, i.e., by considering polygons that refer to city blocks. These latter have been extracted through the free geo-database of the Open Street Map project.

The NDI parameter is defined as

$$NDI_i = \frac{POST_i - PRE_i}{POST_i + PRE_i} \tag{1}$$

where PRE_i and $POST_i$ indicate the mean values of intensity, respectively for pre- and post-seismic images, associated to i-th polygon (see also Figure 2 as example). The intensity values are obtained averaging the Top of Atmosphere (TOA) Red-Green-Blue (RGB) reflectance and corrected by applying the Flat Field procedure for atmospheric correction [27].

The MI index measures the correlation loss between pre- and post-seismic images (see Equation (2)).

$$MI_i = -\ln\left(1 - r_i^2\right) \times 0.5 \tag{2}$$

where r_i^2 is the correlation between the pre- and post-seismic pixels within each polygon. Correlation is obtained from intensity and backscattering values for optical and SAR data, rescpectively. *MI* is inversely proportional to the damage grade.

The KLD parameter is defined as

$$KLD_i = \frac{(PRE_i - POST_i)^2 + Var(PRE_i) + Var(POST_i)}{2} \times \left(\frac{1}{Var(PRE_i)} + \frac{1}{Var(POST_i)}\right) - 2 \tag{3}$$

where PRE_i and $POST_i$ are the same parameters in the Equation (1), and $Var(PRE_i)$ and $Var(POST_i)$ are their variances within the i-th polygon. The *KLD* parameter has the same behavior of NDI: KLD increasing values correspond to increasing damage level. The ICD parameter is calculated on the base of the Pearson Correlation coefficient (ρ_i) estimated on the pre-seismic SAR image pair (*ICpre*) and on the co-seismic SAR image pair (*ICcos*). From these two intermediate outputs, we can compute the ICD

$$ICD_i = ICpre - ICcos \tag{4}$$

3. Dataset and NN Training

The training case study was the Mw 7.0 earthquake that hit Haiti on 12 January 2010. The epicenter was located about 25 km west–southwest of Port-au-Prince city. The disastrous shock caused the collapse of a huge number of buildings and widespread damage.

Table 1 describes satellite dataset used which consists of GeoEye-1 optical images (one pre- and one post-seismic), and three TerraSAR-X SAR images (two pre- and one post-seismic).

Table 1. EO data list for NN training.

Datatype	Satellite	Acquisition
Optical	GeoEye-1	2009/10/01
Optical	GeoEye-1	2010/01/13
SAR	TerraSAR-X	2009/05/01
SAR	TerraSAR-X	2009/10/13
SAR	TerraSAR-X	2010/01/20

Optical images inputs, both for pre- and post-, consisted of TOA reflectances at 2 m spatial resolution for RGB spectral bands; starting from these reflectances, the intensity values are derived (as described in Section 2).

As regards SAR data, intensity images have been obtained by multi-looking 3×3 m TerraSAR-X applying a re-sampling at 10×10 m.

The computation of change indexes based on EO imagery has been performed at object scale by considering a set of polygons, extracted from the open source Open Street Map (OSM) geo-database over Port-au-Prince. A total of 1513 polygons corresponding to city blocks of the affected areas have been considered. All the features (NDI, KLD, MI and ICD) are grouped in a unique dataset necessary for the Neural Network approach.

In addition to satellite data, a Ground Truth (GT) survey for Port-au-Prince town, expressed in terms of collapse ratio (CR), has been used (see Figure 1). The CR has been calculated by using GT information collected during a post-earthquake survey and available from the JRC (Joint Research Center) database, and considering the same polygons (city blocks) used for the satellite features calculation.

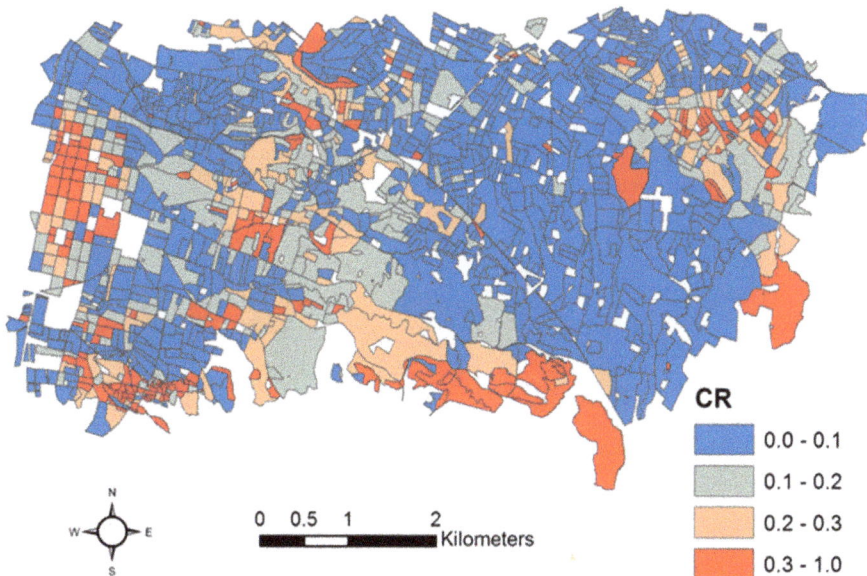

Figure 1. Collapse ratio (CR) for each polygon obtained from JRC survey data over Port-au-Prince.

In this work, Back-Propagation Neural Network (BPNN) has been used [23]. The BPNN to model CR was implemented using, as input, the features extracted from satellite images (see previous section, e.g., NDI in Figure 2), and CR values as target output. The 1513 samples used for NN training were split in Training, Cross Validation, and Test datasets of 65, 20 and 15%, respectively.

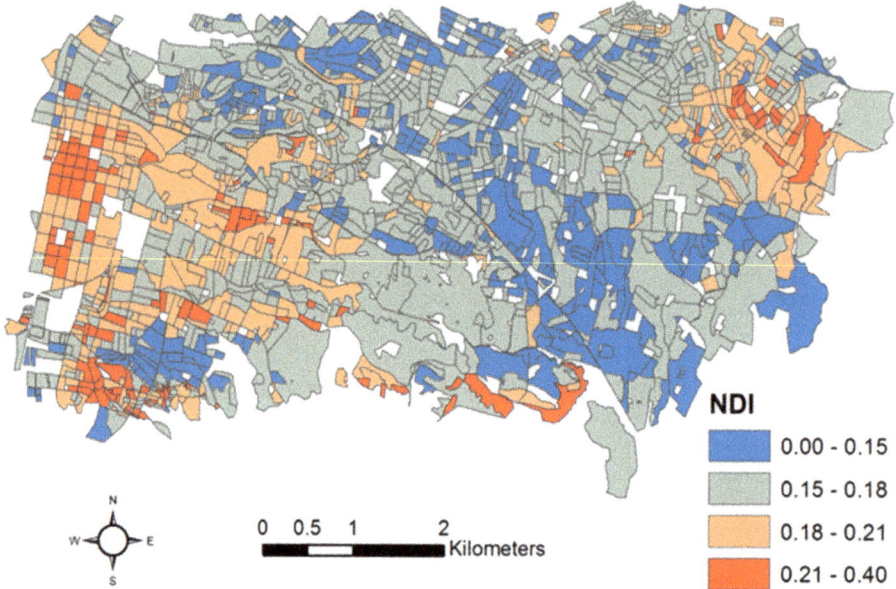

Figure 2. NDI map obtained from pre- and post-seismic GeoEye-1 images over Port-au-Prince.

A first network topology (Exp 1) consisted of only optical features as input (NDI, KLD and MI), five hidden layers with variable number of neurons [5–10–20–10–5], and one output, the CR. Furthermore, in order to perform a data fusion exercise, a different neural network using five inputs was adopted (Exp 2), adding to the optical features also the SAR ones, i.e., IC and KLD.

Figure 3 shows the statistical distributions for CR, train (a), cross-validation (b) and test (c) sets used during NN training phase. Despite of histograms put in evidence that values higher than 0.5 are statistically not well represented, we can consider the dataset a good training ensemble, because it covers the entire range of values.

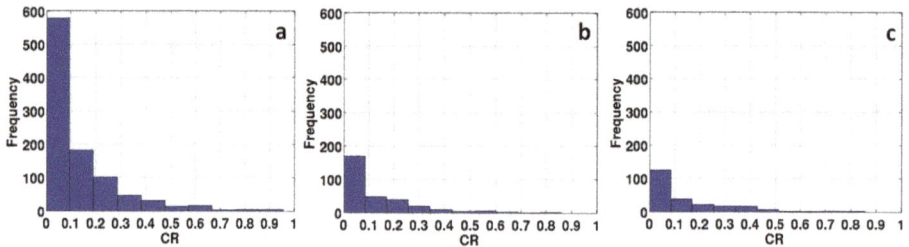

Figure 3. Distribution histograms for (**a**) training; (**b**) cross validation and (**c**) test of Haiti datasets used in the training phase.

Results of NNs training phase for Exp 1 and Exp 2 are depicted in Figures 4 and 5, respectively. As regard the network using only optical features, the regression coefficient obtained applying the network to the test dataset is 0.67 (Figure 4), whilst NN using both and optical features obtains 0.73 (Figure 5). In this phase, it seems that using more features improves the NN retrieval accuracy.

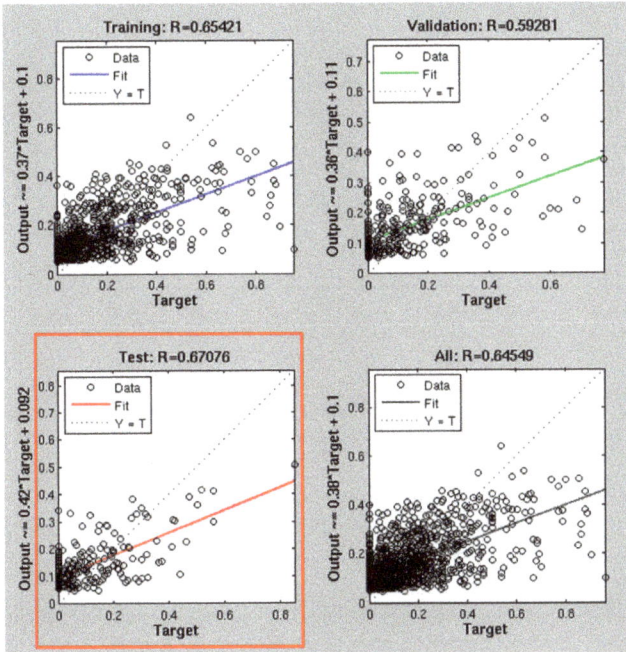

Figure 4. Regression curves for training, cross validation, test and total sets, considering Exp 1. Red squared regression curves describe the result obtained when an independent dataset is applied.

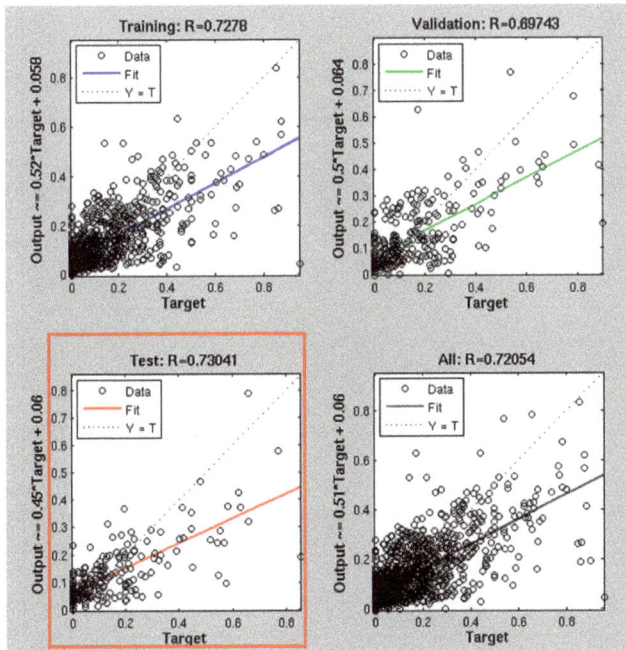

Figure 5. Regression curves for training, validation, test, and total sets, considering Exp 2. Red squared regression curves describe the result obtained when an independent dataset is applied.

4. Application of the Trained NN and Results for the Central Italy Case Study

In order to evaluate the performance of NNs in terms of retrieval accuracy and generalization capability, they were applied to the Amatrice earthquake. The goal is also to evaluate NN capability for modelling the building collapse ratio. The satellite dataset is made up of two Sentinel-2 optical images (1 pre- and 2 post-seismic), three COSMO-Sky SAR images (2 pre- and 1 post-seismic), and a building footprint layer extracted by the Open Street Map service (see Table 2 for EO data characteristics). Both optical and SAR data have a 10 m spatial resolution.

Table 2. EO data list for NN application.

Datatype	Satellite	Acquisition
Optical	Sentinel-2	2016/08/14
Optical	Sentinel-2	2016/09/04
SAR	COSMO-Sky	2016/07/19
SAR	COSMO-Sky	2016/08/20
SAR	COSMO-Sky	2016/08/28

The Amatrice footprint layer extracted from OSM, which is at a single building scale, was modified to obtain polygons surrounding more than one building. In this way, there are more pixels associated to each polygon leading a better estimation of change features over the polygon itself. The resulting layer consists of 112 polygons.

In the Figure 6, histograms for the CR obtained by the NN approach, Exp 1 (a) and Exp 2 (b), are depicted.

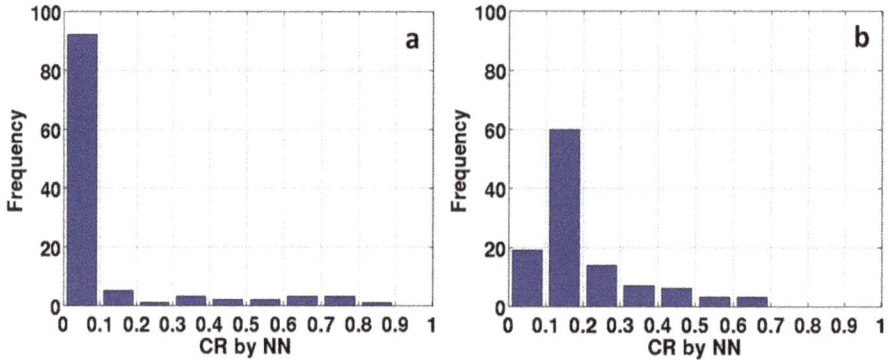

Figure 6. NN Collapse Ratio estimated after the M6.0 earthquake on 24 August 2016. (**a**) NN approach using optical inputs and (**b**) optical and SAR.

The histogram comparison shows that Exp 1 estimates most of polygons with a CR less than 0.1 (92 polygons, 82% of total), which means no damage. Five polygons present a CR between 0.1 and 0.2 and just one between 0.2 and 0.3. Only about 13% of polygons have a CR higher than 0.3, which spans from medium damaging to total collapsing. Specifically, three polygons with CR between 0.3 and 0.4, four between 0.4 and 0.6, six between 0.6 and 0.8, and only one higher than 0.8, indicating total collapse.

The location of polygons is well shown in Figure 7a, which describes the damage CR maps obtained analyzing post 24 August 2016. The map indicates that the severe damage is mainly retrieved in the northeast part of the Amatrice (red blocks), although information from Civil Protection confirms that collapses involved the whole historical center. This is confirmed by looking at an optical very high resolution (VHR) image (Figure 8) acquired after the Amatrice seismic event.

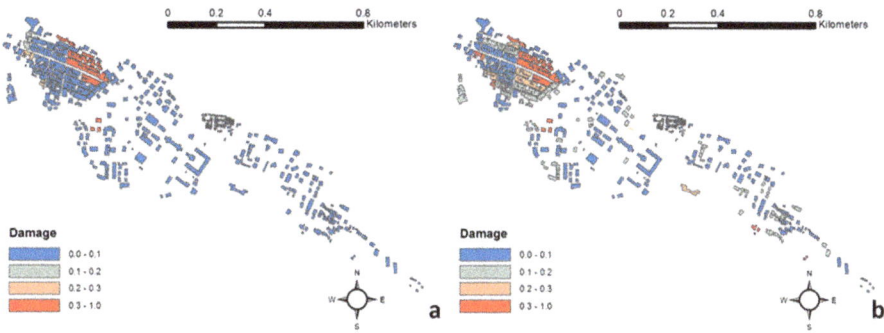

Figure 7. Amatrice CR maps obtained analyzing post 24 August 2016 remote sensing data: (**a**) NN with only optical features as input; (**b**) NN using both optical and SAR features.

Figure 8. Amatrice historical center, post-earthquake damage from RGB VHR DigitalGlobe image.

The fusion of optical and SAR features significantly changes this distribution (Figure 7b), where occurrences of CR less than 0.1 (blue blocks) drastically decrease with respect to the Exp 1 (19 polygons, only 17% of total). About 54% of polygons show a CR between 0.1 and 0.2 (grey blocks), indicating light damage. Looking at map in Figure 7b, these light damages lie partially in the historical center, but most of them are located in the southeast part of the city.

Considering only the historical center and the comparison with the VHR image (Figure 8), we suppose that Exp 2, involving SAR features, is more sensitive to light/medium damage level. Indeed, 14 polygons present a CR between 0.2 and 0.3, which indicate a medium damage level. We also notice that these polygons are located in the southeast part of the historical city center, an area involving severe damage as confirmed by the VHR image depicted in Figure 8. Considering CR higher than 0.3, polygons retrieved by Exp 2 increase by about 27% (19 respects of previous 14 of Exp 1).

The NN employed for the data fusion approach (Figure 7b), better identifies the severely damaged areas in the historical center, also where Exp 1 underestimated the collapses. At the same time, it seems to overestimate the damage in other areas, especially outside of the historical center. Indeed, Exp 2 indicates zones characterized by building collapses also in the southeast and partially in the northwest (Figure 7b). In this sense, the Exp 2 seems to provide a more realistic damage distribution: at first glance the NN regression model using the fusion approach gives the best results better identifying areas affected by collapses, whereas the Exp 1 seems to underestimate them.

5. Conclusions

The present work shows that neural networks, once they have been trained, can be used to rapidly retrieve building collapse ratios from optical and SAR remote sensed data. The implemented ANNs modelled the collapse ratio with quite high accuracy when applied to the post Amatrice earthquake independent dataset. High values of CR (more than 0.5) were retrieved over the historical center that are in agreement with a damage assessment observed by optical VHR imagery. The fusion of optical and SAR derived parameters seem to give a more reliable result using only optical data, even though it probably underestimates the occurrences of collapse ratios higher than 0.5, due mainly to a minor statistical characterization of values during the ANN training phase (only 3%).

Considering that the technique is independent of different types of input data, both for radiometric characteristics and spatial resolution, our work has demonstrated that ANNs are powerful tools able to estimate damaged areas, and they are important in near real time monitoring activities, owing to their fast application.

When looking at using an ANN approach in near real time monitoring, special care has to be taken during the training phase. This is because the neural network needs to be fed and trained continuously also during its operating phase in order to keep phenomena knowledge updated and retrieval performance accurate at the operating stage.

Another aspect is related to the input data characteristics, such as the spatial resolution, which can limit the neural network retrieval accuracy. In this case, a possible improvement could be the use of VHR satellite data, which could provide information at a building scale.

Future work will consider earthquake surveys in order get ground truth datasets to make a quantitative assessment of ANN performance. It will also consider extending the analysis to other areas affected by 2016 Central Italy seismic sequence. Furthermore, because the described approach needs both pre- and post-earthquake images, and in many cases a pre-earthquake image is not available or not up to date, a future goal will be the application of Neural Network approaches only using post-earthquake remote sensed images, available in near real time.

Acknowledgments: The research leading to these results has received funding from the European Union's Seventh Framework Program (FP7/2007-2013) under grant agreement number 606738, APhoRISM project.

Author Contributions: All authors contributed in a substantial way to the writing of the manuscript: Alessandro Piscini conceived the experiments. Alessandro Piscini, Vito Romaniello and Christian Bignami performed the analysis and wrote the manuscript. Salvatore Stramondo supervised the research at all stages.

Conflicts of Interest: The authors declare no conflict of interest.

References

1. Rumelhart, D.E.; Durbin, R.; Golden, R.; Chauvin, Y. Backpropagation: The basic theory. In *Backpropagation: Theory, Architecture, and Applications*; Rumelhart, D.E., Yves, C., Eds.; Lawrence Erlbaum: Hillsdale, NJ, USA, 1995; pp. 1–34.
2. Del Frate, F.; Ferrazzoli, P.; Schiavon, G. Retrieving soil moisture and agricultural variables by microwave radiometry using neural networks. *Remote Sens. Environ.* **2003**, *84*, 174–183. [CrossRef]
3. Tedesco, M.; Pulliainen, J.; Takala, M.; Hallikainen, M.; Pampaloni, P. Artificial neural network-based techniques for the retrieval of SWE and snow depth from SSM/I data. *Remote Sens. Environ.* **2004**, *90*, 76–85. [CrossRef]
4. Verger, A.; Baret, F.; Weiss, M. Performances of neural networks for deriving LAI estimates from existing CYCLOPES and MODIS products. *Remote Sens. Environ.* **2008**, *112*, 2789–2803. [CrossRef]
5. Vilas-González, L.; Spyrakos, E.; Torres-Palenzuela, J.M. Neural network estimation of chlorophyll a from MERIS full resolution data for the coastal waters of Galician rias (NW Spain). *Remote Sens. Environ.* **2011**, *115*, 524–535. [CrossRef]
6. Piscini, A.; Carboni, E.; Del Frate, F.; Grainger, R.G. Simultaneous retrieval of volcanic sulphur dioxide and plume height from hyperspectral data using artificial neural networks. *Geophys. J. Int.* **2014**, *198*, 697–709. [CrossRef]

7. Wu, C.L.; Chau, K.W.; Fan, C. Prediction of rainfall time series using modular artificial neural networks coupled with data-preprocessing techniques. *J. Hydrol.* **2010**, *389*, 146–167. [CrossRef]
8. Chau, K.W.; Wu, C.L. A Hybrid Model Coupled with Singular Spectrum Analysis for Daily Rainfall Prediction. *J. Hydroinform.* **2010**, *12*, 458–473. [CrossRef]
9. Krasnopolsky, V.M.; Breaker, L.C.; Gemmill, W.H. A neural network as a nonlinear transfer function model for retrieving surface wind speeds from the Special Sensor Microwave Imager. *J. Geophys. Res.* **1995**, *100*, 11033–11045. [CrossRef]
10. Ahadzadeh, S.; Valadanzouj, M.; Sadeghian, S.; Ahmadi, S. Detection of damaged buildings after an earthquake using artificial neural network algorithm. In Proceedings of the International Archives of the Photogrammetry, Remote Sensing and Spatial Information Sciences, Beijing, China, 3–11 July 2008.
11. Chakraborty, S.; Kumar, P.; Chakraborty, S. Neural Network Approach to Response of Buildings Due to Earthquake Excitation. *Int. J. Geosci.* **2012**, *3*, 630–639. [CrossRef]
12. Bignami, C.; Chini, M.; Pierdicca, N.; Stramondo, S. Comparing and combining the capability of detecting earthquake damage in urban areas using SAR and optical data. In Proceedings of the IEEE IGARSS, Anchorage, AK, USA, 20–24 September 2004; Volume 1, pp. 55–58.
13. Stramondo, S.; Bignami, C.; Chini, M.; Pierdicca, N.; Tertulliani, A. Satellite radar and optical remote sensing for earthquake damage detection: Results from different case studies. *Int. J. Remote Sens.* **2006**, *27*, 4433–4447. [CrossRef]
14. Romaniello, V.; Piscini, A.; Bignami, C.; Anniballe, R.; Stramondo, S. A multisensor approach for the 2016 Amatrice earthquake damage assessment. *Ann. Geophys.* **2016**, *59*. [CrossRef]
15. Brunner, D.; Lemoine, G.; Bruzzone, L. Earthquake damage assessment of buildings using VHR optical and SAR imagery. *IEEE Trans. Geosci. Remote Sens.* **2010**, *48*, 2403–2420. [CrossRef]
16. Romaniello, V.; Piscini, A.; Bignami, C.; Anniballe, R.; Stramondo, S. Earthquake damage mapping by using remotely sensed data: The Haiti case study. *J. Appl. Remote Sens.* **2017**, *11*. [CrossRef]
17. Kullback, S.; Leibler, R.A. On Information and Sufficiency. *Ann. Math. Stat.* **1951**, *22*, 79–86. [CrossRef]
18. Xie, H.; Pierce, L.E.; Ulaby, F.T. Mutual information based registration of SAR images. In Proceedings of the IEEE International Geoscience and Remote Sensing Symposium, Toulouse, France, 21–25 July 2003; pp. 4028–4031.
19. Tinti, E.; Scognamiglio, L.; Michelini, A.; Cocco, M. Slip heterogeneity and directivity of the ML 6.0, 2016, Amatrice earthquake estimated with rapid finite-fault inversion. *Geophys. Res. Lett.* **2016**, *43*. [CrossRef]
20. Rosenblatt, R. *Principles of Neurodynamics*; Spartan Books: New York, NY, USA, 1962.
21. Hecht-Nielsen, R. *Neurocomputing*; Addison-Wesley: Boston, MA, USA, 1990.
22. Bishop, C. *Neural Networks for Pattern Recognition*; Oxford University Press: Oxford, UK, 1995.
23. Rumelhart, D.E.; Hinton, G.E.; Williams, R.J. Learning internal representation by error propagation. In *Parallel Distributed Processing: Exploration in the Microstructure of Cognition*; Rumelhart, D.E., McClleland, J.L., Eds.; MIT Press: Cambridge, MA, USA, 1986; Volume 1, Chapter 8.
24. Hassoun, M.H. *Fundamentals of Artificial Neural Networks*; MIT Press: Cambridge, MA, USA, 1995.
25. Haykin, S. *Neural Networks: A Comprehensive Foundation*, 2nd ed.; Prentice-Hall: Upper Saddle River, NJ, USA, 1999.
26. Prechelt, L. Automatic Early Stopping Using Cross Validation: Quantifying the Criteria. *Neural Netw.* **1998**, *11*, 761–767. [CrossRef]
27. Rast, M.; Hook, S.J.; Elvidge, C.D.; Alley, R.E. An evaluation of techniques for the extraction of mineral absorption features from high spectral resolution remote sensing data. *Photogramm. Eng. Remote Sens.* **1991**, *57*, 1303–1309.

applied sciences

MDPI

Article

Analysis of the Pyroclastic Flow Deposits of Mount Sinabung and Merapi Using Landsat Imagery and the Artificial Neural Networks Approach

Prima Riza Kadavi [1], Won-Jin Lee [2] and Chang-Wook Lee [1,*]

[1]   Division of Science Education, Kangwon National University, 1 Kangwondaehak-gil, Chuncheon-si,
     Gangwon-do 24341, Korea; rizakadavi@gmail.com
[2]   Earthquake Volcano Research Center, Korea Meteorological Administration, 61 16-Gil, Yeouidaebang-ro,
     Dongjak-Gu, Seoul 07062, Korea; wjleeleo@korea.kr
*   Correspondence: cwlee@kangwon.ac.kr; Tel.: +82-33-250-6731

Received: 25 July 2017; Accepted: 8 September 2017; Published: 11 September 2017

Abstract: Volcanic eruptions cause pyroclastic flows, which can destroy plantations and settlements. We used image data from Landsat 7 Bands 7, 4 and 2 and Landsat 8 Bands 7, 5 and 3 to observe and analyze the distribution of pyroclastic flow deposits for two volcanos, Mount Sinabung and Merapi, over a period of 10 years (2001–2017). The satellite data are used in conjunction with an artificial neural network method to produce maps of pyroclastic precipitation for Landsat 7 and 8, then we calculated the pyroclastic precipitation area using an artificial neural network method after dividing the images into four classes based on color. Red, green, blue and yellow were used to indicate pyroclastic deposits, vegetation and forest, water and cloud, and farmland, respectively. The area affected by a volcanic eruption was deduced from the neural network processing, including calculating the area of pyroclastic deposits. The main differences between the pyroclastic flow deposits of Mount Sinabung and Mount Merapi are: the sediment deposits of the pyroclastic flows of Mount Sinabung tend to widen, whereas those of Merapi elongated; the direction of pyroclastic flow differed; and the area affected by an eruption was greater for Mount Merapi than Mount Sinabung because the VEI (Volcanic Explosivity Index) during the last 10 years of Mount Merapi was larger than Mount Sinabung.

Keywords: Sinabung eruption; Merapi eruption; pyroclastic flow deposits; Landsat imagery; artificial neural network

1. Introduction

Remote sensing research has used multispectral remote sensing imagery to provide additional data, proving to be a valuable source of spatio-temporal data for some applications. Applications that are widely used are: land cover classification, detection catch archeology, extracting spatial features and classification in a residential area, the extraction of the street, the estimation of urban sprawl automatic mapping feature flow, classification and feature extraction transport, mapping snow cover and the evaluation of geomorphological features [1,2].

Land cover assessment (LC) is very important in planning, monitoring and sustaining the utilization of natural resources. LC has a direct impact on water, atmospheric and soil erosion and is therefore directly related to many globally-important environmental issues [3]. Appropriate knowledge, updated and temporal, about LC is very important to address the issue of unplanned development, environmental degradation, loss of wildlife habitat and depletion of primary agricultural and forest land [4]. Therefore, it is important to evaluate and monitor the LC dynamics resulting from anthropogenic activities and natural phenomena to plan, monitor and sustain the utilization of natural resources [5].

In this study, land cover assessment is used to see the distribution of volcanic eruptions' deposits that may be dangerous to the surrounding area when the eruption occurs. The volcano is a naturally-formed entity on the surface of the Earth that occupies an area and displays volcanism. An eruption is the discharge of magma from within the Earth and can be divided into three types: explosive, effusive and hot spot eruptions. The type of eruption that occurs is influenced by many factors such as magma viscosity, the gas content of the magma, the influence of groundwater and the depth of the magma chamber [6]. The products ejected by volcanic eruptions, which are often catastrophic, can be captured by satellite, optical and radar sensors. Remote sensing images can be utilized to detect the spread of eruptive fumes associated with volcanic eruptions that spread in the atmosphere, pyroclastic deposits, incandescent lava, lahar distribution and dome deformation [7].

We analyze the evaluation of volcano hazard, especially pyroclastic deposit, using land classification with Landsat imagery data and the artificial neural network approach. Several different approaches to the evaluation of volcano hazards can be found in the current literature, including direct and indirect heuristic approaches and deterministic, probabilistic and statistical approaches. Lee et al. (2015) summarized the many analyses of volcano activity based on remote sensing and GIS techniques [8]. Recently, studies on volcano activity assessment made use of remote sensing, and many applied probabilistic models such as satellite imagery. We used land classification in satellite imagery using an artificial neural network method to analyze the area affected by an eruption. Lee at al. (2003) showed that the most frequently-used neural network method is the backpropagation learning algorithm, a learning algorithm of a multilayered neural network that consists of an input layer, hidden layers and an output layer [9]. Many experiments have shown that multilayered neural networks are more accurate for land cover classification than traditional statistical methods [10–13].

2. Study Area

Indonesia is a country located in the ring of fire, an area of frequent geological disasters, including earthquakes, volcanoes, flash floods, landslides and tsunamis. These disasters are harmful, destructive and result in a huge loss of life. The most frequent disaster in Indonesia is a volcanic eruption. Indonesia currently has 129 active volcanoes, and 70 volcanoes have erupted in the last 400 years. The area of land threatened by a volcanic eruption in Indonesia is 16,670 km^2, and around 5,000,000 lives are threatened by volcanic eruptions [14].

There have been several volcanic eruptions in Indonesia (Figure 1a) in the last 10 years. The focus of this research is Mount Sinabung (Figure 1b) and Merapi (Figure 1c), the eruptions of which have caused many casualties. The eruption of Merapi in 2010 was the largest eruption in the last 10 years and killed 347 people. Mount Sinabung has erupted two times in the last 10 years in 2010 and 2013–2017. The period of eruption of 2013 is still ongoing and has killed 16 people [14].

Mount Sinabung is in the Karo Highlands, Karo District, North Sumatra Province, Indonesia. The geographical position of the peak of Mount Sinabung is 3°10′ north latitude and 98°23.5′ east longitude. The height of Mount Sinabung is 2460 m above sea level, and its volcano type is a stratovolcano [14]. Mount Sinabung was dormant for 400 years, from ~1600–2010 CE. Mount Sinabung finally erupted in 2010, an activity that was predicted from the three earthquakes that struck Sumatra. The earthquakes measured 8.8, 7.9 and 8.4 on the Richter scale and struck in 2005, 2007 and 2007, respectively [15].

Mount Merapi is located on the border of four districts; Sleman, Yogyakarta and Magelang, Boyolali and Klaten, in Central Java Province. Its geographical position lies at 7°32′30″ north latitude and 110°26′30″ east longitude. Based on its tectonic order, the mountain is located in a subduction zone, where the meeting of the Indo-Australian and Eurasian Plates controls the volcanism in Sumatra, Java, Bali and Nusa Tenggara. The height of Mount Merapi is 2986 m above sea level, and its type is a stratovolcano [14]. Mount Merapi is a very active volcano mountain. Since 1600, Merapi volcano has erupted more than 80 times, and an eruption occurs on average every 4–5 years. Merapi is a volcano cone composed of andesitic-basaltic magma with silica (SiO_2) content ranging from 52–56%. The top morphology is characterized by a horseshoe-shaped crater, in which lava domes grow [16–19].

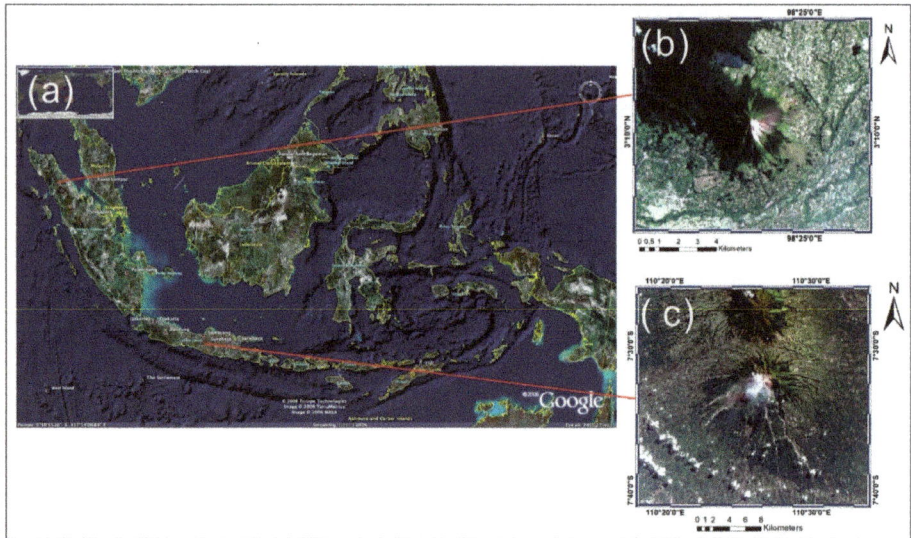

Figure 1. Map of Indonesia using Google Maps (**a**); location of Mt. Sinabung (**b**) and Merapi (**c**) on Sumatra and Java Islands from USGS earth explorer website [20].

3. Data

A pyroclastic flow inundation map can be generated by field surveys from the crater rim to the furthest extent of the pyroclastic flow after a volcanic eruption. However, a field survey in an active volcano can be dangerous because of the exposure to hazardous gases and sudden activity. In contrast, remote sensing techniques are a useful tool for generating pyroclastic flow deposit maps, which provide a safe, cost-effective alternative to field mapping.

A Landsat image is a picture of the surface of the Earth taken from outside the atmosphere at an altitude of approximately 818 km from the Earth's surface, on a scale of 1:250,000. Each Landsat image covers an area of 185 km^2 so that the aspect of a large object can be identified without exploring all of the surveyed areas [21]. Each color in the satellite image has a meaning; a color on the image indicates whether a reflection value corresponds to vegetation, aquatic bodies or a body of the Earth's surface rock [22]. Therefore, geological interpretation of the Landsat image is based on the difference between the reflection values [23].

This study used satellite data from Landsat 7 ETM+ and Landsat 8. Landsat 7 ETM+ was used to see the changes of pyroclastic deposits from 2007 to 2012, while Landsat 8 was used to see changes in pyroclastic deposits from 2013 to 2017. However, the Landsat 7 ETM+ data suffer from the scan line corrector (SLC)-off phenomenon because of the failure of the SLC in the ETM+ instrument. These images can be restored by gap-filling the scan line using one-dimensional interpolation without any other supplementary data with the gap interpolation and filtering technique [24].

The Landsat 7 and Landsat 8 image data from Mount Sinabung and Mount Merapi were taken from USGS earth explorer website [20] using RGB Bands 7, 4 and 2 to see the pyroclastic deposit in Landsat 7 and RGB Bands 7, 5 and 3 to see the pyroclastic deposits in Landsat 8 (Figures 2 and 3). Landsat image data are taken annually to analyze changes in pyroclastic deposit. Landsat images were selected based on their image quality. Several parameters were used to assess the quality of Landsat images, including sunlight and cloud cover in the image of the object of research. Based on these parameters, data selection is performed in different months and dates in each year.

We select the Landsat image data representing data before eruption and after eruption to see the difference in pyroclastic deposit area of Mount Sinabung and Merapi. For Mount Sinabung, we chose Landsat image data taken 7 February 2007 as data before eruption, 17 November 2011 as data after the first eruption, 1 July 2016 as data after the second eruption and 28 July 2017 as current pyroclastic deposit data. For Mount Merapi, we chose Landsat image data taken 16 June 2007 as data before eruption, 10 May 2011 and 2 November 2014 as data after the big eruption in 2010 and 23 May 2017 as current pyroclastic deposit data.

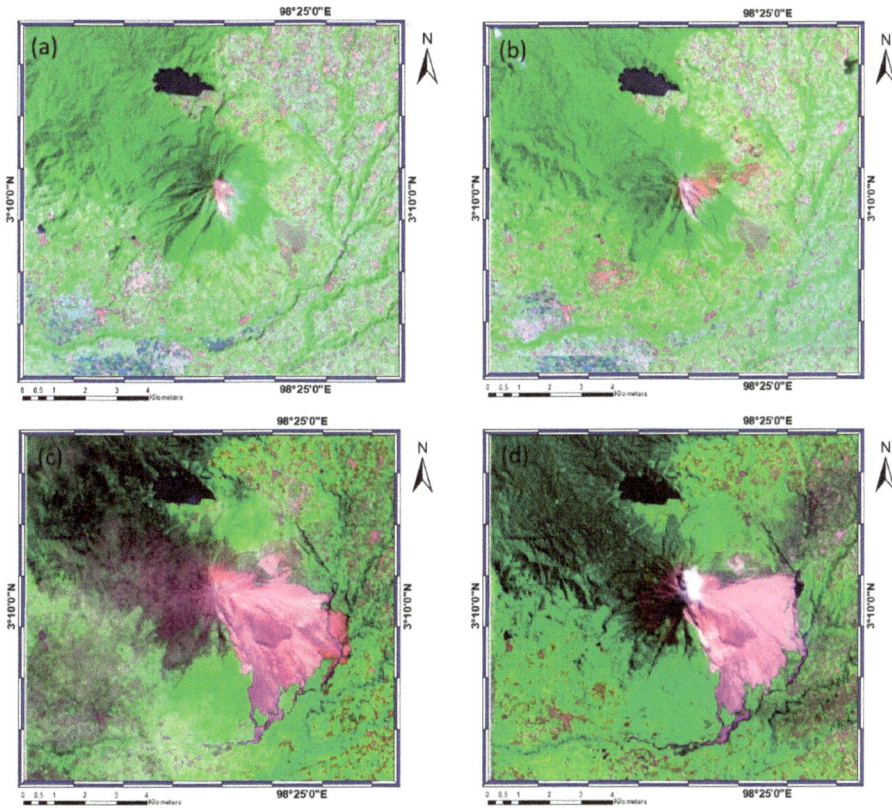

Figure 2. Landsat 7 (Bands 7, 4 and 2) and Landsat 8 (bands 7, 5 and 3) imagery data of Mount Sinabung taken in the year shown below each image. We can see the changes of pyroclastic deposits of Mount Sinabung from before eruption (**a**) 7 February 2007, after the first eruption on (**b**) 17 November 2011, after the second eruption on (**c**) 7 July 2016, and current pyroclastic deposits on (**d**) 28 July 2017, marked with grey color in the middle of the image.

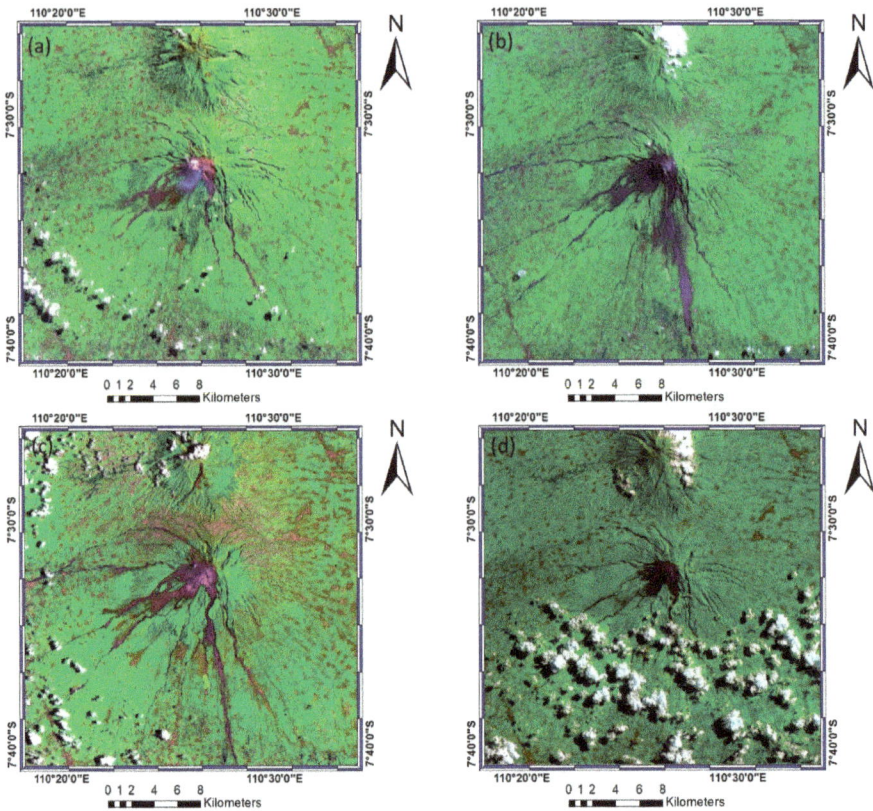

Figure 3. Landsat 7 (Bands 7, 4, 2) and Landsat 8 (Bands 7, 5, 3) imagery data of Mount Merapi taken in the year shown below each image. We can see the changes of pyroclastic deposits of Mount Merapi from before eruption on (**a**) 16 June 2007, and also during the time after the big eruption in 2010, which can see on (**b**) 10 May 2011 and (**c**) 2 November 2014, and current pyroclastic deposits on (**d**) 23 May 2017, marked with dark grey color in the middle of the image.

4. Methodology

A neural network model belongs to the branch of artificial intelligence generally referred to as artificial neural networks (ANNs). ANNs teach a system to execute a task, instead of using a computational programming system to do defined tasks. To perform such tasks, an artificial intelligence system (AI) is generated; a pragmatic model that can quickly and precisely find the patterns buried in data that represent useful knowledge. Neural networks are one example of these AI models. In the area of medical diagnosis relationships with dissimilar data, artificial intelligence is one of the most available techniques [25].

In this method, the analyst first determines some training area (sample area) on the image as a class of the appearance of a particular object. This determination is based on the analyst's knowledge of the region of land cover areas. The pixel values in the sample area are then used by computer software as the key to recognizing other pixels. Areas that have similar pixel values are entered into predefined classes [10]. Therefore, in this method, the system first identifies the class of information, and this class is then used to determine the spectral class that represents the class of information.

An ANN is a mathematical model that has been applied for the identification, modeling, optimization, forecasting, prediction and control of complex systems [26,27]. It can be trained for performing a particular task based on available empirical data [28]. ANN is a nonparametric approach that has advantages over statistical classification techniques and has been widely used in different LC studies in recent years [29,30]. The NN classifier adopted in this study is a nonlinear layered feed-forward model with standard backpropagation for supervised learning. To perform the ANN using ENVI software, the logistic activation method was used, and one hidden layer was selected. The training threshold contribution and training momentum fields were set to a value of 0.9. Finally, the training rate field, training RMS exit criteria field and number of training iterations were set to values of 0.2, 0.1 and 100, respectively.

The links with the neurons located in the so-called hidden neuron layer then take different weights and are trained depending on the required output, thus modeling complex relationships among variables [31]. The system requires feed-forward and backpropagation processes to allow the network to be trained. The visualization of this stage is accomplished through error analysis. If the error becomes smaller and asymptotic, the network will be ready to receive new input data and predict the output [32].

The ANN models used in this study are of the multilayer perceptron ANN type [33]. The architecture is as shown in Figure 4. The input layer is Region 1 (red color as pyroclastic deposits), Region 2 (blue color as water), Region 3 (green color as forest) and Region 4 (yellow color as farmland). In each case, the training of the proposed network was performed with a backpropagation algorithm, which is a supervised learning procedure. It uses the method of gradient descent for minimizing the global quadratic error of the output calculated by the network [32].

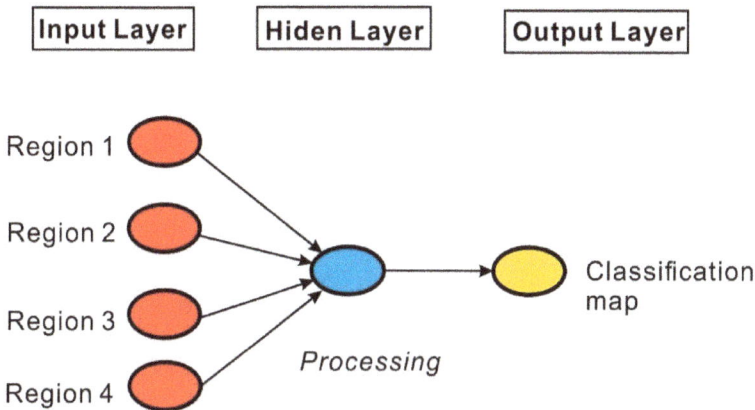

Figure 4. Artificial neural network model used for data analysis. Input layer data Regions 1–4, which are select in ENVI. The output layer is the classification map.

5. Result

5.1. Classification of Landsat Images

The pyroclastic dimension area was analyzed using supervised classification using the artificial neural network approach in the ENVI program. We used the gap-filled Landsat data to delineate the pyroclastic deposit surrounding areas through supervised classification. We established four classes for Mount Sinabung and Mount Merapi. The focus of this study is the pyroclastic deposits, which are marked as red. The classes used on Mt. Sinabung are forest (green), farmland (yellow), water and shadow (blue) and pyroclastic deposit (red). Mt. Merapi is divided into forest (green), farmland (yellow), cloud (blue) and pyroclastic deposit (red).

Land classification on Mount Sinabung using the artificial neural network approach is shown in Figure 5. The land classification is divided into four classes based on its color: green as forest, yellow as farmland, blue as water and shadow and red as pyroclastic precipitate. We selected Landsat image data in this study taken on 7 February 2007 as data before eruption, 17 November 2011 as data after the first eruption, 1 July 2016 as data after the second eruption and 28 July 2017 as current pyroclastic deposit data. Based on the supervised classification maps of Mt. Sinabung (Figure 5), the pyroclastic flow depositional area was primarily in the southern region of Mt. Sinabung before the first eruption in 2010. However, the pyroclastic flow deposit migrated to the eastern part of the volcano after the 2010 eruption (Figure 5a,b). The next eruption was in 2013–2017 (Figure 5b,c), and from the land classification data, it can be seen that after the second eruption on 1 July 2016, there was a very big change in the pyroclastic deposits of Mount Sinabung. The pyroclastic precipitate flows to the southeast and extends downward with a shape resembling a landslide. On 28 July 2017, it was seen that pyroclastic deposits did not change too much from the previous year.

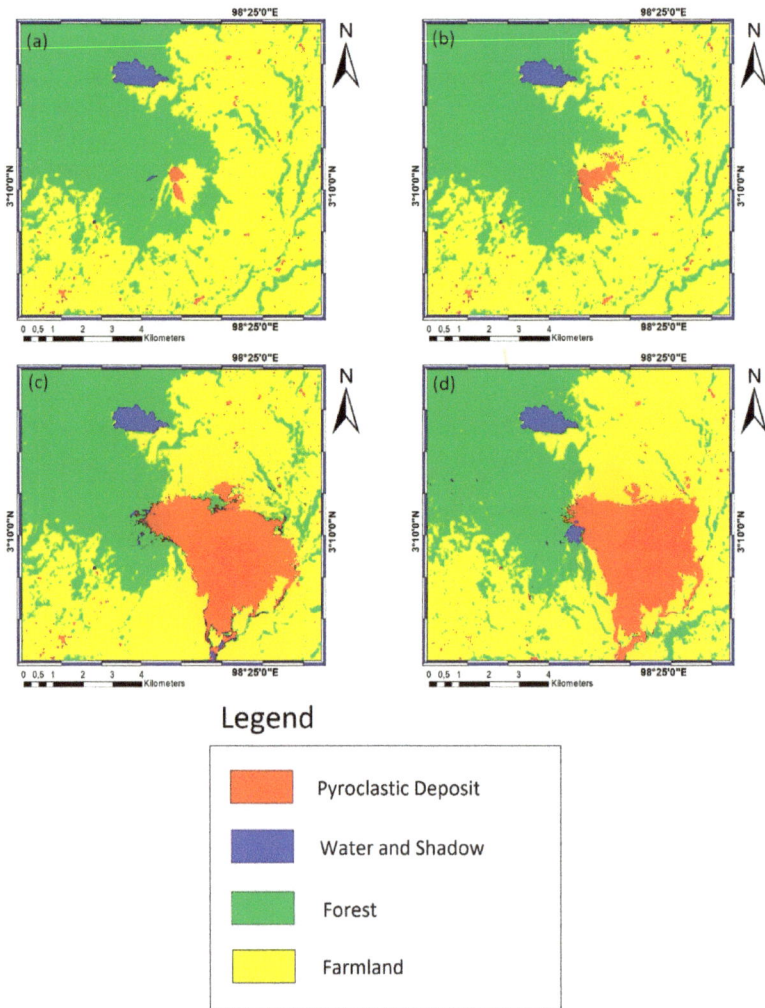

Figure 5. Supervised classification maps from ANN processing in the ENVI program of the Mount Sinabung Landsat Imagery taken on: (**a**) 7 February 2007; (**b**) 17 November 2011; (**c**) 1 July 2016; and (**d**) 28 July 2017.

Land classification on Mount Merapi using the artificial neural network approach is shown in Figure 6. The land classification is divided into four classes based on its color: green as forest, yellow as farmland, blue as cloud and red as pyroclastic precipitate. We selected Landsat image data in this study taken on 16 June Landsat 2007 as data before eruption, 10 May 2011 and 2 November 2014 as data after the big eruption in 2010 and 23 May 2017 as current pyroclastic deposits. Based on the supervised classification maps of Mount Merapi (Figure 6), the pyroclastic flow depositional area primarily existed in the southwestern region before the eruption in 2010. However, the pyroclastic flow deposit migrated to the southern part of the volcano after the 2010 eruption.

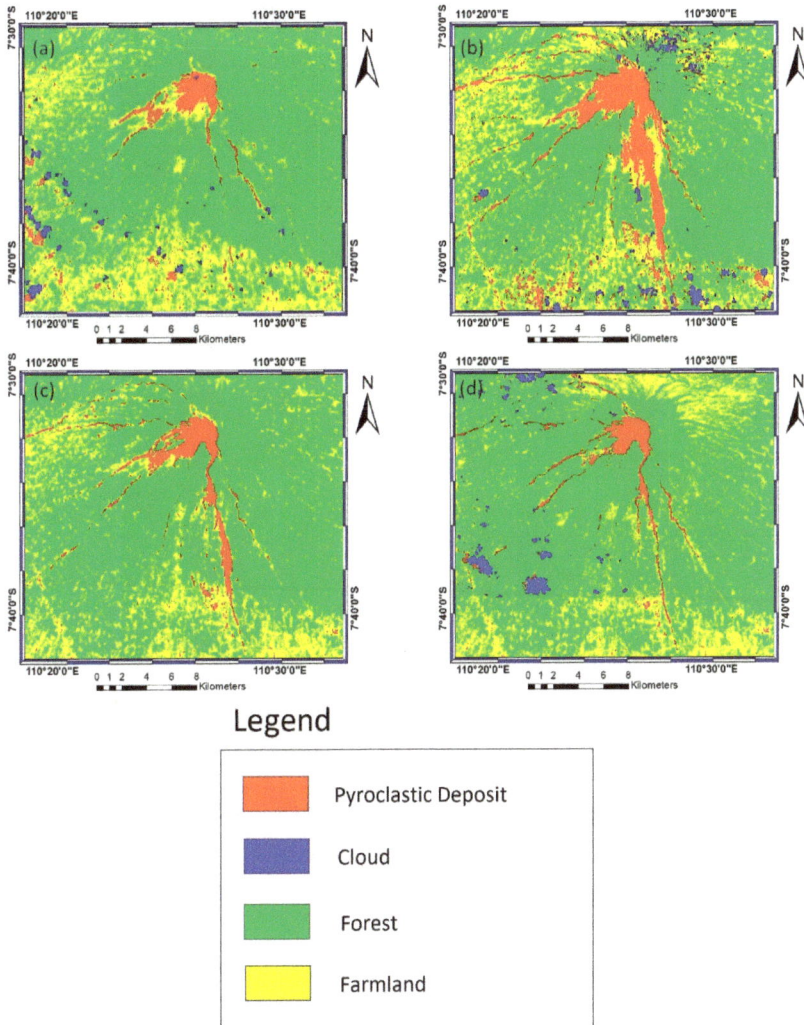

Figure 6. Supervised classification maps using ANN processing in the ENVI program of Mount Merapi Landsat imagery data taken on: (**a**) 16 June 2007; (**b**) 10 May 2011; (**c**) 2 November 2014; and (**d**) 23 May 2017.

5.2. Time Series Analysis of Pyroclastic Deposit

Pyroclastic deposits area are constantly changing after an eruption, and therefore, time series analysis is needed to compare the area of pyroclastic deposits in each year. Time series analysis of pyroclastic deposit land classification with Landsat imagery data is used to calculate the area of pyroclastic deposits. We calculate the area of the pyroclastic deposits using ArcMap. The data of the land classification map with a multilayer format are used to calculate pixels in each class [34], because the concern of this research is the pyroclastic deposit, so we only calculate red pixels, which means pyroclastic deposit on the land classification map. The Landsat image has a resolution of 30 × 30 m, so we just multiply the number of pixels by the resolution of the Landsat image to get the area of pyroclastic deposits.

The pyroclastic deposits on Mt. Sinabung increased after the eruption in 2010. Since Mt. Sinabung has been highly active since 2010, the pyroclastic deposit area changes every year. We can see the change of the pyroclastic deposit of Mt. Sinabung each year based on Table 1. Before the eruption, pyroclastic deposits were calculated as 0.3807 km² in 2007. The area of pyroclastic deposits increased in 2011 due to the eruption in 2010 to 1.5066 km². The increase of pyroclastic precipitate in 2011 was not too big because the eruption that occurred in 2010 was a small eruption. Then, in 2013–2017, the second eruption occurred, and in 2016, the pyroclastic deposit area due to the second eruption of Mount Sinabung in the last ten years increased very significantly to 20.5911 km²; this can be seen in Figure 7. The current pyroclastic deposit area of Mount Sinabung is estimated at 18.4572 km².

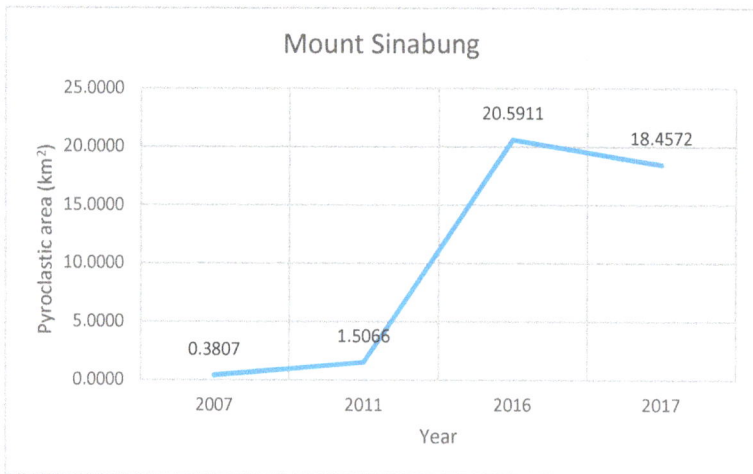

Figure 7. Time series of pyroclastic flow deposits for Mount Sinabung.

Table 1. Time series of pyroclastic flow deposits.

Mount Sinabung No.	Year	Pyroclastic Deposit Area	
		Pixel	km²
1	2007	423	0.3807
2	2011	1674	1.5066
3	2016	22,879	20.5911
4	2017	20,508	18.4572

Mount Merapi erupted in 2006; thus, the pyroclastic deposit of Mount Merapi was still high in 2007, as it was calculated to be 16.443 km² in Table 2. In 2010, Merapi erupted with VEI (Volcanic Explosivity

Index) 4. Therefore, much volcanic material was ejected and caused damage. The pyroclastic deposits caused by the eruption in 2010 were very big, increasing two-fold from previous eruptions (Figure 8). We can calculate the pyroclastic deposit of by the 2010 eruption by 2011 data; it is 38.2536 km². After the big eruption of 2010, the pyroclastic deposits area was decreased, calculated for Landsat 2014 as 20.5911 km². The current pyroclastic deposit area of Mount Sinabung is estimated at 18.4572 km².

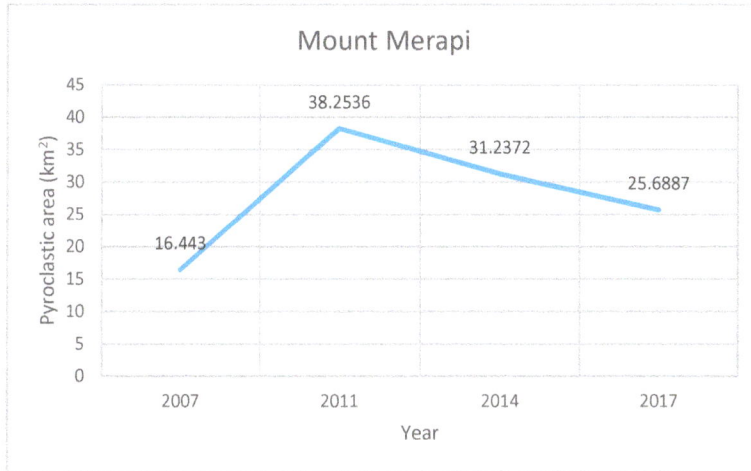

Figure 8. Time series of pyroclastic flow deposit for Mount Merapi.

Table 2. Time series of pyroclastic flow deposit.

Mount Merapi. No	Year	Pyroclastic Deposit Area	
		Pixel	km²
1	2007	18,270	16.443
2	2011	42,504	38.2536
3	2014	34,708	31.2372
4	2017	28,543	25.6887

6. Discussion

The supervised classification neural network analysis was used to view and calculate the area of pyroclastic flow deposits. From the Landsat imagery data, differences in appearance can be seen between pyroclastic flow deposits, forests, lakes and plantations. Landsat image data were used as an object for supervised neural network classification analysis. The supervised neural network classification divides the area based on the pixels that have been selected according to the class. The results of the supervised neural network analysis show that there are differences in the distribution of pyroclastic flow deposits between Mount Merapi and Mount Sinabung. The apparent differences in the Landsat imagery data that have been found by classification by the supervised neural network are the direction of flow, the area of deposit and the distribution of pyroclastic flow deposits. The differences in the distribution of pyroclastic flow deposits between Mount Sinabung and Mount Merapi were caused by several influential factors, such as the formation of volcanoes, regional geological conditions, volcano magmatic properties, volcanic type, lava domes and the Volcanic Explosivity Index.

The geological conditions of Sumatra affect the formation of Mt. Sinabung. The outline of Sumatra Island consists of three tectonic systems, namely the Sumatra Subduction System, the Mentawai Fault system and the Sumatra Fault System. Based on the geological reconstruction by Robert Hall (2000) [35], the initial formation of the Sumatra region began ~50 million years ago (early Eocene). There are at least 19 C-section segments, each with lengths of ±60–200 km, in the Sumatra Fault System, which has a total length of ±1900 km. Lake Toba on the island of Sumatra is evidence of a supervolcano and is the remains of the largest eruptions of Caldera Crescent (scale: 8 VEI).

The eruptions of Mount Sinabung eject many volcanic materials, one of which is pyroclastic flow. The pyroclastic flow deposits of Mount Sinabung spread to the southeast, tend to widen and are not transported far from the crater of Mount Sinabung. These characteristics result from the acid-intermediate (andesitic-rhyolitic) [17] magma of Mount Sinabung, which tends to be thick and prevents the pyroclastic flow deposit from being transported far. The other factor is due to the shape of the slopes of Mt. Sinabung, where the pyroclastic flows and the direction of the river in Sinabung valley are almost perpendicular. Mount Sinabung is a Type B mountain, which means there was no track record of eruption before it erupted in 2010 [14]. The eruption of 2010 was a phreatic-type eruption, as defined by the presence of volcanic dust. A phreatic eruption is the process of magma escaping to the surface because of the influence of steam caused by the direct or indirect interaction between water and magma [36]. The second eruption of Mount Sinabung was larger than that in 2010, which could have been caused by the formation of an old dome on Mount Sinabung after the first eruption. The material ejected in the next eruption mixed with the material on the surface or around the crater, destroying the already formed lava dome and forming a new lava dome. The eruptions of Mount Sinabung are generally rated as VEI 2, which means the eruption is relatively small [37]. However, the eruptions are continuous and interfere with the activity of the residents. A larger eruption is probable to occur since Sinabung is located in the Semangko Fault Zone, the largest fault on Sumatra Island.

Based on seismic data, Mount Merapi is estimated to have a magma pouch because no seismic data from inside the seismic zone exist, indicating that there is a soft material between harder materials in the seismic zone. This soft material is thought to be the magma pocket of Mount Merapi [19]. This hypothesis is supported by the fact that the temperatures around the top of the Woro and Gendol plates can reach 830 °C, which means there is a sufficiently shallow source of heat. It is estimated that these magma pockets arise because of this basic fault, where magma can accumulate. This magma pocket is thought to act as a valve that slows the upward magma migration of the magma chamber. Therefore, the scale of the eruption is reduced [18]. The absence of an earthquake below 5 km strengthens the case for the existence of a magma kitchen at a depth of ~8 km, as proposed by Beaducel (1998) based on tiltmeter and GPS data modeling [6].

Mount Merapi erupts almost every five years because of the existence of magma pockets in the magma chamber of Merapi Volcano, which are likely to have been formed by a major eruption in the past. The major eruptions recorded on Mount Merapi in the 21st Century were in 2006 and 2010, with the 2010 eruption causing more than 350 casualties [14].

Landsat imagery data show that the 2010 eruption issued pyroclastic material from inside Merapi's crater that spread around the Mt. Merapi area. Landsat 8 image data using Bands 4, 3 and 2 in the natural view show that the direction of the pyroclastic flow was to the south and southwest and that Merapi's pyroclastic flow sediment tends to elongate because it is transported far from the peak. The long range of the flow could be caused by the valleys that run in the same direction as the slope, as well as the basaltic-andesitic (intermediate-base) magma type. This type of magma tends to be dilute and can be transported far, along with the outflow of pyroclastic flows from the eruption. Merapi's eruption type can be categorized as a weak volcanic type that is not so explosive, but pyroclastic flow almost always occurs after an eruption. The eruptive activity of Merapi can be seen to undergo a long process, from the initial lava dome formation to incandescent lava and hot clouds (pyroclastic flow). The eruption in 2010 had a strong VEI value of four, and lahar flows occurred in addition to a pyroclastic flow [16]. Lahar flows damage buildings and bridges because of their high speed and because they carry material such as stone and sand. In addition, volcanic ash is widely distributed,

Appl. Sci. **2017**, *7*, 935

the high silica content of which makes it dangerous to breathe. The volcanic ash sediment can be seen in Landsat 7 imagery data from 2010.

7. Conclusions

Mount Sinabung first erupted in 2010 after a long break of over 400 years, and this first eruption was followed by a series of six eruptions between 2013 until 2016. The biggest eruption occurred in 2014 and had a VEI of two, causing ~16 deaths and the destruction of farmland and residents' houses by pyroclastic flows. Mount Merapi erupts roughly every five years because of the high magma activity in its magma chamber, as well as the magma pocket that formed during a past eruption. We analyzed the differences in the pyroclastic deposit flows between Mt. Sinabung and Mt. Merapi, using Landsat 7 and 8 imagery with Bands 7, 4, 2 and 7, 5, 3, respectively, to see the pyroclastic precipitates and deposits of sand and rocks.

An artificial neural network method was used to analyze the pyroclastic precipitates and the pyroclastic precipitate area using Landsat imagery data taken annually from 2007 to 2017 for each volcano. The analysis showed that the pyroclastic sediment of Mount Sinabung spread to the east at the beginning of eruption and then widened to the southeast and south after the series of eruptions took place. The continuous eruption of Mount Sinabung caused the formation of pyroclastic deposits that accumulated around the body of the mountain. The pyroclastic sediment of Mount Sinabung tends to widen and is not transported far because of the direction of river flow on the mountain, which is not in the direction of the mountain's slope. In addition, the andesitic-rhyolitic (intermediate-acid) magma composition of Mount Sinabung prevents pyroclastic flow from being transported far, because the lava tends to be thick. In contrast, the pyroclastic flows of Mount Merapi flow to the south and to the southwest along the slope and river flow direction of Mount Merapi. Thus, the pyroclastic sediment of Mount Merapi does not accumulate much in the body of Merapi Volcano, but flows following the slope of the volcano. The composition of magma on Mount Merapi is also different from Mount Sinabung; Mount Merapi's magma is andesitic-basaltic (intermediate-base), so the flowing lava is thin and easily transported. Although the maximum VEI of Merapi is larger than that of Sinabung (VEIs of four and two, respectively), the pyroclastic precipitates of Mount Sinabung's eruption are wider than those of Mount Merapi (~18 and ~17 km^2, respectively).

Acknowledgments: This research was supported by the Basic Science Research Program through the National Research Foundation of Korea (NRF) funded by the Ministry of Science, ICT & Future Planning (NRF-2015M1A3A3A02013416) and the Korea Meteorological Administration Research and Development Program under Grant KMIPA (2015–3071).

Author Contributions: Prima Riza Kadavi applied the artificial neural networks model for detecting pyroclastic flow deposit area and wrote the paper. Won-Jin Lee designed the volcanic experimental equation corresponding to volcano eruption. Chang-Wook Lee suggested the idea, organized the paper work and interpreted the results of volcanic activities. All of the authors contributed to the writing of each part.

Conflicts of Interest: The authors declare no conflict of interest.

References

1. Elhag, M. Quantification of Land Cover Changes Based on Temporal Remote Sensing Data for Ras Tarout Area, Saudi Arabia. *Indian J. Geo-Mar. Sci.* **2016**, in press.
2. Elhag, M. Evaluation of Different Soil Salinity Mapping Using Remote Sensing Techniques in Arid Ecosystems, Saudi Arabia. *J. Sens.* **2016**, *2016*, 7596175. [CrossRef]
3. Koomen, E.; Stillwell, J. *Modelling Land-Use Change*; Springer: New York, NY, USA, 2007.
4. Anderson, J.R. *A Land Use and Land cover Classification System for Use with Remote Sensor Data Vol 964*; US Government Printing Office: Washington, DC, USA, 1976.
5. Ullah, S.; Farooq, M.; Shafique, M.; Siyab, M.A.; Kareem, F.; Dees, M. Spatial assessment of forest cover and land-use changes in the Hindu-Kush mountain ranges of north ern Pakistan. *J. Mt. Sci.* **2016**, *13*, 1229–1237. [CrossRef]

6. Beauducel, F. Structures Mechanical Behaviour of Merapi Volcano, Java: A Methodological Approach of the Deformation Field. Ph.D. Thesis, University Denis Diderot Paris VII, Paris, France, 1998.

7. Lee, C.W.; Lu, Z.; Won, J.S.; Jung, H.S.; Dzurisin, D. Dynamic deformation of Seguam Volcano, Alaska, 1992–2008, from multi-interferogram InSAR processing. *J. Volcanol. Geotherm. Res.* **2013**, *260*, 43–51. [CrossRef]

8. Lee, C.W.; Kim, J.W.; Jung, H.S.; Lee, J.K.; Degrandpe, K.; Lu, Z. Volcanic activity analysis of Mt. sinabung in Indonesia using remote sensing and GIS techniques. In Proceedings of the Near-Surface Asia Pacific Conference, Waikoloa, HI, USA, 7–10 July 2015; pp. 150–153.

9. Lee, S.; Ryu, J.H.; Min, K.; Won, J.S. Landslide Susceptibility Analysis Using GIS and Artificial Neural Network. *Earth Surf. Process. Landf.* **2003**, *28*, 1361–1376. [CrossRef]

10. Canziani, G.; Perrati, R.; Marinelli, C.; Dukatz, F. Artificial Neural Network and Remote Sensing in the Analysis of the Highly Variable Pampean Shallow Lakes. *Math. Biosci. Eng.* **2008**, *5*, 691–711. [CrossRef] [PubMed]

11. Chen, K.; Yen, S.; Tsay, D. Neural classification of SPOT imagery through integration of intensity and fractal information. *Int. J. Remote Sens.* **1997**, *18*, 763–783. [CrossRef]

12. Saravanan, K.; Sasithra, S. Review on Classification Based On Artificial Neural Networks. *Int. J. Ambient Syst. Appl. (IJASA)* **2014**, *2*. [CrossRef]

13. Serpico, S.B.; Bruzzone, L.; Roli, F. An experimental comparison of neural and statistical non-parametric algorithms for supervised classification of remote sensing images. *Pattern Recognit. Lett.* **1996**, *17*, 1331–1341. [CrossRef]

14. Ministry of Energy and Natural Resources of the Republic of Indonesia. Available online: http://www.vsi. esdm.go.id/ (accessed on 10 July 2017).

15. Lupi, M.; Miller, S.A. Short-lived tectonic switch mechanism for long-term pulses of volcanic activity after mega-thrust earthquakes. *Solid Earth* **2014**, *5*, 13–24. [CrossRef]

16. Hidayati, S.; Ishihara, K.; Iguschi, M.; Ratdomopurbo, A. Focal mechanism of volcano-tectonic earthquakes at merapi volcano, Indonesia. *Indones J. Phys.* **2008**, *19*, 75–82.

17. Iguchi, M.; Ishira, K.; Hendrasto, M. Learned from 2010 Eruptions at Merapi and Sinabung Volcanoes in Indonesia. *Annu. Disaster Prev. Res. Inst. Kyoto Univ.* **2011**, *54*, 185–194.

18. Ratdomopurbo, A.; Poupinet, G. An overview of the seismicity of merapi volcano (Java, Indonesia), 1983–1994. *J. Volcanol. Geotherm. Res.* **2000**, *100*, 193–214. [CrossRef]

19. Surono, M.; Jousset, P.; Pallister, J.; Boichu, M.; Buongiorno, M.F. The 2010 explosive eruption of Java's Merapi volcano—A '100-year' event. *J. Volcanol. Geotherm. Res.* **2012**, *241–242*, 121–135. [CrossRef]

20. Landsat 7 ETM+ and Landsat 8 Imagery. Available online: https://earthexplorer.usgs.gov/ (accessed on 10 July 2017).

21. Richards, J.A. *Remote Sensing Digital Image Analysis*; Springer: Berlin, Germany, 1999; Volume 240.

22. Eom, J.; Lee, C.W. Analysis on the Area of Deltaic Barrier Island and Suspended Sediments Concentration in Nakdong River Using Satellite Images. *Korean J. Remote Sens.* **2017**, *33*, 201–211.

23. Lee, S.; Kim, G.; Lee, C.W. Preliminary Study for Tidal Flat Detection in Yeongjong-do according to Tide Level using Landsat Images. *Korean J. Remote Sens.* **2016**, *32*, 639–645. [CrossRef]

24. Lee, C.W.; Cho, M.; Choi, Y.J. Method and System for Correction of Optical Satellite Image. U.S. Patent US9,117,276 B2, 25 Agusut 2015.

25. Rosenblatt, F. *Principles of Neuro Dynamics*; Spartan: New York, NY, USA, 1962.

26. Ahmed, E.A.; Adam, M.E.-N. Estimate of global solar radiation by using artificial neural network in Qena, Upper Egypt. *J. Clean Technol.* **2013**, *1*, 148–150. [CrossRef]

27. Xiu, L.-N.; Liu, X.-N. Current status and future direction of the study on artificial neural network classification processing in remote sensing. *Remote Sens. Technol. Appl.* **2003**, *18*, 339–345.

28. Kia, M.; Pirasteh, S.; Pradhan, B.; Mahmud, A.; Sulaiman, W.; Moradi, A. An artificial neural network model for flood simulation using GIS: Johor River Basin, Malaysia. *Environ. Earth Sci.* **2012**, *67*, 251–264. [CrossRef]

29. Mishra, V.N.; Rai, P.K.; Mohan, K. Prediction of land use changes based on land change modeler (LCM) using remote sensing: A case study of Muzaffarpur (Bihar), India. *J. Geogr. Inst. Jovan Cvijic* **2014**, *64*, 111–127. [CrossRef]

30. Rokni, K.; Ahmad, A.; Solaimani, K.; Hazini, S. A new approach for surface water change detection: Integration of pixel level image fusion and image classification techniques. *Int. J. Appl. Earth Obs. Geoinf.* **2015**, *34*, 226–234. [CrossRef]

31. McClelland, J.L.; Rumelhart, D.E. *Explorations in Parallel Distributed Processing*; MIT Press: Cambridge, UK, 1988.

32. Rumelhart, D.E.; Hinton, G.E.; Willams, R.J. Learning Internal Representations by Error Propagation. In *Parallel Distributed Processing, Explorations in the Microestruture of Cognition. Vol. 1: Foundations*; Rumelhart, D.E., McClelland, J.L., Eds.; MIT Press: Cambridge, UK, 1986.

33. Minsky, M.L.; Papert, S.A. *Perceptrons, Expanded Edition*; MIT Press: Cambridge, UK, 1969.

34. Lee, S.; Yang, M.; Lee, C.W. Time Series Analysis of Area of Deltaic Barrier Island in Nakdong River Using Landsat Satellite Image. *J. Korean Soc. Surv. Geod. Photogramm. Cartogr.* **2016**, *34*, 457–469. [CrossRef]

35. Robert, H. Cenozoic geological and plate tectonic evolution of SE Asia and the SW Pacific: Computer-based reconstruction, model and animations. *J. Asian Earth Sci.* **2002**, *20*, 353–431.

36. Mulyana, A.R. *Mapping of Disaster Prone Areas Mount Sinabung, Center for Volcanology and Geological Hazard Mitigation*; Ministry of Energy and Mineral Resources Indonesia: Jakarta, Indonesia, 2010.

37. Lee, C.W.; Lu, Z.; Jin, W.K. Monitoring Mount Sinabung in Indonesia Using Multi-Temporal InSAR. *Korean J. Remote Sens.* **2017**, *33*, 37–46. [CrossRef]

applied sciences

MDPI

Article

Classification of Forest Vertical Structure in South Korea from Aerial Orthophoto and Lidar Data Using an Artificial Neural Network

Soo-Kyung Kwon, Hyung-Sup Jung *, Won-Kyung Baek and Daeseong Kim

Department of Geoinformatics, University of Seoul, Seoul 02504, Korea; a99891902@uos.ac.kr (S.-K.K.);
bekwkz@uos.ac.kr (W.-K.B.); kds2991@uos.ac.kr (D.K.)
* Correspondence: hsjung@uos.ac.kr; Tel.: +82-2-6490-2892

Received: 31 August 2017; Accepted: 9 October 2017; Published: 12 October 2017

Abstract: Every vegetation colony has its own vertical structure. Forest vertical structure is considered as an important indicator of a forest's diversity and vitality. The vertical structure of a forest has typically been investigated by field survey, which is the traditional method of forest inventory. However, this method is very time- and cost-consuming due to poor accessibility. Remote sensing data such as satellite imagery, aerial photography, and lidar data can be a viable alternative to the traditional field-based forestry survey. In this study, we classified forest vertical structures from red-green-blue (RGB) aerial orthophotos and lidar data using an artificial neural network (ANN), which is a powerful machine learning technique. The test site was Gongju province in South Korea, which contains single-, double-, and triple-layered forest structures. The performance of the proposed method was evaluated by comparing the results with field survey data. The overall accuracy achieved was about 70%. It means that the proposed approach can classify the forest vertical structures from the aerial orthophotos and lidar data.

Keywords: forestry vertical structure; stratification; forest inventory; aerial orthophoto; lidar (light detection and ranging); ANN (Artificial Neural Network); machine learning

---

1. Introduction

Since forests are important for human life, forest inventories have been investigated for various purposes for centuries. In Europe, the first inventories were carried out in the 14th and 15th century for the purpose of intensive mine development. Since the 1910s, national forest inventories have been carried out in Norway, Sweden, and Finland, with an emphasis on timber production [1]. However, the demands of society have changed rapidly over recent decades. In this context, the principles for the conservation and sustainable management of forests have been newly added by the United Nations General Assembly [2]. This was in response to an increasing interest in non-timber aspects of forest structure and the demand for assessing these aspects [3]. In the Republic of Korea, in the 1970s, when the forest inventory was first investigated, it was aimed at reforestation and forest statistics. The purpose of the forest inventory has changed as the value of forest resources and impacts on the environment have evolved. Currently, forest inventories are developed to provide information that is useful to achieve the following goals: to maintain a healthy forest ecosystem, to preserve and protect the global environment and to promote sustainable development [4–6].

The vertical structure of a forest, which generally has four layers, is one of several elements representing forest vitality. In temperate zones, the forests are divided into layers of canopy, understory, shrub, and grass. The ability of the lower vegetation layer to grow under the canopy layer is determined by the condition of the soil, the species of vegetation, and the quantity of sunlight received by the lower vegetation layer due to the opening and closing rate of the crown [7–9].

In the case of artificial forests, they are usually single- or double-layered forests, as they are planned and managed for the purposes of wood production and agriculture. Natural forests however, have formed from a variety of vegetation communities through natural succession over lengthy periods of time, and have multiple layers. From an ecological point of view, natural forests with multiple-layers are highly resistant to pests, diseases, and environmental stress, and have high quality ecosystem services, such as providing habitats for wildlife. This is not the case with single-layered forests [10–14]. Therefore, vertical structure is a useful measure to evaluate the environmental aspects of a forest.

Typically, forest inventories have been investigated through terrestrial surveys. The Korea Forest Service currently uses aerial orthophotos to survey forests, but it is very difficult to understand forest structure because orthophotos only image the forest canopy. Thus, a field survey is required to understand the stratification of a forest. In Korea, more than 70% of the country is made up of mountainous areas, and a nationwide field survey would be very time- and cost-consuming. Since remote sensing data is advantageous for extensive regional studies [15], we attempt to develop a method of effective forest inventory using remotely sensed data. Such a method could reduce the time and cost of a forest inventory.

Multi-layered mature forests normally have a rough texture in remotely-sensed images, while single-layered young forests have a smoother texture [16–18]. Therefore, the structure of a forest can be estimated through the reflectance differences among the communities. The arrangement of the crown layer (canopy layer) in single-layered artificial forests can be considered consistent, but in natural forests it is inconsistent [17,18]. Since tree height is closely related to forest vertical structure, lidar data used for tree height measurements can be used to classify a forest's vertical structure [7,19–24]. Lidar is an abbreviation for light detection and ranging, and is a device that measures the distance of a target using laser light. The characteristics of remotely-sensed images and lidar data could enable us to classify the vertical structure of a forest.

The objective of this study is to classify forest vertical structure from aerial orthophotos and lidar data using an artificial neural network (ANN) approach. A total of five input layers are generated including: a median-filtered index map, a non-local (NL) means filtered index map, and reflectance texture map generated from the aerial orthophotos, and height difference and height texture maps generated from the lidar data. Since it is difficult to determine the presence of a grass layer in aerial images, it is omitted from our study and the forest vertical structure is classified into three groups for the ANN approach. The groups include: (i) single-layered forest that has only the canopy layer; (ii) double-layered forest that possesses the canopy layer and an understory or shrub layer; and (iii) triple-layered forest that is composed of the canopy, understory, and shrub layers. The red-green-blue (RGB) aerial orthophoto is used to obtain optical image information. The digital surface model (DSM) and digital terrain model (DTM) are extracted from the lidar point cloud. The height information is extracted by subtracting the DTM height from the DSM height. The accuracy of the classification is validated using field survey measurements.

2. Study Area and Dataset

The study area is located in Gongju province, South Korea, as shown in Figure 1. Gongju is 864.29 km^2 in area, about 0.95% of the total area of South Korea (99,407.9 km^2). The area of cultivated land is 185.82 km^2, accounting for 19.76% of the total area. Forests constitute 70% of the area and there are many hilly mountainous areas of 200–300 m above sea level. The average temperatures are around 11.8 °C in spring, 24.7 °C in summer, 14.0 °C autumn, and −0.9 °C in winter. The average annual precipitation is 1466 mm. The soils are loamy soils and clay loams and at the bottom it has sand soils. The image of the study area covers a 3.25 km^2 area that is 2344.8 m in length and 1385.76 m in width.

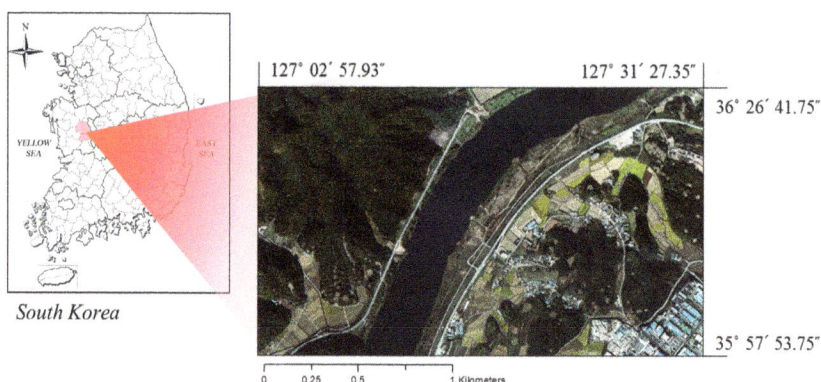

Figure 1. Location of the study area in Gongju province, South Korea. The red-green-blue (RGB) aerial orthophoto shows the spatial distribution of forests in the study area.

The RGB aerial orthophoto shows the spatial distribution of forests in the study area. Most forests are present on steep-sloped mountains and the types of forest include coniferous/broad-leaved forests and artificial/natural forests. Figure 2 shows the three types of forest classification map that are based on vertical structure, dominant species' leaf type, and whether the forest is natural or artificial. These maps were classified based on field survey measurements collected as a part of the Korean 3rd natural environment conservation master plan. The measurements were collected for the whole country over a period of six years from 2006 to 2012. The field survey of the study area was performed in 2009. The fourth masterplan started in 2014 and is due to be completed in 2018. The database construction for the fourth masterplan has not been completed. Figure 2a shows that triple-layered forests were dominant in the study area, and hence the natural forests are shown to be dominant in Figure 2c because the artificial forests are generally single- or double-layered. The study area includes a variety of different forest types including: (i) single-, double-, and triple-layered forests (Figure 2a); (ii) broad-leaved, coniferous, and mixed forests (Figure 2b); and, (iii) natural and artificial forests (Figure 2c).

Figure 3 shows examples of single-, double-, and triple-layered forests. Figure 3a shows an example of a single-layered forest. The forest is a chestnut plantation that the image shows to have been uniformly planted. Thus, it can be understood that it is an artificial forest. In Figure 3b, the image was acquired from a double-layered forest. It is a coniferous forest and the nut pine is dominant. The red reflectance value of coniferous trees was lower than that of broad-leaved trees based on analysis of the RGB image. This confirmed that the brightness is dependent upon species. Considering that coniferous forests are mainly single- or double-layered, species identification could be an important part of the classification of vertical structure. Figure 3c shows an example of a triple-layered forest. The forest is broad-leaved and oak is dominant tree type. Natural forests are a mixture of various species, and may contain coniferous and broad-leaved trees. Therefore, it is expected that it will be more useful to analyze height differences between trees rather than the identification of tree species in the classification of triple-layered forests. As mentioned above, single- and double-layered forests are mostly artificial forests. Artificial forests in the study area include broad-leaved forests, such as the chestnut tree forest, as well as coniferous forests such as the nut pine forest.

The RGB orthophoto was obtained using a DMC II airborne digital camera at an altitude of 1300 m (above sea level) with a redundancy rate of 63% in October 2015. The low-resolution RGB and high-resolution panchromatic (PAN) images were merged to create a high-resolution RGB image. The high-resolution RGB image was then orthorectified and mosaicked. The final mosaicked RGB orthophoto was used for this study. The ground sample distance (GSD) of the orthophoto was 12 cm in both the line and pixel. The lidar point cloud was acquired with an ALTM Gemini167 at an altitude of 1300 m (above sea level) in October 2015 with 2.5 points per square meter in the lowlands and 3~7 points per square meter in the mountain area. The difference in the point density occurred because the overlap rate of the lidar sensors is different between mountainous and non-mountainous areas. Points from terrain was classified from the Lidar point cloud, the DTM with the GSD of 1 m was created from the terrain-derived points. On the other hand, points from trees were extracted, and then the DSM with the GSD of 1 m was produced from the tree-derived points. The terrain- and tree-derived points were extracted by using the commercial software Global Mapper. The height difference map was generated by subtracting DTM from DSM. The height difference map is related with several tree parameters, including the height and density of trees, etc. That is, the pixel values in the height difference map are not tree height, but are closely related with tree height.

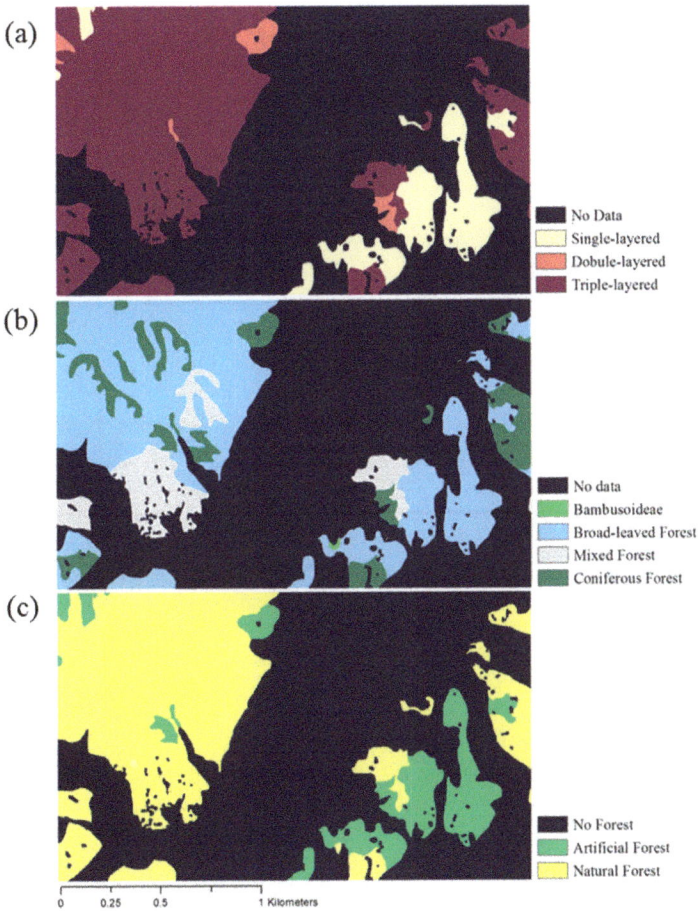

Figure 2. Forest classification maps of the study area produced from the field survey. (**a**) Forest vertical structure; (**b**) dominant species' leaf type; (**c**) natural and artificial forests.

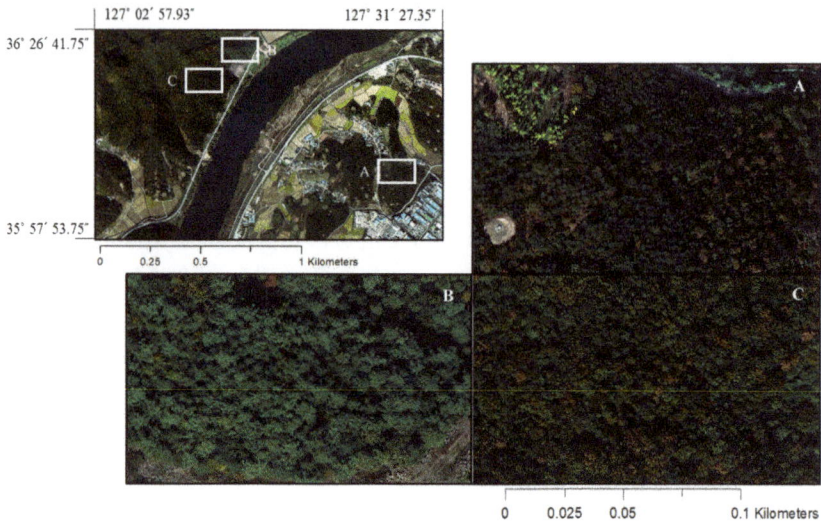

Figure 3. Example images from the study area showing the different forest vertical structures. (**A**) A single-layered forest that is a broad-leaved forest. The chestnut tree (Castanea crenata var. dulcis) is dominant; (**B**) a double-layered forest, which is a coniferous forest in which the nut pine (Pinus koraiensis) is dominant; and (**C**) a triple-layered forest that is a broad-leaved forest. The sawtooth oak (Quercus acutissima) is dominant.

3. Methodology

In this study, we produced five input maps that were resampled to 0.48 m GSD from high-resolution RGB imagery and lidar DSM and DTM data. The three maps from the RGB image include: reflectance values from each tree, reflectance values from each colony, and the variance of reflectance values (i.e., a texture map of reflectance). The two maps from lidar data reflect the tree height of each individual species, and the variance of the height values. We classify forest vertical structure by applying an Artificial Neural Network (ANN) to the five maps. An ANN is a large network of extremely simple computational units that are massively interconnected with nodes and processed in parallel. In this study, a multi-layer artificial neural network was used. It contains three layers: an input layer, a hidden layer, and an output layer, and each layer is composed of nodes. Each node has a weight, which is adjusted from randomly generated initial values through the iterative experiment to obtain the most reasonable output [25,26]. Figure 4 shows the detailed workflow used to classify the forest vertical structure from RGB orthophoto and lidar data using a machine learning approach. The image processing steps such as median, NL-means filtering, texture calculation was implemented by using C language, the MATLAB software was used for the ANN processing, and ER-mapper was used to display input and output maps.

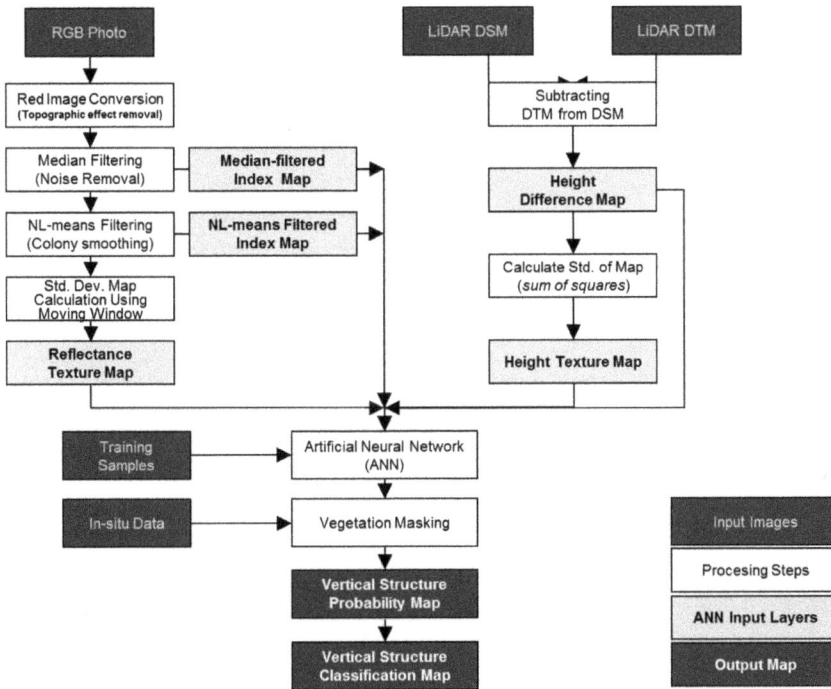

Figure 4. Detailed workflow used to classify the forest vertical structure from RGB orthophoto and lidar data using a machine learning approach.

3.1. Generation of Input Data from RGB Imagery

Vegetation has different reflectance values on remotely sensed imagery because different species have different leaf pigments. When these characteristics were determined from RGB imagery, the needle-leaf forest was brighter than the broad-leaf forest on the converted red image (Median-filtered Index Map). To analyze the optical imaging characteristics of each colony's vegetation, a normalized index map was produced. First, the initial RGB imagery (0.12 m GSD) has a deep shadow due to topographic effects that should be reduced. The shadow effect appears in both the red and green images. Thus, an index map was used instead of the red or green images. The index map is generated in two steps: (i) the red image is converted using the mean and standard deviation of the green image; and (ii) the ratio of the difference between the red and green images and the summation of the red and green images is calculated. This conversion reduces the topographic shadow artefact very effectively. The converted red image ($\overline{R}$) was calculated using:

$$\overline{R}(i,j) = \frac{\sigma_G}{\sigma_R} \times (R(i,j) - \mu_R) + \mu_G \tag{1}$$

where i and j are the line and pixel, respectively, R is the red image, σ_G and σ_R are the means of the green and red images, and μ_G are μ_R are the standard deviation of the green and red images. The normalized index map (NI) was defined using:

$$\text{NI}(i,j) = \left(G(i,j) - \overline{R}(i,j)\right) / \left(G(i,j) + \overline{R}(i,j)\right) \tag{2}$$

where G is the green image.

The index map contains noise from the crown. Since it can degrade the accuracy of the ANN analysis, we applied a median filter with a kernel size of 3 × 3 pixels to reduce the noise. The median filtered index map was used as part of the input data for the ANN. To determine each colony's overall reflectance, we resampled the median-filtered index map to a spatial resolution of 48 cm and applied a NL-means filter with a kernel size of 21 × 21 pixels. The NL-means filter is a powerful smoothing filter that preserves the edges of objects. The NL-means filtered index map represents the overall reflectance of a forest vegetation colony. The third input map from the RGB image indicates the spatial texture of the reflectance values. If a colony has a double- or triple-layered forest vertical structure, it must have several species and the species' reflectance values must be mixed. Thus, if a colony contains various species, the spatial texture of the reflectance would be rough. On the contrary, if only a few species are present such as in an artificial forest, the spatial texture would be smooth. The reflectance texture map is generated by calculating the standard deviation of the difference between the median-filtered and NL-means filtered maps using a moving window of 5 × 5 pixels. Figure 5 shows the three input layers used for the ANN analysis. To ensure that the topographic effect was mitigated in the median-filtered index map from Figure 5a, it was compared with Figure 1. As shown in Figure 5b, the NL-means filtered index map was smoothed with well preserved object edges. The smooth or rough texture of the surface reflectance map can be recognized from Figure 5c. The water surface had lower values while the forest areas had higher texture values.

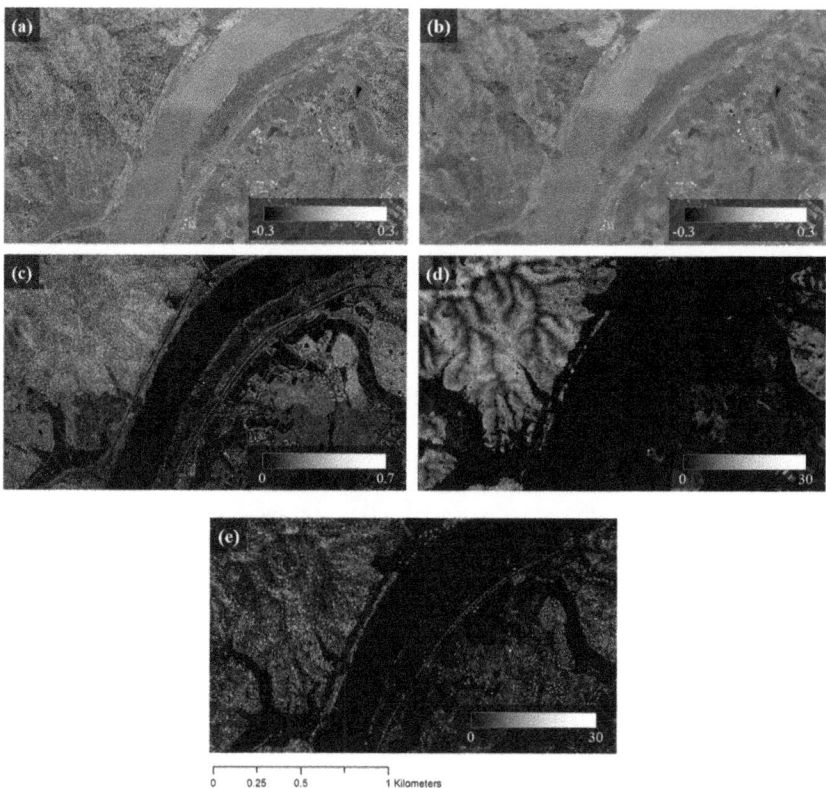

Figure 5. The five input layers used for the artificial neural network (ANN) analysis. (**a**) median-filtered index map; (**b**) non-local (NL)-means filtered index map; (**c**) reflectance texture map; (**d**) height difference map; and (**e**) height texture map.

3.2. Generation of Input Data from Lidar Data

Forest vertical structure is highly related to tree height. This is because the criteria for classification of canopy, understory, and shrub layers in a forest inventory are tree species and tree height. Two input layers were generated from the lidar data (initially 1 m GSD, but resampled to 0.48 m GSD): a height difference map and a height texture map. The height difference map was generated by subtracting the DTM from the DSM. The DSM is a kind of digital elevation model (DEM) that represents the Earth surface including features such as building, trees, and houses. The DTM is also a kind of DEM, but it only represents the Earth's terrain and excludes the natural and manmade features. Thus, the height difference map shows the height of the natural and manmade features, including the tree height Pixel values in the height difference map are not direct measurements of tree height, but a close approximation.

The height difference map was used for the ANN analysis. Since the triple- or double-layered forests have various tree species, the variance of the tree height in a colony would be uneven and the standard deviation of height difference in a colony can be large. The calculation of the height texture map from the lidar data is performed by two processing steps. In the first step, the height difference map is smoothed using a NL-means filter with the kernel size of 11 × 11 pixels. The second step involves calculating the standard deviation of the difference between the height difference map and the NL-means filtered height difference map using a moving window of 5 × 5 pixels. This map is used for the ANN classification processing. Figure 5d shows the height difference map. The relative height difference among trees can be recognized from Figure 5d. Figure 5e shows the height texture map that indicates the spatial variance in tree height.

Figure 6 shows the characteristics of the five input layers used for the ANN. The areas in the boxes labeled A, B, and C in Figure 6 are representative of single-, double-, and triple-layered forests, respectively. The dominant species in the area under box A is chestnut tree and the area is a broad-leaved and artificial forest. The area under box B includes nut pine trees as the dominant species and is a coniferous and artificial forest. The dominant species in the area under box C is oak tree and the area is a broad-leaved and natural forest. The nut pine (box B in Figure 6) was the brightest of the forests among the reflectance index maps. The reflectance texture maps in the single- and double-layered forests were smoother than the triple-layered forest. In box A of Figure 6d, there are some individual trees in the artificial forest that appear as bright patches. These trees show different reflectance values on the RGB and median-filtered index maps as compared with other trees in the same community and appear to be a different species. They show odd values on the maps from the lidar data as well. The brightest parts of the Height Texture Map (Figure 6e) are where the standard deviation was high.

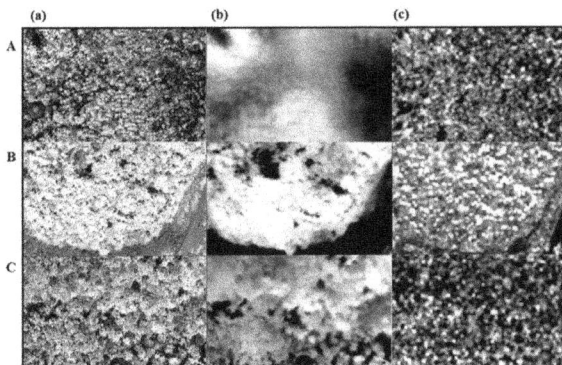

Figure 6. *Cont.*

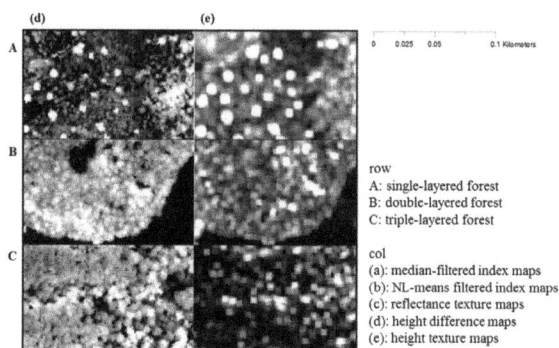

Figure 6. Characteristics of the five input layers used for the ANN approach in boxes A, B, and C of Figure 3: (**a**) median-filtered index maps; (**b**) NL-means filtered index maps; (**c**) reflectance texture maps; (**d**) height difference maps; and (**e**) height texture maps. The areas in boxes A, B, and C are the single-, double-, and triple-layered forests, respectively.

4. Results and Discussion

Figure 7 shows probability maps of the single-, double-, and triple-layered forests generated using the ANN and the five input layers. The ANN is a pixel-based analysis that calculates a probability of an event happening, and the brightness values represent probabilities. Training samples of ten thousand pixels were selected for the single-, double-, and triple-layered forests from field survey measurements. The non-vegetation areas and the road (red-dotted box in Figure 8) were masked out. The ANN determined the weighting factors from the five input layers. Table 1 summarizes the weighting factor for each input layer.

From the single-, double-, and triple-layered forests shown in Figure 7a–c, a final forest structure classification map was produced by selecting the highest value from the probability maps of the single-, double-, and triple-forests. Figure 8 shows the ANN-classified forest structure map and the field forest structure map for comparison. The red-dotted box in Figure 8b was classified as single-layered forest using the ANN approach, but the field measurements recorded it as triple-layered forest (e.g., compare Figure 8a,b). The field measurements have low spatial sampling because most forests in Korea occur in high mountainous regions. This discrepancy was caused by the spatial resolution differences. The forest vertical structure classes between the ANN approach and field measurement were compared and the results are summarized in Table 2. In Table 2, the number denotes the number of pixels. The classification accuracies of the single-, double-, and triple-layered forests were 82.84%, 56.56%, and 69.03%, respectively. The overall accuracy was 71.10%.

Table 1. Estimated weighting factor of the input layers.

Input Layers	Weighting Factor		
	Single-Layered	Double-Layered	Triple-Layered
Median-filtered Index Map	0.13	0.09	0.22
NL-means Filtered Index Map	0.17	0.47	0.23
Reflectance Texture Map	0.10	0.07	0.14
Height Difference Map	0.46	0.16	0.24
Height Texture Map	0.15	0.21	0.17

Summation of the weighting factors is 1.

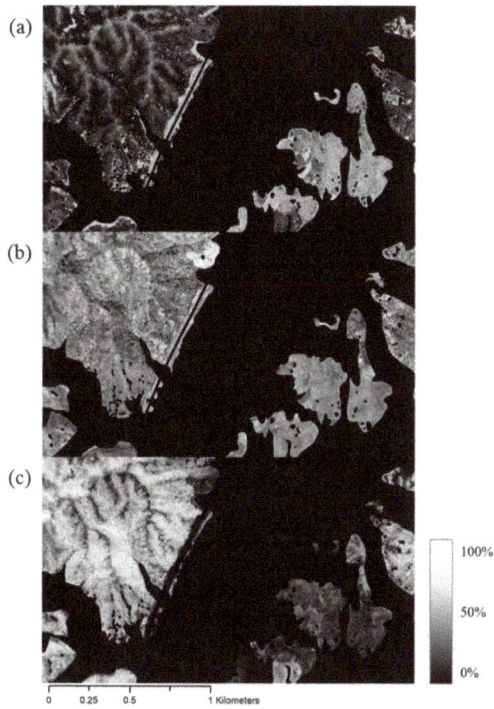

Figure 7. Probability maps of the (**a**) single-; (**b**) double-; and (**c**) triple-layered forests.

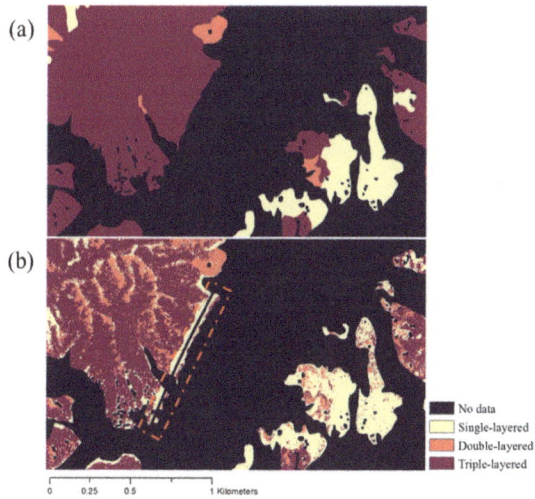

Figure 8. (**a**) Field survey classification map of forest vertical structure (same as Figure 2a); (**b**) final classification map of forest vertical structure using the ANN with red box -road way-.

Table 2. Validation of forest vertical structure classified by using the ANN approach.

ANN \ Field	Single-Layered Forest	Double-Layered Forest	Triple-Layered Forest
Single-layered Forest	849,983 *	42,380	785,080
Double-layered Forest	50,920	105,795 *	611,421
Triple-layered Forest	125,118	38,862	3,112,900 *
total	1,026,021	187,037	4,509,401

* Correctly classified pixels.

For the sections with low classification accuracy, the reasons for the misclassification need to be investigated. However, the overall accuracy of the classification is relatively high, considering that it is difficult to classify the vertical structure of a mixed-species forest using remotely-sensed data [17]. Our results show that it is possible to classify the forest vertical structure using remotely-sensed data. Several discrepancies were identified and the causes determined. The triple-layered coniferous colonies on the ridge of the mountain were mostly classified as double-layered coniferous forest. This was due to the ANN classification being strongly influenced by the dominant species. The red-dotted box in Figure 8 is a roadway, which was classified as triple-layered forest in the field measurement data collected in 2009. However, a single-layered forest exists in the roadway when we checked it in the field in 2015 RGB imagery. The road was constructed before 2009. In this case, the ANN approach performed better than the field measurements due to a low spatial resolution of the field survey sampling. However, we should be aware that the time lag between airborne images and field survey data can result in differences such as boundary changes or forest succession. The discrepancy between the ANN and field measurement classifications was very high in the areas of double-layered forest (Table 2). This is because the double-layered forest is composed of a canopy layer and an understory or shrub layer, and it was very difficult to distinguish from the other vertical structures using the aerial orthophoto and lidar data. Even though using remotely-sensed data and an ANN has some limitations, the approach could be very useful to classify forest vertical structure, because it has a better spatial resolution and is more time- and cost-effective. One other issue was that the RGB image used for this study was created through image tiling by a vendor. The left side of the image (Figure 8b), is more comparable to the field survey data (Figure 8a) than the right side. The reason may be due to a slight difference in color when tiling several raw images to process one orthophoto. Therefore, in future studies, it may be better to use raw image data or high resolution satellite image, which have the same photographic conditions over a single image area.

There is much research on the internal structure of forests using dense lidar data [27], but this method is limited due to the high cost of investigating large areas. Since the forests of Korea comprise a variety of species, it has been difficult to apply the methods of previous studies that identified stratification in coniferous forests composed of similar species [28]. Therefore, the novelty of the approach used in this study is that it can estimate and classify the complex inner structure of a forest with simple remotely-sensed data.

5. Conclusions

Forest vertical structure one element that represents the vitality of a forest. In general, the forest inventory has been investigated through field surveys. Korea Forest Service currently uses the aerial orthophotos to survey the forests, but it is very difficult to understand the infrastructure of the forests using the remote sensing images because of the cost. Intensive lidar data and aerial orthophotos were used because the orthophotos can just image the forest canopy. Thus, the field survey is inevitable in order to understand the stratification of forest.

In this study, forest vertical structure was classified by using the ANN approach from aerial orthophotos, and lidar DTM and DSM. A total of five input layers were generated from these datasets: (i) a median-filtered index map; (ii) a NL-means filtered index map; (iii) a reflectance texture map;

(iv) a height difference map; and (v) a height texture map. Using these maps, a forest structure classification map was generated with the ANN. The classification accuracies of the single-, double-, and triple-layered forests were 82.84%, 56.56% and 69.03%, respectively. The overall accuracy was 71.10%. The accuracy seems good considering that it is not easy to detect the under layers of a mixed broad-leaf and conifer forest from the remotely sensed data. The ANN approach has a better spatial resolution, is a more time- and cost-effective procedure for mapping forest vertical structure. Future studies should consider the effect of time gaps between datasets, and include maps with more detailed information related plant growth such as soil, sunlight, and species characteristics.

Acknowledgments: This work was supported by "Public Technology Development Project based on Environmental Policy" (2016000210001) provided by Korea Environmental Industry and Technology Institute.

Author Contributions: H.-S.J. conceived and designed the experiments. S.-K.K. performed the experiments. H.-S.J. and S.-K.K. analyzed the data. W.-K.B. and D.K. helped with data processing for the imagery. S.-K.K. and H.-S.J. wrote the paper.

Conflicts of Interest: The authors declare no conflict of interest.

References

1. Lund, H.G. *IUFRO Guidelines for Designing Multipurpose Resource Inventories: A Project of IUFRO Research Group 4.02.02;* International Union of Forest Research Organizations (IUFRO): Vienna, Austria, 1998.
2. UN General Assembly. United Nations Sustainable Development. In Proceedings of the United Nations Conference on Environment and Development, Rio de Janeiro, Brazil, 3–14 June 1992; Available online: https://sustainabledevelopment.un.org/content/documents/Agenda21.pdf (accessed on 11 October 2017).
3. Kenning, R.; Ducey, M.; Brissette, J.; Gove, J. Field efficiency and bias of snag inventory methods. *Can. J. For. Res.* **2005**, *35*, 2900–2910. [CrossRef]
4. Kim, E.S.; Kim, C.M.; Kim, K.M.; Ryu, J.H.; Lim, J.S.; Kim, J.C. *The Change of Korean National Forest Inventory System (1971~2010);* Korea Forest Institute: Seoul, Korea, 2015; ISBN 978-89-8176–158-5 93520.
5. Kim, S.H.; Kim, J.C. *Guide Book for the Sixth Korean National Forest Inventory and Fieldwork for Forest Health and Vitality;* Korea Forest Institute: Seoul, Korea, 2011; ISBN 978-89-8176-805-8.
6. Park, S.G.; Kang, H.M. Characteristics of Vegetation Structure in Chamaecyparis Obtusa Stands1. *Korean J. Environ. Ecol.* **2015**, *29*, 907–916. [CrossRef]
7. Morsdorf, F.; Marell, A. Discrimination of vegetation strata in a multi-layered Mediterranean forest ecosystem using height and intensity information derived from airborne laser scanning. *Remote Sens. Environ.* **2010**, *114*, 1403–1415. [CrossRef]
8. Lee, K.J.; Han, S.S. *Forest Ecology,* 2nd ed.; Hyangmunsa: Seoul, Korea, 1999; ISBN 8971871377.
9. Kang, S.S. *Biology,* 3rd ed.; Academy Books: Seoul, Korea, 2000; ISBN 9788976161987.
10. Korea Forest Conservation Movement. Available online: http://www.kfca.re.kr/ (accessed on 16 August 2017).
11. Jeon, S.W.; Kim, J. A Study on the Forest Classification for Ecosystem Services Valuation. *Korean Environ. Res. Technol.* **2013**, *16*, 31–39. [CrossRef]
12. Isbell, F.; Calcagno, V. High plant diversity is needed to maintain ecosystem services. *Nature* **2011**, *477*, 199–202. [CrossRef] [PubMed]
13. Fraf, R.F.; Mathys, L. Habitat assessment for forest dwelling species using LiDAR remote sensing: Capercaillie in the Alps. *For. Ecol. Manag.* **2009**, *257*, 160–167. [CrossRef]
14. Onaindia, M.; Dominguez, I. Vegetation diversity and vertical structure as indicators of forest disturbance. *For. Ecol. Manag.* **2004**, *195*, 341–354. [CrossRef]
15. Seong, N.; Seo, M.; Lee, K.S.; Lee, C.; Kim, H.; Choi, S.; Han, K.S. A water stress evaluation over forest canopy using NDWI in Korean peninsula. *Korean J. Remote Sens.* **2015**, *31*, 77–83. [CrossRef]
16. Hay, G.J.; Niemann, K.O.; McLean, G.F. An object-specific image-texture analysis of H-resolution forest imagery. *Remote Sens. Environ.* **1996**, *55*, 108–122. [CrossRef]
17. National Forestry Cooperative Federation. Available online: http://iforest.nfcf.or.kr/ (accessed on 16 August 2017).
18. Korea Forest Service. Available online: http://www.forest.go.kr/ (accessed on 16 August 2017).

19. Joe, H.G.; Lee, K.S. Comparison between Hyperspectral and Multispectral Images for the Classification of Coniferous Species. *Korean J. Remote Sens.* **2014**, *30*, 25–36. [CrossRef]

20. Zimble, D.A.; Evans, D.L. Characterizing vertical forest structure using small-footprint airborne LiDAR. *Remote Sens. Environ.* **2003**, *87*, 171–182. [CrossRef]

21. Sun, G.; Ranson, K.J. Forest vertical structure form GLAS: An evaluation using LVIS and SRTM data. *Remote Sens. Environ.* **2008**, *112*, 107–117. [CrossRef]

22. Mund, J.P.; Wilke, R. Detecting multi-layered forest stands using high density airborne LiDAR data. *J. Geogr. Inf. Sci.* **2015**, *1*, 178–188. [CrossRef]

23. Fernadez-Ordonez, Y.; Soria-Ruiz, J.S.R. Forest Inventory using Optical and Radar Remote Sensing. In *Advances in Geoscience and Remote Sensing*; Jedlovec, G., Ed.; InTech: Rijeka, Yugoslavia, 2009; ISBN 978-953-307-005-6.

24. Yoon, J.S.; Lee, K.S.; Shin, J.I.; Woo, C.S. Characteristics of Airborne Lidar Data and Ground Points Separation in Forested Area. *Korean J. Remote Sens.* **2006**, *22*, 533–542.

25. LG CNS Blog 'Creative N Smart'-'what is the 'Artificial Neural Network?''. Available online: http://www.blog.lgcns.com/1359/ (accessed on 20 September 2017).

26. Gopal, S.; Curtis, W. Remote Sensing of Forest Change Using Artificial Neural Networks. *IEEE Trans. Geosci. Remote Sens.* **1996**, *34*, 398–404. [CrossRef]

27. Jayathunga, S.; Owari, T.; Tsuyuki, S. Analysis of forest structural complexity using airborne LiDAR data and aerial photography in a mixed conifer–broadleaf forest in northern Japan. *J. For. Res.* **2017**, 1–15. [CrossRef]

28. Falkowski, M.J.; Evans, J.S.; Martinuzzi, S.; Gessler, P.E.; Hudak, A.T. Characterizing forest succession with lidar data: An evaluation for the Inland Northwest, USA. *Remote Sens. Environ.* **2009**, *113*, 946–956. [CrossRef]

applied
sciences

MDPI

Article

Impacts of Sample Design for Validation Data on the Accuracy of Feedforward Neural Network Classification

Giles M. Foody

School of Geography, University of Nottingham, University Park, Nottingham NG7 2RD, UK;
giles.foody@nottingham.ac.uk; Tel.: +44-115-951-5430

Received: 20 July 2017; Accepted: 21 August 2017; Published: 30 August 2017

Abstract: Validation data are often used to evaluate the performance of a trained neural network and used in the selection of a network deemed optimal for the task at-hand. Optimality is commonly assessed with a measure, such as overall classification accuracy. The latter is often calculated directly from a confusion matrix showing the counts of cases in the validation set with particular labelling properties. The sample design used to form the validation set can, however, influence the estimated magnitude of the accuracy. Commonly, the validation set is formed with a stratified sample to give balanced classes, but also via random sampling, which reflects class abundance. It is suggested that if the ultimate aim is to accurately classify a dataset in which the classes do vary in abundance, a validation set formed via random, rather than stratified, sampling is preferred. This is illustrated with the classification of simulated and remotely-sensed datasets. With both datasets, statistically significant differences in the accuracy with which the data could be classified arose from the use of validation sets formed via random and stratified sampling ($z = 2.7$ and 1.9 for the simulated and real datasets respectively, for both $p < 0.05\%$). The accuracy of the classifications that used a stratified sample in validation were smaller, a result of cases of an abundant class being commissioned into a rarer class. Simple means to address the issue are suggested.

Keywords: cross-validation; multi-layer perceptron; remote sensing; classification error; sample design; machine learning

1. Introduction

Artificial neural networks are widely used for supervised classification applications. In these applications, cases of known class membership are used to train the neural network in order to allow it to predict the class membership of previously unseen and unlabeled cases. This type of analysis is common in, for example, the production of thematic maps, such as those depicting land cover, from remotely-sensed imagery [1–3]. The imagery contain data on the remotely-sensed response of the land surface that is converted into information on land cover class via the classification analysis and a wide variety of approaches and applications have been investigated, e.g., [2,4,5]. Neural networks have become a popular method for image classification as numerous studies have shown that they can yield more accurate maps than a variety of other alternative approaches to classification [6–8]. The relative performance of neural network classifiers in relation to a range of alternative methods, including standard statistical classifiers, machine learning methods, and decision trees, is discussed in the literature (e.g., [1]). As with all classifiers, the quality of the final classification is, however, in part a function of the classifier and the nature, notably the size and quality, of the ground reference dataset on the class membership used [9–12]. Ground reference data are used to provide data to train the neural network and to evaluate the quality of its predictions. The latter is typically expressed in terms of the overall accuracy of the classification output from the network.

In remote sensing applications, the class membership of the cases in the training set is typically determined by ground-based observation or interpretation of very fine spatial resolution imagery [1,13]. Clearly, the nature and quality of the ground reference data used will impact upon the accuracy of the predictions obtained from the neural network. Classification accuracy is, for example, influenced by the size, composition, and quality of the ground reference dataset used in training [2,9,10,14–16]. The way the ground reference data are used is also important, especially if part of it is used for the purpose of validation [10]. The use of some reference data for validation purposes, to indicate the quality of the trained model generated, is commonplace in defining a supposedly optimal neural network. In common remote sensing applications, the optimal approach would be the one that yielded the most accurate thematic map when applied to the image data.

In a typical remote sensing application, the aim of the classification is to accurately map the land cover classes in the region of study. Throughout there are a set of basic assumptions made in classification analyses. These include the need for the set of classes to be mutually exclusive and exhaustively defined. Failure to satisfy these assumptions will result in errors. Cases of an untrained class will, for example, typically have to be allocated erroneously to one of the defined classes [17]. Although there are instances when a neural network may be used to obtain a non-standard classification, such as soft classification when classes intergrade or are mixed together [17], the standard hard classification in which a case belongs fully to a single class is the focus on this article. In all cases, the ground reference data used in training, validating, and testing the neural network have a key role to play in the production of an accurate map. Hence, the design of these datasets is important.

Good practices for classification accuracy assessment for the evaluation of the quality of thematic maps derived via remote sensing have been defined and include guidance on the construction of the testing sample upon which the assessment is to be based. Typically, for example, the use of a reference dataset acquired following a probabilistic sampling design is recommended [18,19]. This allows a rigorous design-based assessment of classification accuracy, typically based upon the analysis of an error or confusion matrix in which the predicted class label obtained from the neural network is cross-tabulated against the label in the reference dataset for the sample of cases under consideration [1,18].

The goal of training a classifier is different to that of testing its output predictions and, hence, the nature of the ideal training set may be very different to that of the testing set. In training, the aim is, essentially, to guide the neural network to learn the identity of the classes from their remotely-sensed response. Thus, the observed remotely-sensed response for the training sample of known class membership is used in network learning to ultimately form decision rules to accurately label cases of unknown class membership in order to map the region of interest [1]. Much conventional guidance on the design of the training set in remote sensing applications is based upon historical work undertaken with conventional statistical classifiers, such as the maximum likelihood classification. This advice typically calls for the sample size to be estimated following basic sampling theory in order to derive a representative and unbiased description of each class to allow cases of unknown class membership to then be allocated the label of the class they had greatest similarity to. Alternatively, a simple heuristic, such as the use of a sample of cases for each class, the size of which is at least 10 times the number of discriminating variables, such as the spectral wavebands, are used [10,20,21]. In essence, this type of approach is calling for the use of a stratified sample design in the formation of the training set. Despite the development of new classifiers, such approaches are still widely used even though the nature of the ideal training set varies between classifiers.

With neural networks and machine learning methods such as the support vector machine (SVM) and relevance vector machine (RVM) attention in training is focused more on individual cases than broad statistical summary statistics that are central to statistical classifiers, such as the maximum likelihood classification [1]. The individual training cases in the training set can vary greatly in value to an artificial neural network classification [22,23]. In addition, different classifiers may ideally focus upon different cases in the training set. For example, a SVM may require only a very small training sample and, ideally, cases that lie in the boundary region between classes, while a RVM might also require only a small number of training cases, but these are anti-boundary in nature [12]. Considerable research has, therefore, focused on how the properties of the training dataset impact on a classification, typically with a desire to maximize the final mapping accuracy (e.g., [9,10,24–26]). Sometimes it may be possible to predict the location of useful candidate training cases for a classification [27] or, contrary to conventional approaches, even deliberately use mixed cases in training and the analyst can seek to form a training set intelligently for a given application scenario [22,28,29]. As such, there is no single ideal way to define a training set that is universally applicable. Given the popularity of consensus or ensemble methods that use a variety of different classifiers [7,30] it is common to see relatively large training sets acquired following conventional guidance. While this approach may sometimes be inefficient, notably in that the sample may be larger than needed, it has the capacity to provide useful training data for a wide range of classifiers. It is, therefore, common to see either a training sample designed explicitly for a specific task, which could be a small and highly unrepresentative sample [29,31], or the use of a stratified design that seeks to ensure each class can be described well [1]. The design details are, however, important. The nature of the training set has a significant impact on the accuracy of predictions by a neural network [2,15,32]. For example, the size of the training set, notably in relation to the complexity of the network, can have a marked effect on classification by a neural network [9,16,33]. The composition of the training set in terms of relative class abundance is also important [15,34]. Variations in class abundance can yield imbalanced datasets that, as in other classifiers, can have substantial impacts on the final classification.

In a supervised classification with a conventional feedforward neural network it is common for part of the training sample to be used for validation purposes [1,2,10,33–35]. In this data splitting approach part of the training set is used in the normal way to provide examples of the classes upon which the classifier may learn to form rules to classify cases of unknown membership. The remaining part of the training set forms the validation set and is used to evaluate the performance of the network in terms of the accuracy with which the validation set is classified, as well as help determine when to stop network learning [10,36–39]. A variety of approaches exist for the splitting of the training cases to form the training set and the validation set. If reference data are plentiful the training set and validation set could be completely separate and independent samples but if this is not the case other approaches to cross-validation, such as the leave one out approach, may be used [34]; for simplicity, the focus here is on the use of a completely independent validation set. The use of a validation set is important in neural network-based approaches to classification as there is a desire to avoid overfitting to the training data and there are a variety of network parameters that require definition. The latter includes the basic structure of the network (e.g., number of hidden layers and units) and the learning algorithm and parameters (e.g., momentum, learning rate and number of iterations). There is an extensive literature on this topic [2,10]. The basic idea is that a range of different networks can be defined and one that is optimal for the task at hand, defined with the aid of the validation dataset. Thus, for example, a range of different networks may be generated and the one that classifies the validation set to the highest accuracy is selected for the final classification analysis to produce a thematic map. This analysis might also suggest ways to further enhance the classification by, for example, indicating redundant discriminating variables that could be deleted in order to allow more rapid computation [28,40]. It is also useful as some approaches used in training may artificially inflate class separability and be unhelpful [23]. Given that the ultimate aim of the analysis in a typical remote sensing application is the production of an accurate thematic map via a classification analysis the nature of the validation set can be important. Commonly, with classifiers that use validation data, the validation set is formed by simply taking cases randomly from the entire set of data acquired for training activity [38,41] or a separate, often stratified, sample of cases is obtained (e.g., [42]). The size of the validation set is important. As in other aspects of the analysis, the literature contains guidance on the way the reference data should be divided up. For example, Mas et al. [43,44] suggests that half of the labelled cases be used for training, a quarter for validation and the final quarter for testing.

The composition of the sample of cases forming the validation set, notably in terms of the relative abundance of the cases of the classes may vary as a function of the design used to acquire the training dataset. If the validation sample is formed with a simple random sample design it is likely to be imbalanced in composition, with the number of cases of a class reflecting its relative abundance in the region being mapped. The use of a stratified sample design in the generation of the validation sample will act to give a balanced dataset but this need not be an ideal approach. Indeed, the use of a stratified design could result in the selection of a network that was sub-optimal for the task if the classes vary in abundance and separability. For example, standard approaches to the assessment of overall classification accuracy weight errors equally and could inflate the importance of rare classes while deflating that of abundant classes. Sometimes it may be possible to account for the sample design if there is information on class abundance [18]. Alternatively, if the study aim is focused on a single class, which is often the case, the accuracy assessment used in the validation could be focused on that class at the expense of the others. However, for a general purpose map, it may be more appropriate to follow the guidance on sampling that is typically used in the formation of the testing set as this typically allows for variations in class abundance to be accounted for.

The effect of different sample designs for the formation of the validation dataset is explored in this paper. Specifically, the focus is on the use of samples acquired by simple random and stratified random sampling designs (both without replication). Due to the way overall accuracy is typically estimated with a validation sample, it is hypothesized that the overall accuracy of the final classification, evaluated using the testing set, will be larger for a neural network trained using a validation sample generated via simple random as opposed to stratified random sampling. Indeed a series of outcomes may be predicted as having the potential to arise as a function of the validation dataset used. For example, in relation to two classes that overlap, it would be expected that the hyperplane to best separate the classes fitted when a balanced validation set was used would migrate away from the more abundant class if an imbalanced training set was used. As a consequence of this, there is an opportunity for the accuracy with which the more abundant class is classified to rise as fewer cases of it will be omitted from it and fewer of its cases commissioned by the rarer class (es). These trends arise as classification errors are weighted equally in standard assessments of overall accuracy. The overall trend expected would be for the accuracy with which the abundant class is classified to increase while the accuracy of the classification of the rarer class would decline. As such it is hypothesized that the use of a stratified sample design may not be ideal as its use relative to a randomly-defined validation dataset would be associated with a decrease in overall accuracy, arising noticeably through a decrease in the accuracy for the abundant class (es) as a result of an increase in the commission of cases of abundant class (es) by the set of rarer classes.

2. Data and Methods

Two datasets were used. First, a simulated dataset was used to illustrate the issues and also to facilitate, if desired, replication. Second, a real dataset consisting of remotely-sensed data and associated ground reference data on land cover class labels was used.

A very simple multi-class classification scenario was simulated. This comprised data on four classes acquired in two dimensions, or bands. The data for each class were formed using a random number generator using analyst provided values for the class mean and standard deviation on the assumption that the data for each class were normally distributed. For each class the standard deviation was set equal to 5 and the mean values used to generate the data in each band are shown in Table 1. In the scenario generated, class 1 was the most abundant class. Specifically, class 1 was five times more abundant than each of the other classes. Most attention was focused on class 1 and class 2, which exhibited a degree of overlap in their distributions with class 3 and, especially, class 4 was highly separable (Figure 1).

For the analyses of the simulated dataset, training validation and testing sets were generated (Table 2). These were used in two series of analyses, one using a validation dataset formed via simple random sampling and the other formed via a stratified random sample design. In brief, a single training set was generated using a stratified random sample of 400 cases per-class. Similarly, a single testing set was used to evaluate the accuracy of the classifications from the neural networks selected as optimal in each series of analyses undertaken. This testing set was simulated to represent a sample acquired by simple random sampling and comprised 800 cases. Due to the way the scenario was designed, class 1 was five times more abundant in it than the other classes. The two validation datasets used each contained 800 cases, but one was formed via a simple random sample in which class abundance varied as in the testing set, while the other was formed with a stratified random sample in which each class was equally represented. Each series of neural network analyses used a software package that sought to generate an optimal network, with optimality defined as the maximization of the overall accuracy with which the validation set was classified.

The remotely-sensed data were acquired by an airborne thematic mapper (ATM) sensor for a test site near the village of Feltwell, Norfolk, UK. The latter is located approximately 58 km to the northeast of the city of Cambridge. The land around the village was topographically flat and its land cover mosaic was characterized mainly by large agricultural fields. At the time of the ATM data acquisition these fields had also typically been planted with a single crop type (Figure 2).

The ATM used was a basic multispectral scanning system that acquired data in 11 spectral wavebands from blue to thermal infrared wavelengths (Table 3). Given the relatively low altitude of airborne data acquisition (~2000 m), the spatial resolution of the imagery was very much smaller than the typical field size, approximately 5 m. As a result, image pixels tended to represent an area composed of a single class (i.e., pure pixels) and, hence, were appropriate for hard image classification analysis; boundary pixels were ignored. Attention focused here on the six crop classes that dominated the region at the time of the ATM data acquisition. These classes and their approximate coverage (%) of the study area at the time were: sugar beet (S, 30.3%), wheat (W, 30.0%), barley (B, 16.0%), carrot (C, 10.3%), potato (P, 7.8%), and grass (G, 5.4%).

Table 1. The classes in the simulated dataset; note: units are arbitrary.

Band	Class 1	Class 2	Class 3	Class 4
Band 1	50	35	83	20
Band 2	50	60	50	20

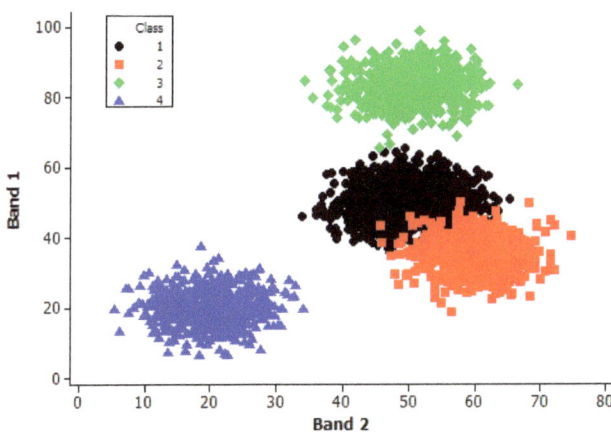

Figure 1. Simulated data (training, validation (random), and testing). Note: class 2 is shown overlaid on top of part of class 1 and the area of overlap may be inferred.

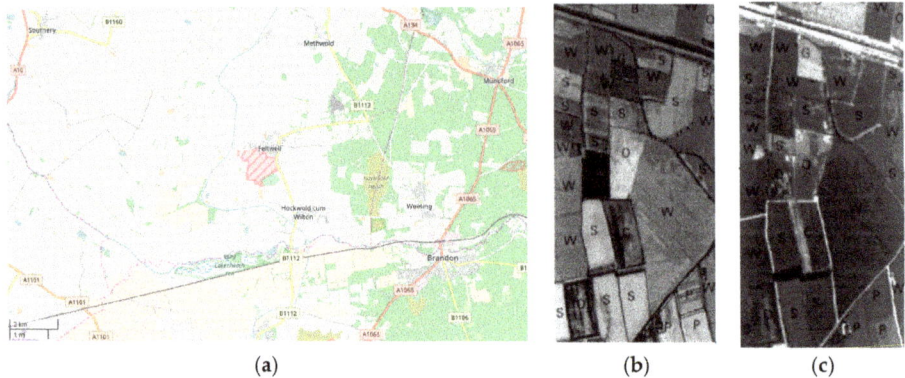

(a)   (b)   (c)

Figure 2. Data for Feltwell. (**a**) The location of Feltwell, © OpenStreetMap contributors; (**b**) airborne thematic mapper (ATM) image extract in waveband 4 and (**c**) ATM image extract in waveband 7.

Table 2. Class composition of the datasets used for analyses of the simulated data.

Dataset	Size and Class Composition
Training	400 cases of each class; total = 1600 cases
Validation (random)	500 class 1 100 class 2 100 class 3 100 class 4; total = 800 cases
Validation (stratified)	200 cases of each class; total = 800 cases
Testing	500 class 1 100 class 2 100 class 3 100 class 4; total = 800 cases

Table 3. The 11 spectral wavebands of the ATM sensor used.

Waveband	Wavelength (µm)
1	0.42–0.45
2	0.45–0.52
3	0.52–0.60
4	0.60–0.63
5	0.63–0.69
6	0.69–0.75
7	0.76–0.90
8	0.91–1.05
9	1.55–1.75
10	2.08–2.35
11	8.50–13.00

Ground reference data for the purposes of network learning (training) and evaluation were acquired. Following [43,44], the ground reference dataset was partitioned such that 50% was used for training, 25% for validation, and 25% for testing. The composition of these datasets, however, sometimes varied (Table 4).

Conventional guidance on the design of the training set was followed with 110 cases of each class obtained for training purposes; meeting the often stated requirement of a sample of at least 10 times the number of discriminating variables (wavebands) used as input. The training set, therefore, contained 660 cases acquired by a stratified random sampling design. This training set was used throughout.

Testing sets should, ideally, be acquired using a probability sampling design [18,19]. Here, a simple random sample (without replication) was used to acquire 330 cases to use for testing. Note that this sample size exceeds that required for accuracy estimation of a map with an accuracy of 85%, a standard if contentious target accuracy in remote sensing, with an allowable error of 4%. This testing set was used in all analyses of the ATM data. This latter issue impacts on comparisons of accuracy estimates and requires the use of a technique suited for related samples [45].

Table 4. Class composition of the datasets used for analyses of the remotely sensed data.

Dataset	Size and Class Composition
Training	110 cases of each class; total = 660 cases
Validation (random)	100 sugar beet 99 wheat 53 barley 34 carrot 26 potato 18 grass; total = 330 cases
Validation (stratified)	55 cases of each class; total = 330 cases
Testing	100 sugar beet 99 wheat 53 barley 34 carrot 26 potato 18 grass; total = 330 cases

As with the analyses of the simulated dataset, the search for an optimal neural network was undertaken twice. In each case the training and testing sets were the same, the only difference was the composition of the validation set used to identify the optimal network from a set of candidate networks generated for the task. In one set of analyses, the validation set was generated by simple random sampling and thus the number of cases of each class tended to reflect the actual abundance of the classes in the region to be mapped. Indeed, here, the sample was selected to ensure that the class composition of the validation set equalled that of the testing set. This sample comprised 330 cases. In the other set of analyses, a validation sample of the same size, but acquired following a stratified random sample such that each class was equally represented was used. The nature of the datasets used is summarized in Table 4.

In the search for an optimal neural network to classify the ATM data, optimality was defined in relation to the maximum overall accuracy of the classification of the validation data. The accuracy of each classification was calculated from the confusion or error matrix generated for it which shows a cross-tabulation predicted and actual class label for each case in the dataset analysed [1,18]. Using the layout and notation defined for the confusion matrix shown in Table 5, which is used throughout the article, overall accuracy, O, was calculated using Equation (1):

$$O = \frac{\sum n_{ii}}{n} \tag{1}$$

In addition to the global estimate of classification quality conveyed by overall accuracy, the producer's, P, and user's accuracy, U, were calculated with reference to each class [1,18]. These were obtained for class i from Equations (2) and (3) respectively:

$$P_i = \frac{n_{ii}}{n_{.i}} \tag{2}$$

$$U_i = \frac{n_{ii}}{n_{i\cdot}} \tag{3}$$

Note that, as is common, the calculations of these measures of accuracy are based on the raw counts of cases shown in the elements of the confusion matrix. These approaches to accuracy assessment were used to support all analyses, whether based on the training, validation, or testing datasets. Most focus is, however, on the accuracy values arising from analyses of the validation and testing datasets.

To determine if the use of different validation samples impacted on the accuracy of the final thematic classification, the statistical significance of differences in the accuracy of the classifications of the testing set were assessed. Standard approaches for the comparison of accuracy values that are popular in remote sensing projects are unsuitable here as the same sample of testing cases was used throughout. To accommodate for this situation, the statistical significance of differences in accuracy was assessed using the McNemar test [45,46]. The latter is a non-parametric test that is based on a binary confusion matrix which shows the cross-tabulation of the cases that have been labelled correctly and incorrectly by the two classifications being compared. The test focuses on the discordant cases, those which were classified correctly by one classifier, but incorrectly by the other. Without continuity correction, the test is based on the normal curve deviate, z, as expressed as:

$$z = \frac{n_{CI} - n_{IC}}{\sqrt{n_{CI} + n_{IC}}} \tag{4}$$

where n_{CI} indicates the number of cases in the relevant element of the matrix with the subscript C indicating if the classification was correct in its labelling or I if it was incorrect and order of the subscripts indicates the specific classification from the pair under study. For a standard two-tailed test at the 95% level of confidence, the null hypothesis of no significant difference is rejected if the calculated z exceeds the critical value of $|1.96|$. Similarly, for a one-tailed test, if the hypothesis under test has a directional component, the direction (sign) needs consideration and the magnitude of the critical value of z to indicate that a significant difference exists at the 95% level of confidence is 1.645.

Table 5. The confusion matrix based on raw counts of cases for a classification of q classes. Matrix columns show the label in the reference data and rows the label in the classification.

Class	1	2	...	q	Total
1	n_{11}	n_{12}	...	n_{1q}	$n_{1\cdot}$
2	n_{21}	n_{22}	...	n_{2q}	$n_{2\cdot}$
:	:	:		:	:
q	n_{q1}	n_{q2}	...	n_{qq}	$n_{q\cdot}$
Total	$n_{\cdot 1}$	$n_{\cdot 2}$	...	$n_{\cdot q}$	n

3. Results and Discussion

With the simulated dataset, two sets of analyses were undertaken to identify optimal neural networks, one using the validation set formed via simple random sampling and the other formed using a stratified random sample. The key properties of the selected networks and their ability to classify the datasets are defined in Table 6 with a full set of confusion matrices for each set of analyses shown in Tables 7 and 8.

Attention focused especially on the accuracy with which the testing set was classified, as this reflects the accuracy of the final product obtained. It was evident that the accuracy of classification obtained with the use of the simple random sample (98.37%) was slightly higher than that arising from the use of the validation set formed via stratified sampling (96.62%). Although small, this difference was significant ($p < 0.05$), with the calculated value of z from Equation (4) given the 26 discordant cases observed being 2.745 (Table 9). It was also evident that the predictions of what might happen as moving from a balanced stratified to random validation sample outlined in the introduction occurred.

Note, for example, that relative to the classification obtained with a stratified validation sample, the use of the validation set formed by random sampling resulted in a higher overall accuracy of the testing set. This arose because of a higher accuracy with which the abundant class was classified due to a reduction in omission error (from 27 to 7 cases) and the accuracy of the rarer class it was confused with declined (Tables 7 and 8). Critically, the results suggest that the use of a balanced training set acquired via a stratified random sample may produce a sub-optimal final output. As hypothesized, the use of a stratified sample resulted in a lower accuracy than was achievable because the accuracy with which the abundant class was classified declined, associated with a large omission error due to cases being commissioned into a rarer class.

A series of analyses was undertaken to define an optimal neural network using the two validation datasets in the analyses of the remote sensing data. The core focus here is on the classification results from the network determined to be optimal from each series of analyses. Although the precise details of the neural networks are relatively unimportant as the classifications that arise from them are the focus of attention, key details on the networks selected are summarized in Table 10 and the confusion matrices for the testing set are shown in Tables 11 and 12.

Table 6. Key characteristics of the networks selected for the analyses of the simulated data; note: network architecture is expressed as input:hidden:output.

Validation Dataset	Architecture	Algorithm and Iterations	Classification Accuracy (%) Training Validation Testing		
Random	2:15:2	Backpropagation 48	97.25	97.62	98.37
Stratified	2:8:2	Conjugate gradient 55	98.25	98.75	96.62

Table 7. Confusion matrices for the network selected using a validation set formed by random sampling.

			Training			
Class	1	2	3	4	Total	User's (%)
1	398	40	2	0	440	90.45
2	2	360	0	0	362	99.44
3	0	0	398	0	398	100
4	0	0	0	400	400	100
Total	400	400	400	400	1600	
Producer's (%)	99.50	90.00	99.50	100		

			Validation			
Class	1	2	3	4	Total	User's (%)
1	490	9	0	0	499	98.19
2	10	91	0	0	101	90.09
3	0	0	100	0	100	100
4	0	0	0	100	100	100
Total	500	100	100	100	800	
Producer's (%)	99.00	91.00	100	100		

			Testing			
Class	1	2	3	4	Total	User's (%)
1	493	6	0	0	499	98.79
2	7	94	0	0	101	93.06
3	0	0	100	0	100	100
4	0	0	0	100	100	100
Total	500	100	100	100	800	
Producer's (%)	98.60	94.00	100	100		

Table 8. Confusion matrices for the network selected using a validation set formed by stratified sampling.

(a)	Training					
Class	1	2	3	4	Total	User's (%)
1	386	14	0	0	400	96.50
2	14	386	0	0	400	96.50
3	0	0	400	0	400	100
4	0	0	0	400	400	100
Total	400	400	400	400	1600	
Producer's (%)	96.50	96.50	100	100		

(b)	Validation					
Class	1	2	3	4	Total	User's (%)
1	195	5	0	0	200	96.50
2	5	195	0	0	200	96.50
3	0	0	200	0	200	100
4	0	0	0	200	200	100
Total	200	200	200	200	800	
Producer's (%)	97.50	97.50	100	100		

(c)	Testing					
Class	1	2	3	4	Total	User's (%)
1	473	0	0	0	473	100
2	27	100	0	0	127	78.74
3	0	0	100	0	100	100
4	0	0	0	100	100	100
Total	500	100	100	100	800	
Producer's (%)	94.60	100	100	100		

Table 9. Cross-tabulation of labelling from classifiers using random (columns) and stratified (rows) validation sets for the simulated data.

	Correct	Incorrect	Total
Correct	767	6	773
Incorrect	20	7	27
Total	787	13	800

Table 10. Key characteristics of the networks selected for the analyses of the remotely-sensed data.

Validation Dataset	Architecture	Algorithm and Iterations	Classification Accuracy (%) Training Validation Testing		
Random	9:11:6	Conjugate gradient, 435	97.27	98.18	97.87
Stratified	11:16:6	Conjugate gradient, 205	96.21	97.27	96.36

The selected neural networks were able to classify the data accurately. In each case, however, it was evident that classification accuracy was slightly, no more than 1.51%, less accurate when the stratified, rather than random, validation dataset had been used in network selection. Although small, these differences can still be significant. Indeed, a key result was that the accuracy of the classification of the testing dataset, which indicates the accuracy of the land cover map obtainable, was higher when the validation set formed using random (overall accuracy = 97.87%), rather than stratified, sampling (overall accuracy = 96.36%). Although small, a test of the significance of the difference in overall accuracy, using a one-sided McNemar test to recognize the directional nature of the test and the use of the same testing set, revealed it to be statistically significant at the 95% level of confidence (Table 13). Specifically, given the seven discordant cases observed (Table 13), Equation (4) yields $z = 1.889$.

The difference in accuracy between the classifications of the testing sets by the two selected neural networks (Tables 11 and 12) was attributable mostly to seven cases of sugar beet being commissioned into the potato class when the validation set formed using a stratified sample was used. As a result of these errors the producer's accuracy for the sugar beet class declined from 99.00% when the validation sample acquired with a random sample was used to 93.00% when the validation sample had been generated with a stratified sample. The user's accuracy for the potato class also differed for the classifications of the testing set obtained when using the validation set defined with random and stratified sampling, the accuracies being 96.00% and 77.42%, respectively.

Given that the testing set had been generated using a simple random sample design the variation in the number of cases per-class reflected the relative abundance of the classes in the region to be mapped. Critically, the size of the sample of cases for the potato class was approximately one quarter of that for the sugar beet class. As hypothesized, the overall accuracy of the testing set decreased when the stratified rather than random sample was used in validation because cases of an abundant class (sugar beet) were commissioned by a relatively rare class (potato). Thus, even in a situation such as that encountered here, in which the classes are very highly separable, the sample design used in the formation of the validation dataset can have a statistically significant effect on the overall accuracy of the final land cover map, as reflected in the accuracy of the classifications of the testing set (Tables 11 and 12).

Table 11. Confusion matrix for the testing set from the network selected using a validation set formed by random sampling.

Class	S	W	B	C	P	G	Total	User's (%)
S	99	0	1	0	0	0	100	99.00
W	0	96	0	0	1	0	97	98.97
B	0	2	52	0	0	0	54	96.30
C	0	0	0	34	1	0	35	97.14
P	1	0	0	0	24	0	25	96.00
G	0	1	0	0	0	18	19	94.74
Total	100	99	53	34	26	18	330	
Producer's (%)	99.00	96.97	98.11	100	92.31	100		

Table 12. Confusion matrix for the testing set from the network selected using a validation set formed by stratified sampling.

Class	S	W	B	C	P	G	Total	User's (%)
S	93	0	1	0	0	0	94	98.94
W	0	97	0	0	1	0	98	98.98
B	0	2	52	0	0	0	54	96.30
C	0	0	0	34	1	0	35	97.14
P	7	0	0	0	24	0	31	77.42
G	0	0	0	0	0	18	18	100
Total	100	99	53	34	26	18	330	
Producer's (%)	93.00	97.98	98.11	100	92.31	100		

Table 13. Cross-tabulation of labelling from classifiers using random (columns) and stratified (rows) validation sets for the remotely sensed data.

	Correct	Incorrect	Total
Correct	317	1	318
Incorrect	6	6	12
Total	323	7	330

Finally, it was also evident that the networks selected for the analyses of the remotely-sensed data differed, most notably in terms of architecture (Table 10). When the validation set constructed with simple random sampling had been used, the data acquired in wavebands 1 and 8 were deemed unnecessary and, hence, only nine input units used. In addition this latter network also had fewer hidden units than the network selected when the validation set had been formed with a stratified sample. Overall, the network formed with the use of the validation sample acquired by random sampling was smaller and less complex than that selected when the validation set formed with a stratified sample was used. As such the network might be expected to be less likely to over-train and have a higher ability to generalize than that selected from the use of the validation sample formed via stratified sampling. Slightly different trends were observed for the simulated dataset. Here, however, it should be noted that a large number of networks of very different size, but very similar performance in terms of ability to classify the data, were generated, limiting the ability to comment on the issue; note, for example, that some candidate networks that yielded the same accuracy for the classification of the testing set after use of the random validation set had the smallest number of hidden units.

The results show that the design of the validation sample has a significant effect on classification by a neural network. Thus, the sample design used to form the validation set should be considered carefully when using neural networks. If, for example, there are constraints that limit design possibilities, it may be possible to make simple adaptations to standard practice. For example, if a stratified sample must be used for the validation sample, then this feature of the dataset should be accounted for in the assessment of the accuracy of the classification of the validation set. Thus, rather than use a standard confusion matrix, as indicated in Table 5, the elements of the matrix could be converted from raw counts to proportions via $p_{ij} = W_i \frac{n_{ij}}{n_{i.}}$ where W_i is the proportion of the area mapped as class i [18]. The use of this approach, for example, shows that the estimated accuracy with which the validation set formed by stratified sampling of the simulated data was slightly less (98.12%) than the naïve assessment of the matrix (98.75%) and may indicate that the network it is associated with is, therefore, less attractive as a candidate for the task at hand than it first appears from the naïve assessment. Alternatively, if the focus of the application is on a subset of the classes it may be sensible to weight errors differentially or focus on only on the classes of interest rather than use overall accuracy. Critically, consideration needs to be given to the design of the validation sample in classification by a neural network. It should also be noted that this is only one small part of a set of broader validation issues that should be considered in the use of neural networks [47].

4. Conclusions

Feedforward neural networks are often constructed with the aid of a validation dataset. The latter data are typically used to indicate the accuracy of the neural network on a dataset independent of that used in its training phase. Commonly, the optimal network is selected on the basis of the overall accuracy with which the validation dataset is classified based on the analysis of a raw confusion matrix. A problem with this approach is that all classification errors are typically treated equally and the magnitude of the overall accuracy can be distorted by the sample design used to form the validation sample. Here, it was shown that the use of a stratified, rather than random, sample of cases as a validation set resulted in a statistically significant reduction of the accuracy with which an independent test set was classified. This difference in accuracy arose because of the commission of cases of an abundant class by a relatively rare class. Moreover, these results were obtained for analyses of a datasets in which the classes were very highly separable within the feature space of both the simulated and the remotely-sensed datasets used. Assuming that the desired aim is the production of a map with high overall accuracy, simple ways to address issues connected with the design of the validation dataset are to the use of a validation sample acquired by simple random sampling or to ensure the accuracy of its classification is based on proportions calculated on the basis of class abundance rather than basic raw counts.

Acknowledgments: The ATM data were acquired through the European AgriSAR campaign and the research benefits from earlier work with colleagues, notably Manoj Arora, which is gratefully acknowledged. The neural network analyses were undertaken with the Trajan package. The map in Figure 2 was obtained from OpenStreetMap which are available under the Open Database Licence; further details at http://www.openstreetmap.org/copyright.

Conflicts of Interest: The author declares no conflict of interest. The founding sponsors had no role in the design of the study; in the collection, analyses, or interpretation of data, in the writing of the manuscript, and in the decision to publish the results.

References

1. Tso, B.; Mather, P.M. *Classification Methods for Remotely Sensed Data*, 2nd ed.; Taylor & Francis: London, UK; New York, NY, USA, 2001.

2. Mas, J.F.; Flores, J.J. The application of artificial neural networks to the analysis of remotely sensed data. *Int. J. Remote Sens.* **2008**, *29*, 617–663. [CrossRef]

3. Jensen, R.R.; Hardin, P.J.; Yu, G. Artificial neural networks and remote sensing. *Geogr. Compass* **2009**, *3*, 630–646. [CrossRef]

4. Yue, J.; Zhao, W.; Mao, S.; Liu, H. Spectral-spatial classification of hyperspectral images using deep convolutional neural networks. *Remote Sens. Lett.* **2015**, *6*, 468–477. [CrossRef]

5. Li, L.; Chen, Y.; Xu, T.; Huang, C.; Liu, R.; Shi, K. Integration of Bayesian regulation back-propagation neural network and particle swarm optimization for enhancing sub-pixel mapping of flood inundation in river basins. *Remote Sens. Lett.* **2016**, *7*, 631–640. [CrossRef]

6. Peddle, D.R.; Foody, G.M.; Zhang, A.; Franklin, S.E.; LeDrew, E.F. Multi-source image classification II: An empirical comparison of evidential reasoning and neural network approaches. *Can. J. Remote Sens.* **1994**, *20*, 396–407. [CrossRef]

7. Lu, D.; Weng, Q. A survey of image classification methods and techniques for improving classification performance. *Int. J. Remote Sens.* **2007**, *28*, 823–870. [CrossRef]

8. Serpico, S.B.; Bruzzone, L.; Roli, F. An experimental comparison of neural and statistical non-parametric algorithms for supervised classification of remote-sensing images. *Pattern Recognit. Lett.* **1996**, *17*, 1331–1341. [CrossRef]

9. Paola, J.D.; Schowengerdt, R.A. A review and analysis of backpropagation neural networks for classification of remotely-sensed multi-spectral imagery. *Int. J. Remote Sens.* **1995**, *16*, 3033–3058. [CrossRef]

10. Kavzoglu, T.; Mather, P.M. The use of backpropagating artificial neural networks in land cover classification. *Int. J. Remote Sens.* **2003**, *24*, 4907–4938. [CrossRef]

11. Pal, M.; Foody, G.M. Evaluation of SVM, RVM and SMLR for accurate image classification with limited ground data. *IEEE J. Sel. Top. Appl. Earth Obs. Remote Sens.* **2012**, *5*, 1344–1355. [CrossRef]

12. Foody, G.M.; Pal, M.; Rocchini, D.; Garzon-Lopez, C.X.; Bastin, L. The sensitivity of mapping methods to reference data quality: Training supervised image classifications with imperfect reference data. *ISPRS Int. J. Geo Inf.* **2016**, *5*, 199. [CrossRef]

13. Antoniou, V.; Fonte, C.C.; See, L.; Estima, J.; Arsanjani, J.J.; Lupia, F.; Minghini, M.; Foody, G.; Fritz, S. Investigating the feasibility of geo-tagged photographs as sources of land cover input data. *ISPRS Int. J. Geo Inf.* **2016**, *5*, 64. [CrossRef]

14. Kavzoglu, T. Increasing the accuracy of neural network classification using refined training data. *Environ. Model. Softw.* **2009**, *24*, 850–858. [CrossRef]

15. Foody, G.M.; McCulloch, M.B.; Yates, W.B. The effect of training set size and composition on artificial neural network classification. *Int. J. Remote Sens.* **1995**, *16*, 1707–1723. [CrossRef]

16. Zhuang, X.; Engel, B.A.; Lozano-Garcia, D.F.; Fernandez, R.N.; Johannsen, C.J. Optimisation of training data required for neuro-classification. *Int. J. Remote Sens.* **1994**, *15*, 3271–3277. [CrossRef]

17. Foody, G.M. Hard and soft classifications by a neural network with a non-exhaustively defined set of classes. *Int. J. Remote Sens.* **2002**, *23*, 3853–3864. [CrossRef]

18. Olofsson, P.; Foody, G.M.; Herold, M.; Stehman, S.V.; Woodcock, C.E.; Wulder, M.A. Good practices for estimating area and assessing accuracy of land change. *Remote Sens. Environ.* **2014**, *148*, 42–57. [CrossRef]

19. Stehman, S.V. Basic probability sampling designs for thematic map accuracy assessment. *Int. J. Remote Sens.* **1999**, *20*, 2423–2441. [CrossRef]

20. Piper, J. Variability and bias in experimentally measured classifier error rates. *Pattern Recognit. Lett.* **1992**, *13*, 685–692. [CrossRef]

21. Garson, G.D. *Neural Networks: An Introductory Guide for Social Scientists*; Sage: London, UK, 1998.

22. Ahmad, S.; Tesauro, G. Scaling and generalisation in neural networks: A case study. In *Proceedings 1988 Connectionist Models Summer School*; Touretzky, D., Hinton, G., Sejnowsjki, T., Eds.; Morgan Kaufmann: San Mateo, CA, USA, 1989; pp. 3–10.

23. Foody, G.M. The significance of border training patterns in classification by a feedforward neural network using back propagation learning. *Int. J. Remote Sens.* **1999**, *20*, 3549–3562. [CrossRef]

24. Li, C.; Wang, J.; Wang, L.; Hu, L.; Gong, P. Comparison of classification algorithms and training sample sizes in urban land classification with Landsat thematic mapper imagery. *Remote Sens.* **2014**, *6*, 964–983. [CrossRef]

25. Silva, J.; Bacao, F.; Dieng, M.; Foody, G.M.; Caetano, M. Improving specific class mapping from remotely sensed data by cost-sensitive learning. *Int. J. Remote Sens.* **2017**, *38*, 3294–3316. [CrossRef]

26. Ma, X.; Tong, X.; Liu, S.; Luo, X.; Xie, H.; Li, C. Optimized sample selection in SVM classification by combining with DMSP-OLS, Landsat NDVI and GlobeLand30 products for extracting urban built-up areas. *Remote Sens.* **2017**, *9*, 236. [CrossRef]

27. Foody, G.M.; Mathur, A. Toward intelligent training of supervised image classifications: Directing training data acquisition for SVM classification. *Remote Sens. Environ.* **2014**, *93*, 107–117. [CrossRef]

28. Chang, E.I.; Lippmann, R.P. Using genetic algorithms to improve pattern classification performance. In *Advances in Neural Information Processing Systems*; Lippmann, R.P., Moody, J.E., Touretzky, D.S., Eds.; Morgan Kaufmann: San Mateo, CA, USA, 1991; Volume 3, pp. 797–803.

29. Mathur, A.; Foody, G.M. Crop classification by support vector machine with intelligently selected training data for an operational application. *Int. J. Remote Sens.* **2008**, *29*, 2227–2240. [CrossRef]

30. Du, P.; Xia, J.; Zhang, W.; Tan, K.; Liu, Y.; Liu, S. Multiple classifier system for remote sensing image classification: A review. *Sensors* **2012**, *12*, 4764–4792. [CrossRef] [PubMed]

31. Foody, G.M.; Mathur, A. The use of small training sets containing mixed pixels for accurate hard image classification: Training on mixed spectral responses for classification by a SVM. *Remote Sens. Environ.* **2006**, *103*, 179–189. [CrossRef]

32. Mueller, A.V.; Hemond, H.F. Statistical generation of training sets for measuring NO^{3-}, NH^{4+} and major ions in natural waters using an ion selective electrode array. *Environ. Sci. Process. Impacts* **2016**, *18*, 590–599. [CrossRef] [PubMed]

33. Bishop, C.M. *Neural Networks for Pattern Recognition*; Oxford University Press: Oxford, UK, 1995.

34. Lek, S.; Giraudel, J.L.; Guegan, J.-F. Neuronal networks: Algorithms and architectures for ecologists and evolutionary ecologists. In *Artificial Neuronal Networks. Application to Ecology and Evolution*; Lek, S., Guegan, J.-F., Eds.; Springer: Berlin, Germany, 2000; pp. 3–27.

35. Fardanesh, M.T.; Ersoy, O.K. Classification accuracy improvement of neural network classifiers by using unlabeled data. *IEEE Trans. Geosci. Remote Sens.* **1998**, *36*, 1020–1025. [CrossRef]

36. Twomey, J.M.; Smith, A.E. Bias and variance of validation methods for function approximation neural networks under conditions of sparse data. *IEEE Trans. Syst. Man Cybern. Part C* **1998**, *28*, 417–430. [CrossRef]

37. Prechelt, L. Automatic early stopping using cross validation: Quantifying the criteria. *Neural Netw.* **1998**, *11*, 761–767. [CrossRef]

38. Setiono, R. Feedforward neural network construction using cross validation. *Neural Comput.* **2001**, *13*, 2865–2877. [CrossRef] [PubMed]

39. Huynh, T.Q.; Setiono, R. Effective neural network pruning using cross-validation. In Proceedings of the IEEE International Joint Conference on Neural Networks, Montreal, QC, Canada, 31 July–4 August 2005; Volume 2, pp. 972–977.

40. Lee, C.; Landgrebe, D.A. Decision boundary feature extraction for neural networks. *IEEE Trans. Neural Netw.* **1997**, *8*, 75–83. [CrossRef] [PubMed]

41. Zhang, G.; Hu, M.Y.; Patuwo, B.E.; Indro, D.C. Artificial neural networks in bankruptcy prediction: General framework and cross-validation analysis. *Eur. J. Oper. Res.* **1999**, *116*, 16–32. [CrossRef]

42. Pal, M.; Mather, P.M. Support vector machines for classification in remote sensing. *Int. J. Remote Sens.* **2005**, *26*, 1007–1011. [CrossRef]

43. Mas, J.F. Mapping land use/cover in a tropical coastal area using satellite sensor data, GIS and artificial neural networks. *Estuar. Coast. Shelf Sci.* **2004**, *59*, 219–230. [CrossRef]

44. Mas, J.F.; Puig, H.; Palacio, J.L.; Sosa-López, A. Modelling deforestation using GIS and artificial neural networks. *Environ. Model. Softw.* **2004**, *19*, 461–471. [CrossRef]

45. Foody, G.M. Thematic map comparison: Evaluating the statistical significance of differences in classification accuracy. *Photogramm. Eng. Remote Sens.* **2004**, *70*, 627–633. [CrossRef]

46. Agresti, A. *Categorical Data Analysis*, 2nd ed.; Wiley: New York, NY, USA, 2002.

47. Humphrey, G.B.; Maier, H.R.; Wu, W.; Mount, N.J.; Dandy, G.C.; Abrahart, R.J.; Dawson, C.W. Improved validation framework and R-package for artificial neural network models. *Environ. Model. Softw.* **2017**, *92*, 82–106. [CrossRef]

applied
sciences

MDPI

Article

Real-Time Transportation Mode Identification Using Artificial Neural Networks Enhanced with Mode Availability Layers: A Case Study in Dubai

Young-Ji Byon [1,*], Jun Su Ha [2], Chung-Suk Cho [1], Tae-Yeon Kim [1] and Chan Yeob Yeun [3]

[1] Department of Civil Infrastructure and Environmental Engineering,
Khalifa University of Science and Technology, P.O. Box 127788, Abu Dhabi L2017E, UAE;
chung.cho@kustar.ac.ae (C.-S.C.); taeyeon.kim@kustar.ac.ae (T.-Y.K.)
[2] Department of Nuclear Engineering, Khalifa University of Science and Technology, P.O. Box 127788,
Abu Dhabi L2017E, UAE; junsu.ha@kustar.ac.ae
[3] Department of Electrical and Computer Engineering, Khalifa University of Science and Technology,
P.O. Box 127788, Abu Dhabi L2017E, UAE; cyeun@kustar.ac.ae
* Correspondence: youngji.byon@kustar.ac.ae; Tel.: +971-02-501-8336

Received: 30 July 2017; Accepted: 6 September 2017; Published: 8 September 2017

Abstract: Traditionally, departments of transportation (DOTs) have dispatched probe vehicles with dedicated vehicles and drivers for monitoring traffic conditions. Emerging assisted GPS (AGPS) and accelerometer-equipped smartphones offer new sources of raw data that arise from voluntarily-traveling smartphone users provided that their modes of transportation can correctly be identified. By introducing additional raster map layers that indicate the availability of each mode, it is possible to enhance the accuracy of mode detection results. Even in its simplest form, an artificial neural network (ANN) excels at pattern recognition with a relatively short processing timeframe once it is properly trained, which is suitable for real-time mode identification purposes. Dubai is one of the major cities in the Middle East and offers unique environments, such as a high density of extremely high-rise buildings that may introduce multi-path errors with GPS signals. This paper develops real-time mode identification ANNs enhanced with proposed mode availability geographic information system (GIS) layers, firstly for a universal mode detection and, secondly for an auto mode detection for the particular intelligent transportation system (ITS) application of traffic monitoring, and compares the results with existing approaches. It is found that ANN-based real-time mode identification, enhanced by mode availability GIS layers, significantly outperforms the existing methods.

Keywords: artificial neural network; traffic monitoring; GPS; GIS; mode detection

1. Introduction and Related Works

Assessments of level of service (LOS) measures for various modes of transportation are crucial in monitoring and managing the performance of a transportation system network that consists of multiple available modes of transportation via inter-modal connection points. Traditionally, departments of transportation (DOTs) have used either fixed-point sensors or probe vehicles for traffic monitoring purposes. Fixed-point sensors include loop detectors and video cameras. Loop detectors embedded under the road surfaces sense fluctuations in electric currents as vehicles pass over it and estimate their speeds based on the time it took for vehicles to travel the distance between their two axles. Video cameras are either monitored by dedicated personnel at DOTs or enhanced with motion detection algorithms in efforts to estimate their speeds on screen that are often challenged by weather conditions affecting the accuracy, such as rainfall, making the road surface darker, or snowfall, making the road too bright. Fixed-point sensors are expensive, at first, with high capital costs associated with their installations, yet provide near-permanent raw data at economically feasible costs continuously afterwards. However, they are only capable of monitoring traffic conditions at those fixed points instead of throughout the route which can often result in misleading local traffic condition estimations. In order to overcome the issue, DOTs have regularly dispatched dedicated probe vehicles [1] with dedicated drivers and data logging personnel who are later replaced by conventional GPS data loggers in efforts to reduce labor costs. The collected data have been typically post-processed to estimate the traffic conditions throughout the route after the probes have run. This after-the-fact approach provides insights on near future traffic conditions for planning purposes which is not suitable for real-time traffic monitoring.

With an emergence and rapid market penetration rates of smartphones, there are new opportunities for collecting traffic data as smartphone users voluntarily travel throughout all sectors of transportation networks in all modes of transportation. The majority of smartphones are typically equipped with multiple sensors including assisted GPS (AGPS) chips, accelerometers, and magnetometers connected on an economically feasible data plan that continues to lower in costs. AGPS is an enhanced version of traditional GPS sensors that is typically assisted by cellular tower signals that improve the startup performance. If the mode of transportation a smartphone user is in can reliably be identified, technically speaking, the users can be considered as traffic probes themselves that are hovering over an entire transportation network and can provide enormous amounts of raw data that greatly benefit traffic monitoring and management applications in the field of intelligent transportation system (ITS).

Privacy issues associated with such an approach in data collection is beyond the scope of this paper. However, the users may still give consent to reliable smartphone apps to collect such data with expectations of valuable real-time traffic information or tax-related benefits from the government in return, for example. In fact, many users of Internet service providers, such as Google™, and various other social network services (SNS) have given such consents to them in exchange of different forms of benefit.

Among all modes of transportation, an "auto" mode is especially useful for traffic monitoring purposes as it refers to standard private vehicles that have movement patterns that account for the majority of vehicles on the roads. The auto mode users are also the major audiences of real-time traffic information. A transit bus, for example, would not be a good candidate to be categorized as an auto mode. A transit bus on an uncongested road would still frequently stop and typically operates at a lower maximum speed than private vehicles. The traffic conditions the bus experiences are not generally considered as proper representations of current traffic conditions.

An artificial neural network (ANN) simulates a human brain that consists of neurons and excels at pattern recognition rather than analytically processing and formulating for a given phenomenon. Conventional GPS receivers are only capable of directly measuring location information from which acceleration data can only be post-estimated. Recent smartphone models have accelerometers that directly measure the acceleration or deceleration values. In addition, smartphones are also equipped with magnetometers that can measure the strength of electromagnetic fields in the vicinity. A well-trained ANN could determine the mode of transportation a user is in by observing the user's GPS, accelerometer, and magnetometer related information such as speed, location, quality of GPS data, acceleration, and electromagnetic field measurements. As an analogy, a man with his eyes and ears covered may still be able to "feel" and determine whether he is in a typical car (auto mode) or a bus by sensing the physical characteristics around him, such as acceleration and deceleration patterns.

A geographic information system (GIS) is a platform on which traffic conditions can be spatially managed and is an important urban analysis tool in geoinformatics. Applying GIS-based spatial analyses on traffic-related data can enhance mode detection accuracies by considering available modes near the location where a mode estimation is made.

Dubai is a city with unique environmental factors that can affect mode detection accuracies. Due to the high density of high-rise buildings typically with a ratio of a building height to a street-canyon-width of above 100 to 1, GPS signals are prone to multi-path errors. The extremely hot outdoor temperature has introduced indoor walk modes at pathways connecting their metro and major destinations including the Dubai Mall, known as a Metro Link Bridge (https://thedubaimall.com/en/plan-your-visit/getting-here). It is interesting to see how the unique geographic factors play their roles as far as the ANN-based mode detection rates are concerned. For example, the high-rise buildings may tend to introduce noise in the GPS signals making location determination and associated speed information less accurate. The indoor pathways for pedestrians may block the direct line of sight to GPS satellites, which may also deteriorate the strength of the signal.

1.1. Artificial Intelligence in Civil Engineering

The field of civil engineering is considered as one of the oldest engineering disciplines which can also be seen as one of the most matured fields of engineering. With the advancements in computer technology accompanied by emerging artificial intelligence algorithms, there have been various attempts to utilize the artificial intelligence in various sectors in civil engineering. French et al. [2] have applied neural networks for forecasting rainfalls in the field of hydrology. Goh and Jeng et al. [3,4] have assessed seismic liquefaction potentials with neural networks in the field of geotechnical engineering. Tsai and Lee [5] and Mizumura [6] have used back-propagation neural networks and Kalman filtering for tidal-level and ocean data forecasting in the field of ocean engineering. Bassuoni and Nehdi [7] have developed a neuro-fuzzy based prediction model for the durability of self-consolidating concretes. Prasad et al. [8] have used ANNs for predicting compressive strength of various structural materials in the field of structural engineering. In the field of transportation engineering, traffic related forcasts often deal with classification problems and ITS applications, including traffic monitoring problem, can greatly benefit from using the ANNs.

1.2. Traffic Monitoring

Traffic monitoring requires remotely sensing the traffic conditions on desired locations on the roads. Smith et al. [9], in their pioneering work with wireless location technology (WLT)-based traffic monitoring, conducted multiple operational tests and concluded that there are many advantages, including the fact that the WLT offers a new platform for collecting traffic data spatially. The authors point out that there are associated sampling challenges that need to be addressed with concerted efforts in the future. Byon et al. [10] developed a GPS and GIS-integrated traffic monitoring system, named GISTT, whose main aim is to replace traditional traffic probes with dedicated data logging personnel with GPS data loggers. The study assumes that the GPS data loggers are always in the auto mode. Uno et al. [11] used GPS data for WLT-based transit performance monitoring services. Cathey and Dailey [12] attempted to utilize public transit vehicles that voluntarily travel on their routes regularly, for monitoring the general traffic by correlating the behaviors of transit buses to the general traffic. Due to the intrinsic nature of the bus mode that frequently stops even on uncongested roads for serving bus stops, the proposed method, at its best, is an indirect estimation of the actual general traffic conditions. Bar-Gera [13] implemented a WLT-based traffic monitoring system by solely using the triangulation of cell phone towers without the use of GPS signals on major highways. This approach results in relatively lower accuracies (100–300 m) and needs sufficiently large sample sizes in order to overcome the accuracy issues. In addition, the accuracy also depends on the density of cell tower distributions, which happen to be sufficiently high only near major roads which limits the method's expandability into arterial roads. A GPS-based WLT can provide more accurate location information especially when AGPS chips are assisted with cell phone signals. (Zahradnik, F., Assisted GPS, A-GPS, AGPS, https://www.lifewire.com/assisted-gps-1683306) Zhong et al. [14] estimated passenger traffic flows of a transportation hub using mobile phone data using AGPS-enabled phones.

1.3. Transportation Mode Detection

Transportation mode detection is crucial for categorizing raw data collected from WLT-based methods into desired modes for specific ITS applications. Tsui and Shalaby [15] developed a fuzzy logic-based mode identification methodology for raw data collected from personal travel-survey GPS data loggers. The conventional GPS data loggers were carried by study participants for multiple days as they traveled in various modes of transportation. Then the collected data were post-processed in an attempt to estimate their travelled modes retroactively. This after-the-fact approach is suitable for transportation planning purposes rather than real-time applications, including traffic monitoring. Zong et al. [16] also used a similar GPS dataset for identifying travel modes for transportation planning purposes. Chen and Bierlaire [17] proposed a probabilistic method that predicted trip paths taken and their associated modes by using raw data from smartphone sensors including GPS and Bluetooth sensors. The method adopts a post-processing of data and is useful for ITS applications with longer time horizons. Gonzalez et al. [18] and Byon et al. [19] developed real-time mode identification methods with a conventional GPS receiver with ANNs. Their methods used both directly-measured and indirectly-processed data. Directly-measured GPS data are location and speed values, while the indirectly-processed data are acceleration values that are estimated by observing differential speed values with known sampling intervals. The studies find that an ANN-based mode detection approach is effective with reasonable accuracies. Byon et al. [20] enhanced the ANN-based mode detection by utilizing directly-measured acceleration values from smartphones and found that accelerometer values from smartphones did improve the mode detection performances. Recently, there are more specialized applications of mode detection. Maghrebi et al. [21] attempted to estimate travel modes from social media in hopes to utilize social network services as additional input data for mode detection. Cardoso et al. [22] focused on detecting the transportation mode for elderly care while Lan et al. [23] used kinetic energy-harvesting wearables for detecting the mode of transportation.

1.4. Statistical Considerations

For real-time traffic monitoring to be considered reliable, there needs to be a certain guideline for required sample sizes. Li et al. [24] investigated the sample size requirement for GPS-based traffic monitoring and asserted that existing methods either overestimate [25] or underestimate the required sample sizes. The authors found that, for reliably estimating travel times, delay and work-zone conditions require minimum sample sizes of 50, 20, 10, and eight for road sections with lengths of 0.5, 1.5, 2.5, and 3.3 km, respectively. The findings imply that, from the perspectives of GPS-based real-time traffic monitoring, there needs to be at least those minimum number of GPS receivers simultaneously on a particular section of the road. It could also be interpreted that the traffic monitoring procedure should wait until the minimum sufficient number of samples are collected prior to attempting to estimate the current traffic conditions when there are not enough simultaneous GPS probes. Assemi et al. [26] developed and validated a statistical model for travel mode identifications using smartphones.

2. Study Objective

The objective of this study is to develop an ANN-based real-time mode detection methodology that utilizes spatial analysis techniques and improves mode detection accuracies that would support various ITS applications in transportation geoinformatics. This paper assesses the feasibility of using proposed mode availability GIS layers for enhancing performances of transportation mode detection ANNs with data from smartphone sensors in Dubai. The GIS layers contain spatial information of available modes of transportation in a 2D space on a map of the study region. The developed ANNs are tested with three different experimental scenarios with varying choice sets of transportation modes, monitoring duration, device querying frequency, data collection time periods, and routes. The experimental scenarios are:

1. Universal mode detection with a mode availability layer;
2. Auto mode detection with mode availability layers;
3. Peak (rush hour) versus non-peak (non-rush hour) period comparisons.

In all scenarios, performance of the newly-proposed method with mode availability GIS layers are compared with existing methods using conventional GPS data loggers and smartphone sensors.

3. Data Collection

Currently, Dubai has various modes of transportation available including; auto; metro; bus; tram; truck; walk (outdoor and indoor); bike; ferry and boat modes. (Road and Transport Authority of Dubai; https://traffic.rta.ae/trfesrv/public_resources/service-catalogue.do).

Ideally speaking, if all of these modes can be reliably estimated, many ITS applications can benefit from the results in the future. Figure 1 shows all currently-available modes of transportation in Dubai. This paper first develops a universal mode detection ANN that classifies the multiple modes of transportation. Then, a specific ITS application of traffic monitoring based on private vehicles, known as an auto mode, with ANNs is developed which can be useful, especially for transportation practitioners.

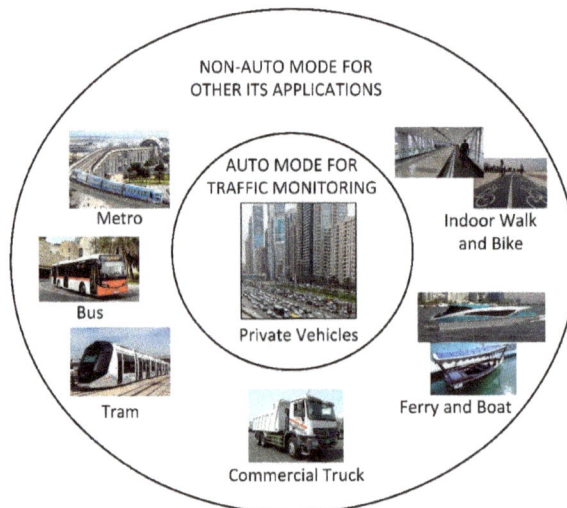

Figure 1. Available modes of transportation in Dubai.

Conventional GPS data loggers (GL-750FL by Canmore Electronics Co. LTD, Hsinchu City, Taiwan, http://www.canmore.com.tw/product.php) and smartphones (Galaxy S8 by Samsung Electronics, Seoul, South Korea, with GPS, accelerometer, and magnetometer sensors, http://www.samsung.com/global/galaxy/galaxy-s8/) are dispatched along major routes of Dubai on different modes of transportation: auto, metro, bus, tram, truck, walk, bike, and water transit vehicles (ferries and boats). The devices have been attached on either side of the waist belt of data collection participants. The sampling rate is set at 1 Hz (once per second) for both devices. In the spring of 2017, in total, 110 weekday hours of data are collected including morning peak (7–9 a.m.), afternoon peak (5–7 p.m.), and non-peak periods (10 a.m.–2 p.m.). The morning peak, afternoon peak, and non-peak periods account for 25%, 25%, and 50% of all data, respectively. Sixty percent of the data are used to train the ANNs and 40% of the data are used as unseen data for evaluating the performance of the developed ANNs. The ratio between the training and validating data are arbitrarily chosen in such a way that the training phase is given with more data for building the ANNs. Analyses on the effect of varying ratios between the two datasets are beyond the scope of this paper. Ten people in total with 10 GPS data loggers and 10 smartphones have collected similar amounts of data from different modes of transportation.

The following criteria are considered for choosing routes and time slots for the data collection in Dubai.

1. Representative routes of Dubai should share as many modes as possible in order for the different modes to experience similar physical surrounding conditions nearby so that the proposed ANN's distinguishing power can focus on the characteristic variances among the different modes.
2. The data should be collected from different types of roads (highways, arterial roads) from different regions of the city that would fairly represent the overall road conditions throughout the city.
3. The data should be collected from both peak periods (rush hour) and non-peak periods (non-rush hour) because the traffic conditions are typically different between those time periods.

Figure 2 shows the study area with different routes for different modes. For the auto mode, major highways (E11, E44, and D63) and a loop of nearby arterial roads (318th, 8th, Al Thanya St., Al Marabea Rd., Al Waha St., and Umm Al Sheif Rd.) are chosen. For the metro mode, which operates on an elevated rail that is about 20 m above the ground, a section between Nakheel station and Al Rigga station is selected. For the bus mode, a transit Route 8 is chosen for its wide coverage throughout the city while Route C9 is chosen because it travels through the central business district (CBD) which is notorious for extremely high-rise buildings that may cause serious multi-path errors with GPS signals. For the tram mode, the Dubai Tram route is chosen, which is the only tram route available in Dubai. For the truck mode, highway E311, also known as a "truck road" is chosen which has intentionally been built in parallel with a major highway, E11, in order to reduce the load on E11. (Zaman, S., New highway connects Abu Dhabi to Dubai, Gulf News, http://gulfnews.com/news/uae/transport/new-highway-connects-abu-dhabi-to-dubai-1.1937213) For the walk and bike modes, the data are collected from Jumeira Beach due to the relatively higher density of the population using the two modes in the city. Due to the extremely hot weather, major metro stations are connected with indoor pathways that are elevated roughly 20 m above the ground to major shopping malls that often stretch over a few kilometers which introduces a new indoor-walk mode in Dubai. This enables collecting data from an indoor walk mode based on sufficiently-long indoor pathways that can provide consistent streams of raw data to the mode detection ANNs in this study. Dubai is located by the Persian Gulf and its waterways play a major role as a part of the overall transportation infrastructure. Ferries (Route F1 and F2) and "Abra" boats (B1 and B2) are chosen for the data collection. Abra boats refer to small motorized water taxis that operate in the Dubai Creek area.

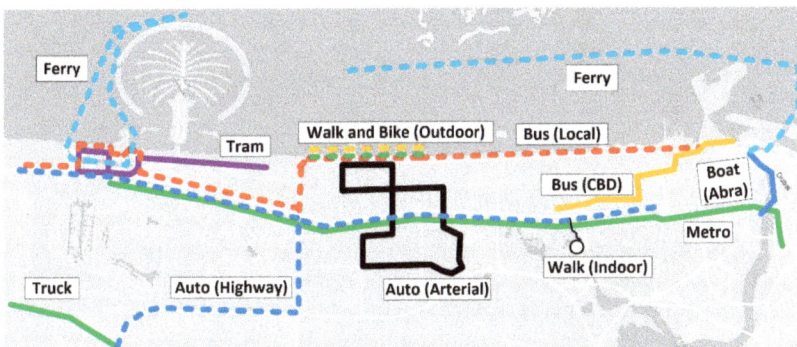

Figure 2. Data collection area in Dubai in various modes of transportation.

4. Implementation of Mode Detection Artificial Neural Networks

A commercial package, Neurosolutions 7 (http://www.neurosolutions.com) is used for implementing the ANNs for various scenarios, training the ANNs, and computing the mode detection rates. Table 1 shows the common components of the software package and Figure 3 shows one of the most popular ANNs consisting of a single hidden layer, an input layer, and an output layer with a back-propagation algorithm. The particular design of ANN in Figure 3 is described in more detail in Section 7. If a transportation mode can correctly be identified with one of the simplest ANNs as adopted in this research with reasonable accuracies, more complex versions of it with sufficient fine-tuning would only improve the accuracies further.

Table 1. Common components of Neurosolutions.

Icon	Name	Description
	Axon	Layer of processing elements
	TanhAxon	Hyperbolic nonlinear transfer function. M-P PE can be built by attaching TanhAxon after Axon
	FullSynapse	Connects Axons from the left to the right
	L2Criterion	Cost function (J)
	BackAxon	Placed together with Axon and manages the back-propagation algorithm
	BackFullSynapse	Connects BackAxons from the right to the left
	Momentum	Updates weights with momentum learning algorithm

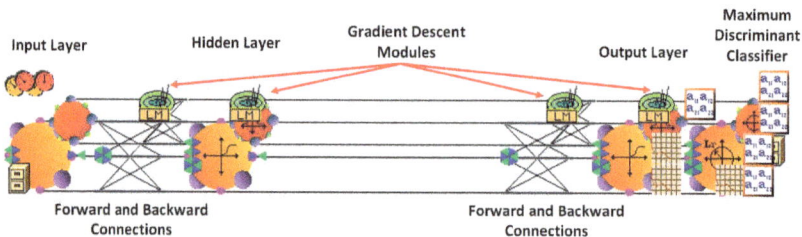

Figure 3. Setup of a single-hidden-layer artificial neural network (ANN).

5. Empirical Analysis on Input Factors

Emerging smartphones are generally equipped with AGPS and magnetometer sensors. They can provide additional assistance in classifying the different modes. Byon et al. [20] find that those sensors are beneficial in detecting the transportation modes by producing more raw data in addition to the speed and acceleration values. For example, a boat mode, which usually operates with a wider view of the sky with less influences from buildings that would cause multi-path errors, can often see a greater number of satellites in view. Tram vehicles that run on electric motors with associated electronic parts generally produce higher electromagnetic fields. By utilizing the newly-available sensors from the smartphones, it is possible to further assist with classifying the mode with higher mode detection rates. Table 2 shows the summary of data collection in terms of the average number of satellites in view and average magnitudes of electromagnetic fields for each mode. The values of average number of satellites in view and average magnitude of magnetic field, are computed from the entire set of collected data in the field.

Table 2. Distinctive input variables for each mode.

Modes	Average Number of Satellites in View	Average Magnitude of Magnetic Field (µT)
Auto	8	49
Bus	6	55
Truck	9	45
Tram	6	138
Walk	8	38
Bike	7	33
Boat	10	25

6. Building Mode Availability Layer

On a particular section of parallel roads, there are only a certain number of modes available and ANNs can utilize this additional input which can further aid the ANNs classifying capabilities. In this paper, a newly-developed type of GIS map layer in raster format, which contains binary (1 for available and 0 for not-available state) information is created. Mode-specific vector map layers are generated by selecting a particular subset of road networks corresponding to the availability of the mode throughout the map. Then, the newly-generated vector map is transformed into a raster map with 10 m × 10 m cells, where cells within a 100-m buffer of the road networks available with the particular mode are turned on. Once the newly-formed mode availability (MA) layers are prepared and ready to be accessed for each mode, raw data associated with any particular coordinates of GPS data from the field can be matched against those MA layers for checking the availability of each mode at that location (i.e., geocentric coordinates of x, y, z). This would form a new set of binary input factors to further aid the mode detection processes with ANNs. Figure 4 illustrates the concept of MA layers for each mode in a raster format. An auto mode is enabled in all sections in Figure 4 while the transit mode is available via points A-B-D-C and the metro mode is available via points C-D-E.

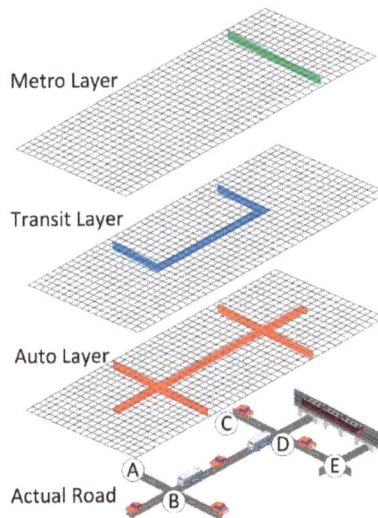

Figure 4. Mode availability (MA) layers in a raster format.

ANNs are generally categorized as a "black box" approach, referring to the fact that there are no well-specified analytical relationships between input and output as mathematical functions. The method is also data-driven. To some, those facts may be viewed as major flaws as the approach is missing the understanding of phenomenon from the perspectives of intuitive mathematical reasoning. However, this very black-box approach indeed offers pattern recognition power that explorers search spaces that analytically approaching minds would not have even recognized they existed. In brief, the proposed MA approach produces binary (1 or 0) values for each cell on a raster map. When raw data are collected from within those "enabled" regions for a certain available mode, it gives additional stimulation to the ANNs and forces its mode classifying estimations towards the actual mode during the training phase. In other words, the extra stimulation would still help ANNs to estimate the actual mode more often than the case without the additional input, even when the output layer lists other non-available dummy modes among its output mode options.

7. Design of an Artificial Neural Network Classifier for Mode Detection

The characteristics of GPS data streams vary depending on the transportation mode in which the smartphones are located. In this paper, a pattern classification approach is adopted to identify the mode based on the characteristics of data streams from various sensors and mode availability layer. In its most general form, a typical neural network (NN) consists of multiple layers of neurons, connection weights among neurons, and associated nonlinear transfer functions within neurons. Multi-layer perceptrons (MLP) [27] have been used in the proposed ANNs. More specifically, one-hidden-layer MLP ANNs with McCulloch-Pitts (M-P) (Marsalli, M., McCulloch-Pitts Neurons, http://www.mind.ilstu.edu/curriculum/modOverview.php?modGUI=212) processing elements are used. The MLP is the most common supervised learning ANN. Supervised learning refers to the case where there are input and output pairs of data for training the network. For the mode identification, the input of GPS data used include speed, acceleration, and the average number of satellites in view, while the output includes the corresponding mode of transportation. If the mobile device or the server can store (and transmit) both current and a few recent GPS readings, the additional past speed and acceleration values can also be the input to the ANN to possibly improve the detection accuracy. The ANN maps each set of input data to the most probable mode. The training process adjusts connection weights between neurons as an entire set of training input data, also known as an epoch, is repeatedly passed through the network. All ANNs used in this paper consist of three layers of neurons connected in a feed-forward fashion, trained with the well-established, error-back-propagation algorithm. Depending on the specifics of each scenario, the number of inputs ranges from six (DL with two pings) to 220 (MAGIS with 20 pings) and the number of output classes is either seven (auto, bus, truck, tram, walk, bike, boat) or two (auto or non-auto). Seven output classes are used for classifying individual modes; i.e., auto, bus, truck, tram, walk, or bike. Two output classes are used when the classification is for auto vs. non-auto modes. As shown in Figure 5, the number of input variables varies depending on the monitoring duration and number of pings (n) per unit of time (as indicated as "Ping n" in Figure 5) in each scenario. For example, if the monitoring duration is 10 min and the number of pings or queries is 10 in that duration (one ping per minute), there are 10 speed values, 10 acceleration values, and 10 values for the number of satellites in view. Hyperbolic tangent functions are used as the nonlinear transfer functions in the M-P PEs (McCulloch-Pitts processing elements). The number of neurons (nodes) in the hidden layer is varied as 5, 10, 20, and 30. For different scenarios, different numbers of hidden nodes have been used, depending on the number of input variables and output classes. As a rule of thumb, having too many hidden nodes forces the NN to memorize the input-output training data, and leads to poorer generalization. Having too few hidden nodes, on the other hand, gives the NN difficulties in identifying and isolating the different classes. Therefore, it is desirable to keep the number of hidden nodes to a minimum without sacrificing the discriminating power. From the 110 h of collected data, for each scenario of data logger (DL), smartphone (SP), and smartphone with mode availability GIS (MAGIS), the entire set of data are shuffled randomly first. Then, 60% of the data are used to train the network and the remaining 40% are used for the validation process as unseen data. For example, in the case of the two pings –5 min scenario of DL, 10 different ANNs are formed with a randomly-shuffled set of data of which 60% of the data is used to train/develop the ANNs. Then, the average detection rate of those 10 runs of an identical scenario of two pings –5 min is considered as the resulting detection rate for the scenario. The ratio of training and validating data is set arbitrarily while assigning more data towards the training phase. The final number of hidden nodes used is determined after varying the number of hidden nodes over 10 to 20 trials. The number of nodes was kept to a minimum so long as performance of the network did not degrade compared with networks with a greater number of nodes. During the training processes, classification performance seemed to stabilize after 3000 epochs. If the error did not stabilize or continued to fluctuate after 3000 epochs, the number of hidden nodes is increased.

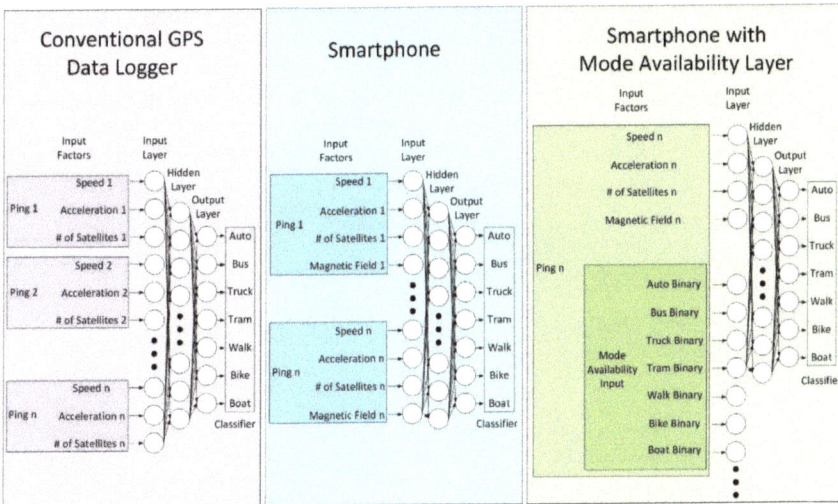

Figure 5. Artificial neural network layout for universal mode detection with data logger (DL), smartphone (SP), and smartphone with mode availability GIS (MAGIS).

8. Experimental Scenarios and Results

This paper considers a few scenarios that effectively test the feasibility of the proposed ANN-based mode detection method with the developed MA layers. In order to compare the performances of the proposed method against existing methods, two existing methods are also carried on in addition to the proposed one. The two existing techniques are the conventional GPS data logger (DL) method and smartphone (SP) method. The DL approach uses a conventional GPS data logger that is capable of directly measuring its current location and speed. Indirectly-estimated acceleration values are produced from the speed values. In the case of the SP approach, the GPS location values are fixed quicker with the aid of signals from cell phone towers with the smartphones' embedded AGPS chips. In addition, directly measured acceleration values from the accelerometers are collected in forms of absolute magnitude of 3D acceleration vectors. Finally, another measurement from a magnetometer further aids the mode detection.

The newly-proposed method involves using the smartphone approach with a mode availability GIS layers, namely the MAGIS method. The MAGIS requires pre-processing of a city's road networks filtered for available mode routes information. However, once such GIS layers are built, the database can be easily maintained in the form of a live GIS database and can be easily updated as the routes for each mode can be modified in real-time in the database.

8.1. Universal Mode Detection with Smartphone with Mode Availability GIS (MAGIS)

Figure 5 shows the designs of ANNs for the universal mode detection for all modes of transportation based on their physical characteristics. Auto mode would typically experience the highest maximum speed, while transit buses would frequently accelerate and decelerate even during free-flow periods for serving transit stops. As Table 2 suggests, a boat mode would have a clearer view of the sky and "see" a greater number of satellites. In all of the above cases, observing a single particular set of data is not sufficient for reliably detecting the mode, as Li et al. [24] suggest. By making observations for a longer duration of time and sampling the data more frequently, it would give more opportunities for ANNs to capture particular sets of patterns for each mode and make more reliable mode estimations. In this scenario, 10 min of monitoring duration with five data samplings, also known as "pings", are used for all three approaches of DL, SP, and MAGIS.

Table 3 and Figure 6 show the results from the universal mode detection using DL, SP, and MAGIS approaches. All three results show the highest values in its main diagonals, meaning that the proposed ANNs are detecting the correct modes the most among other modes. However, the mode detection rates of DL are the lowest among the three methods, probably due to its estimated acceleration values and lacking newly-available emerging input values from smartphones. The MAGIS performs significantly better than DL and SP methods due to its additional input factors from smartphone sensors and the binary input variables prepared from MA layers.

Table 3. Results of universal mode detection rates (%) with DL, SP, and MAGIS.

A. Universal Mode Detection with Conventional GPS Data Logger							
Actual Modes	**Predicted Modes**						
	Auto	**Bus**	**Truck**	**Tram**	**Walk**	**Bike**	**Boat**
Auto	42	12	18	11	4	5	8
Bus	8	48	10	26	3	2	3
Truck	25	13	42	11	4	4	1
Tram	11	12	10	42	4	11	10
Walk	4	3	2	4	52	23	12
Bike	7	8	5	7	15	53	5
Boat	2	5	8	17	14	9	45

B. Universal Mode Detection with Smartphone							
Actual Modes	**Predicted Modes**						
	Auto	**Bus**	**Truck**	**Tram**	**Walk**	**Bike**	**Boat**
Auto	61	9	13	7	3	4	3
Bus	7	63	9	14	4	2	1
Truck	18	9	57	10	3	2	1
Tram	9	11	9	53	3	7	8
Walk	1	2	3	4	74	11	5
Bike	5	6	4	5	11	67	2
Boat	1	3	4	13	11	7	61

C. Universal Mode Detection with Mode Availability Layer							
Actual Modes	**Predicted Modes**						
	Auto	**Bus**	**Truck**	**Tram**	**Walk**	**Bike**	**Boat**
Auto	78	5	7	4	2	3	1
Bus	4	82	4	7	2	1	0
Truck	8	2	82	4	2	1	1
Tram	4	7	5	76	2	5	1
Walk	2	3	4	5	74	7	5
Bike	3	4	5	4	8	72	4
Boat	1	2	2	5	3	5	82

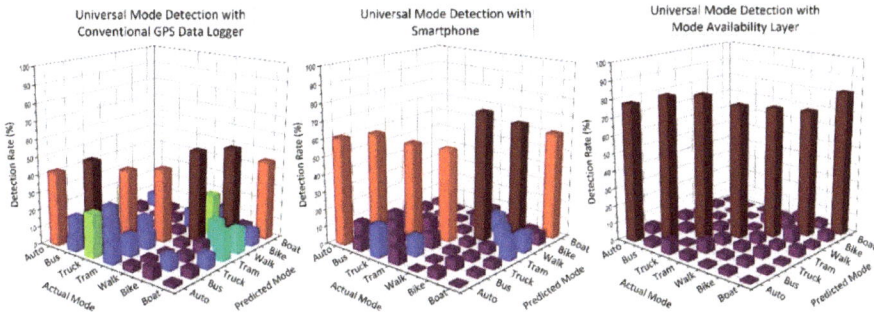

Figure 6. Bar graphs of the results from universal mode detection with DL, SP, and MAGIS.

8.2. Auto Mode Detection with MAGIS

A particular ITS application of traffic monitoring is considered in this scenario. The traffic conditions experienced by private auto mode vehicles are usually considered as the general traffic flow because other modes of transportation, such as public transit or trucks, behave differently due to their specific roles, such as serving passengers at bus stops or carrying heavy loads of goods. Therefore, in the case of traffic monitoring, the mode detection among auto vs. non-auto modes are sufficient. This implies that all modes, other than the auto modes, are considered as non-auto modes altogether. In this section, raw data from non-auto modes are randomly shrunk down in size to match the size of the data from the auto mode with 50/50 ratio, in order to truly see the classifying power of the ANNs. Figure 7 shows the design layout of the ANNs for auto mode detection with DL, SP, and MAGIS approaches.

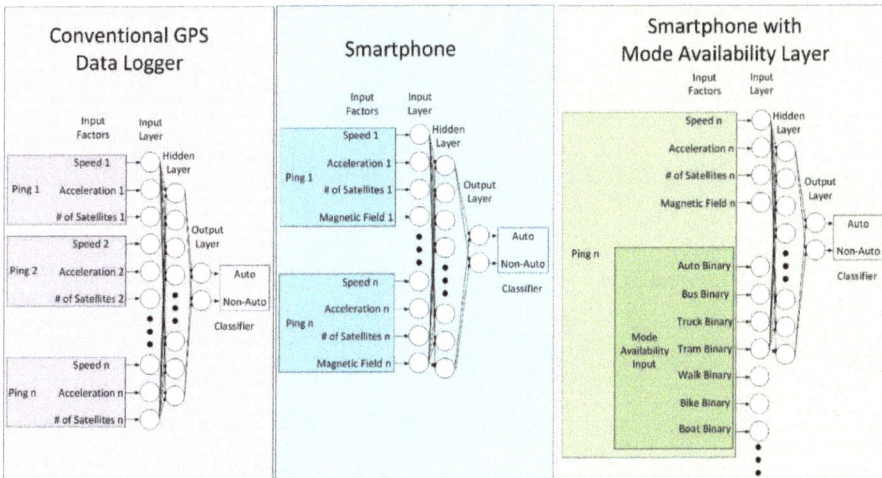

Figure 7. Artificial neural network layout for auto mode detection with DL, SP, and MAGIS.

Table 4 and Figure 8 show the results of auto mode detection with DL, SP, and MAGIS approaches. When the actual mode of transportation is auto mode and its mode is correctly estimated as auto mode, it is denoted as an A-A detection rate. Similarly, if the actual mode is non-auto mode and the estimated mode is non-auto mode, it is denoted as an N-N detection rate. Ideally, if the A-A and N-N detection rates are both 100%, the mode detection is considered perfect.

Table 4. Results of auto mode detection rates (%) with DL, SP, and MAGIS.

A. Auto Mode Detection Rates with Conventional GPS Data Logger

# of Pings	Monitoring Duration							
	5 min		10 min		15 min		20 min	
	A-A	N-N	A-A	N-N	A-A	N-N	A-A	N-N
2	49	68	52	72	54	75	55	78
5	53	70	59	72	60	77	67	80
10	58	74	62	74	67	81	71	84
15	63	79	65	78	71	83	74	86
20	67	81	69	84	73	87	77	89

B. Auto Mode Detection Rates with Smartphone

# of Pings	Monitoring Duration							
	5 min		10 min		15 min		20 min	
	A-A	N-N	A-A	N-N	A-A	N-N	A-A	N-N
2	54	77	53	77	56	79	58	81
5	63	83	64	83	63	85	65	85
10	66	82	69	86	73	88	77	89
15	72	85	74	89	77	91	82	94
20	75	88	76	90	84	94	87	92

C. Auto Mode Detection Rates with Mode Availability Layer

# of Pings	Monitoring Duration							
	5 min		10 min		15 min		20 min	
	A-A	N-N	A-A	N-N	A-A	N-N	A-A	N-N
2	62	82	65	83	69	86	73	88
5	68	85	70	86	78	89	80	90
10	77	88	75	89	82	91	89	93
15	82	91	84	92	89	92	92	94
20	84	92	88	95	93	94	95	97

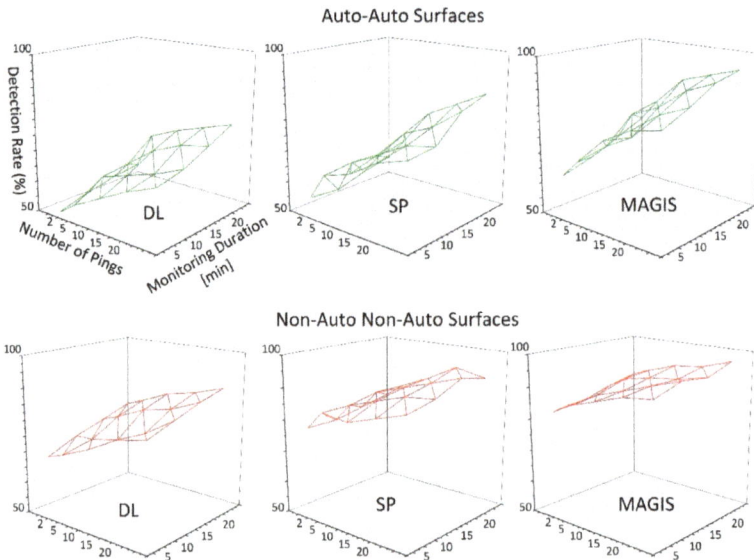

Figure 8. A-A surface and N-N surface of auto mode detection with DL, SP, and MAGIS.

The results show the DL approach, while showing the main diagonal with the highest values for both A-A and N-N are all lower than the SP or MAGIS-based results. The MAGIS approach shows the highest rates for both A-A and N-N. The results are significantly higher than DL or SP approaches and it shows that the mode availability binary input does improve the detection rates for both A-A and N-N rates.

8.3. Peak Versus Nonpeak Auto Mode Detection with MAGIS

During rush hour, (7–9 a.m. and 5–7 p.m.) with 10 pings with 10 min duration scenario, the A-A rate is found to be 68%, 73%, 81%, for DL, SP, and MAGIS approaches, respectively. During the non-rush hour period (10 a.m.–2 p.m.) the A-A rate is found to be 56%, 67%, 74%, for DL, SP, and MAGIS approaches, respectively. It is interesting to note that the mode detection rates are better during the peak period. It seems that due to congestion, the vehicles tend to experience more fluctuations in accelerations and decelerations unique to each mode and it actually helps the ANNs to detect the intrinsic patterns embedded in each mode.

9. Future Research

The mode detection accuracies may be further improved by incorporating more complex ANNs with more hidden layers and varying the number of neurons. There is a need for a consistent guideline for optimizing the complexity of the designs and layout of the ANNs for the mode detection.

The MA layers can be implemented as an online database and be available for real-time access in the case of certain sections being unavailable due to accidents or temporarily paralyzed by grid-lock situations in the CBD.

Routes of certain modes of transportation are only available at particular ranges of elevations. For example, a boat mode usually operates at sea level, while elevated rails for the metro is always at higher elevations from the ground. Different transit bus routes may operate within different ranges of elevations. When combined with digital elevation models (DEM) it is possible to identify the mode of transportation more easily by incorporating the operating height information of each mode.

Currently in Dubai, where this study has been conducted, the metro trains operate above the ground. However, the majority of metro systems internationally operate underground. It is a challenge to enhance the proposed approach with underground applications without GPS data due to the blocked line of sight from GPS satellites to the metro users. It would be interesting to see how other sensors in smartphones, combined with mode availability layers, perform together for mode detection.

This paper briefly develops universal mode detection ANNs, followed by focusing on a particular ITS application of traffic monitoring. In the future, unique and specific ITS applications can be built for each mode. For example, for bike mode-related applications, the ANNs can further be optimized for detecting the bike mode followed by conducting slope analyses of the ground by processing the DEM with slope-determining GIS operations in 3D for the flattest route instead of the shortest route computations for the bike mode.

10. Conclusions

With advancements in computing technology and artificial intelligence algorithms, the field of civil engineering, which is often seen as a classic field of engineering, can benefit from the newly-available tools.

Traditionally, traffic monitoring requires fixed sensors with high capital costs that can only monitor a few fixed locations or probe vehicles with a dedicated vehicle and driver with an advantage of monitoring traffic conditions throughout the route, yet with high labor costs. There are existing methods that attempt to replace the probe vehicles with conventional GPS data loggers or smartphones with additional sensors that can automatically collect data as long as their mode of transportation can be reliably detected as an "auto" mode. This paper develops an artificial neural network

(ANN)-based mode detection procedure enhanced with a mode availability (MA) GIS map layer, named as the MAGIS approach, which can detect various modes of transportation for potential future ITS applications and an auto mode for traffic monitoring purposes. A new approach of constructing MA layers in a raster format is proposed, traffic data associated with certain locations are queried against the MA layers, and the results in binary form are fed into the mode detection ANNs. A thorough data collection is carried out in Dubai where extremely tall buildings and indoor pathways offer unique environments to conduct this study. It is found that ANN-based mode detection enhanced with mode availability layers improve the detection accuracies significantly.

In the case of universal mode detection among seven different modes of transportation, including auto, bus, truck, tram, walk, bike, and boat, the proposed MAGIS approach has resulted in correct detection rates that range from 72 to 82% while the existing DL and SP methods resulted in correct detection rates from 42 to 53% and from 53 to 74%, respectively, for all modes of available transportation. In the case of the auto mode detection, the MAGIS approach produces correct detection rates ranging between 62% and 97% across all scenarios, while DL and SP methods have produced correct detection rates ranges between 49% and 89% and between 54% and 92%, respectively. Figure 9 shows the trends from the results. The proposed MAGIS approach can be adapted for cities with different sets of available modes as long as their MAGIS layers can be prepared and managed in parallel with the real-time operation of ANNs.

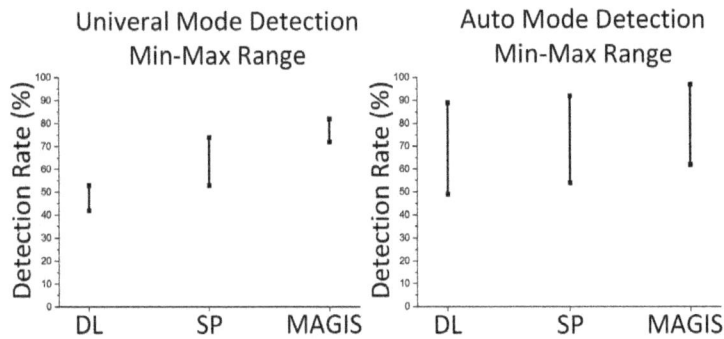

Figure 9. Minimum and maximum mode detection rates of DL, SP, and MAGIS methods.

Acknowledgments: This research was conducted by the Transport-GIS Lab at the Khalifa University of Science and Technology (KUST) with an interdisciplinary team of faculties from departments of Civil Infrastructure and Environmental Engineering, Industrial and Systems Engineering, and Electrical and Computer Engineering funded by KUIRF Level 1 (2016) at KUST, and an ADEC Award for Research Excellence 2016–2017 in Abu Dhabi, UAE.

Author Contributions: Young-Ji Byon and Jun Su Ha conceived, designed, and analyzed the experiments; and Young-Ji Byon, Chung-Suk Cho, Tae-Yeon Kim, and Chan Yeob Yeun collected data, contributed the tools and wrote the paper.

Conflicts of Interest: The authors declare no conflict of interest.

References

1. Tarnoff, P.J.; Bullock, D.M.; Young, S.E. Continuing Evolution of Travel Time Data Information Collection and Processing. In Proceedings of the Transportation Research Board 88th Annual Meeting, Washington, DC, USA, 1–15 January 2009.
2. French, M.N.; Krajewski, W.F.; Cuykendall, R.R. Rainfall Forecasting in Space and Time Using a Neural Network. *J. Hydrol.* **1992**, *137*, 1–31. [CrossRef]
3. Goh, A.C.T. Seismic Liquefaction Potential Assessed by Neural Networks. *J. Geotech. Eng.* **1994**, *120*, 1467–1480. [CrossRef]

4. Jeng, D.S.; Lee, T.L.; Lin, C. Application of Artificial Neural Networks in Assessment of Chi-Chi Earthquake-induced Liquefaction. *Asian J. Inf. Technol.* **2003**, *2*, 190–198.

5. Tsai, C.P.; Lee, T.L. Back-propagation Neural Network in Tidal-level Forecasting. *J. Waterw. Port Coast. Ocean Eng.* **1999**, *125*, 195–202. [CrossRef]

6. Mizumura, K. Application of Kalman Filtering to Ocean Data. *J. Waterw. Port Coast. Ocean Eng.* **1984**, *110*, 334–343. [CrossRef]

7. Bassuoni, M.T.; Nehdi, M.L. Neuro-fuzzy based Prediction of the Durability of Self-consolidating Concrete to Various Sodium Sulfate Exposure Regimes. *Comput. Concr.* **2008**, *5*, 573–597. [CrossRef]

8. Prasad, B.K.R.; Eskandari, H.; Reddy, B.V.V. Prediction of Compressive Strength of SCC and HPC with High Volume Fly Ash using ANN. *Constr. Build. Mater.* **2009**, *23*, 117–128. [CrossRef]

9. Smith, B.L.; Zhang, H.; Fontaine, M.; Green, M. Cellphone Probes as an ATMS Tool. In *Research Project Report for the National ITS Implementation Research Center and US DOT University Transportation Center*; Center for Transportation Studies at the University of Virginia: Charlottesville, VA, USA, 2003.

10. Byon, Y.J.; Shalaby, A.; Abdulhai, B. GISTT: GPS-GIS Integrated System for Travel Time Surveys. In Proceedings of the Transportation Research Board 85th Annual Meeting Preprint CD-ROM, Washington, DC, USA, 22–26 January 2006.

11. Uno, N.; Kurauchi, F.; Tamura, H.; Iida, Y. Using Bus Probe Data for Analysis of Travel Time Variability. *J. Intell. Transp. Syst.* **2009**, *13*, 2–15. [CrossRef]

12. Cathey, F.W.; Dailey, D.J. Transit Vehicles as Traffic Probe Sensors. *Transp. Res. Rec. J. Transp. Res. Board Natl. Acad.* **2002**, *1804*, 23–30. [CrossRef]

13. Bar-Gera, H. Evaluation of a Cellular Phone-Based System for Measurements of Traffic Speeds and Travel Times: A Case Study from Israel. *J. Transp. Res. Board Natl. Acad. Part C Emerg. Technol.* **2007**, *15*, 380–391. [CrossRef]

14. Zhong, G.; Wan, X.; Zhang, J.; Yin, T.; Ran, B. Characterizing Passenger Flow for a Transportation Hub Based on Mobile Phone Data. *IEEE Trans. Intell. Transp. Syst.* **2017**, *18*, 1507–1518. [CrossRef]

15. Tsui, A.; Shalaby, A. An Enhanced System for Link and Mode Identification for GPS-based Personal Travel Surveys. *J. Transp. Res. Board Natl. Acad.* **2006**, *1972*, 38–45. [CrossRef]

16. Zong, F.; Yuan, Y.; Liu, J.; Bai, Y.; He, Y. Identifying travel mode with GPS data. *Transp. Plan. Technol.* **2017**, *40*, 242–255. [CrossRef]

17. Chen, J.; Bierlaire, M. Probabilistic Multimodal Map-matching with Rich Smartphone Data. *J. Intell. Transp. Syst.* **2015**, *19*, 134–148. [CrossRef]

18. Gonzalez, P.A.; Weinstein, J.S.; Barbeau, S.J.; Labrador, M.A.; Winters, P.L.; Georggi, N.L.; Perez, R. Automating mode detection for travel behavior analysis by using global positioning systems enabled mobile phones and neural networks. *IET Intell. Transp. Syst.* **2010**, *4*, 37–49. [CrossRef]

19. Byon, Y.J.; Abdulhai, B.; Shalaby, A. Real-Time Transportation Mode Detection via Tracking Global Positioning System Mobile Devices. *J. Intell. Transp. Syst.* **2009**, *13*, 161–170. [CrossRef]

20. Byon, Y.J.; Liang, S. Real-Time Transportation Mode Detection using Smartphones and Artificial Neural Networks: Performance Comparisons between Smartphones and Conventional Global Positioning System sensors. *J. Intell. Transp. Syst.* **2014**, *18*, 264–272. [CrossRef]

21. Maghrebi, M.; Abbasi, A.; Waller, S.T. Transportation Application of Social Media: Travel Mode Extraction. In Proceedings of the 2016 IEEE 19th International Conference on Intelligent Transportation Systems (ITSC), Rio de Janeiro, Brazil, 1–4 November 2016.

22. Cardoso, N.; Madureira, J.; Pereira, N. Smartphone-based Transport Mode Detection for Elderly Care. In Proceedings of the 2016 IEEE 18th International Conference on e-Health Networking, Applications and Services (Healthcom), Munich, Germany, 14–17 September 2016.

23. Lan, G.; Xu, W.; Khalifa, S.; Hassan, M.; Hu, W. Transportation Mode Detection Using Kinetic Energy Harvesting Wearables. In Proceedings of the 2016 IEEE International Conference on Pervasive Computing and Communications, Sydney, Australia, 14–18 March 2016.

24. Li, S.; Zhu, K.; van Gelder, B.H.W.; Nagle, J.; Tuttle, C. Reconsideration of Sample Size Requirements for Field Traffic Data Collection with Global Positioning System Devices. *J. Transp. Res. Board Natl. Acad.* **2002**, *1804*, 17–22. [CrossRef]

25. Quiroga, C.A.; Bullock, D. Travel Time Information Using GPS and Dynamic Segmentation Techniques. *J. Transp. Res. Board Natl. Acad.* **1999**, *1660*, 48–57. [CrossRef]

26. Assemi, B.; Safi, H.; Mesbah, M.; Ferreira, L. Developing and Validating a Statistical Model for Travel Mode Identification on Smartphones. *IEEE Trans. Intell. Transp. Syst.* **2016**, *17*, 1920–1931. [CrossRef]

27. Principe, J.C.; Euliano, N.R.; Lefebvre, W.C. *Neural and Adaptive Systems: Fundamentals through Simulations*; Wiley: New York, NY, USA, 2000.

applied
sciences

MDPI

Article

Severity Prediction of Traffic Accidents with Recurrent Neural Networks

Maher Ibrahim Sameen [†] and Biswajeet Pradhan [†,*]

Department of Civil Engineering, Geospatial Information Science Research Center (GISRC),
Faculty of Engineering, University Putra Malaysia, UPM, Serdang 43400, Malaysia; maherrsgis@gmail.com
* Correspondence: biswajeet24@gmail.com or biswajeet@upm.edu.my; Tel.: +60-3-8946-6383
† These authors contributed equally to this work.

Academic Editor: Saro Lee
Received: 14 March 2017; Accepted: 28 April 2017; Published: 8 June 2017

Abstract: In this paper, a deep learning model using a Recurrent Neural Network (RNN) was developed and employed to predict the injury severity of traffic accidents based on 1130 accident records that have occurred on the North-South Expressway (NSE), Malaysia over a six-year period from 2009 to 2015. Compared to traditional Neural Networks (NNs), the RNN method is more effective for sequential data, and is expected to capture temporal correlations among the traffic accident records. Several network architectures and configurations were tested through a systematic grid search to determine an optimal network for predicting the injury severity of traffic accidents. The selected network architecture comprised of a Long-Short Term Memory (LSTM) layer, two fully-connected (dense) layers and a Softmax layer. Next, to avoid over-fitting, the dropout technique with a probability of 0.3 was applied. Further, the network was trained with a Stochastic Gradient Descent (SGD) algorithm (learning rate = 0.01) in the Tensorflow framework. A sensitivity analysis of the RNN model was further conducted to determine these factors' impact on injury severity outcomes. Also, the proposed RNN model was compared with Multilayer Perceptron (MLP) and Bayesian Logistic Regression (BLR) models to understand its advantages and limitations. The results of the comparative analyses showed that the RNN model outperformed the MLP and BLR models. The validation accuracy of the RNN model was 71.77%, whereas the MLP and BLR models achieved 65.48% and 58.30% respectively. The findings of this study indicate that the RNN model, in deep learning frameworks, can be a promising tool for predicting the injury severity of traffic accidents.

Keywords: severity prediction; GIS; traffic accidents; deep learning; recurrent neural networks

1. Introduction

Traffic accidents are a primary concern due to many fatalities and economic losses every year worldwide. In Malaysia, recent statistics show that there are nearly 24 deaths per 100,000 people for all road users [1]. Expressways are potential sites of fatal highway accidents in Malaysia. Better accident severity prediction models are critical to enhancing the safety performance of road traffic systems. According to recent literature, driver injury severity can be classified into a few categories such as property damage, possible/evident injury, or disabling injury/fatality [2]. Therefore, modeling accident severity can be addressed as a pattern recognition problem [3], which can be solved by deep learning, statistical techniques and sometimes by physical modelling approaches [4–7]. In a deep learning model, an input vector is often mapped into an output vector through a set of nonlinear functions. In the case of accident severity, the input vectors are the characteristics of the accident, such as driver behavior and highway, vehicle and environment characteristics. The output vector is the corresponding classes of accident severity. Deep learning allows computational models to learn hierarchal representations of data with multiple levels of abstraction at different processing layers. The advantage of deep learning neural networks over statistical techniques is that they involve a more general mapping procedure i.e., a specific function is not required in model building [8]. However, these techniques can be treated as black box methods, if the network architecture is not carefully designed and its parameters are not optimized.

Several studies have investigated Neural Network (NN) and deep learning methods in transportation related applications [9–12]. For example, Abdelwahab and Abdel-Aty [8] used NN to predict driver injury severity from various accident factors (i.e., driver, vehicle, roadway and environment characteristics). In their study, the NN model developed performed better than the ordered probit model. In another paper, Delen et al. [10,13] applied NN to model injury severity of road accidents using 17 significant parameters. The NN model was used to predict the injury severity levels, resulting in a low overall accuracy of 40.71%. More recently, Hashmienejad and Hasheminejad [14] proposed a novel rule-based technique to predict traffic accident severity based on users' preferences instead of conventional data mining methods. The proposed method outperformed the NN and support vector machine methods. In a recent paper, Alkheder et al. [11] used NN methods to predict the injury severity (minor, moderate, severe, death) of traffic accidents in Abu Dhabi. Their analysis was based on 5973 traffic accident records that had occurred over a 6-years period. The overall accuracy of the model for the training and testing data were 81.6% and 74.6%, respectively. In addition, Zeng and Huang [15] proposed a training algorithm and network structure optimization method for crash injury severity prediction. Their results indicated that the proposed training algorithm performed better than the traditional back-propagation algorithm. The optimized NN, which contained less nodes than the fully connected NN, achieved reasonable prediction accuracy. Also, the fully connected and optimized NN models outperformed the ordered logit model. Also, their results showed that optimization of the NN structure could improve the overall performance of model prediction.

In recent years, deep learning has become a popular technique in image [16] and natural language processing applications [17]. Many researchers have applied deep learning based techniques in transportation related applications such as traffic flow [18] and accident hotspots prediction [19]. A detailed literature review showed that optimization of network structures is critical for crash severity prediction. Generally speaking, deep learning allows compositionality and the design of flexible network structures with multilayer modules, and therefore it is expected that the performance of deep learning models will be better than the traditional NN models. A deep learning architecture is a multilayer stack of simple modules, most of which are subjected to learning computing nonlinear input–output mappings [20]. Each module in the stack transforms its input to increase both selectivity and invariance of the representation. A Recurrent Neural Network (RNN) is a type of deep learning module, which is more appropriate for sequential data (e.g., traffic accident data with temporal correlations). It is effective for a wide range of applications including text generation [21] and speech recognition [22]. The application of RNNs in various fields motivated the authors to evaluate these models in deep learning frameworks for severity prediction of traffic accidents.

The main novelty of this work is the development of a RNN model that accurately predicts injury severity of traffic accidents utilizing the temporal structure of accident data. The specific objective is to design a deep learning model using a RNN for severity prediction of traffic accidents. The hyperparameters of the model will be selected through a systematic grid search technique. In addition, the sensitivity of model parameters and configurations will be assessed, and then will be compared with the well-known Multilayer Perceptron (MLP) and Bayesian Logistic Regression (BLR) models to understand its advatange. Finally, the impacts of accident-related factors on injury severity outcome will be determined using the profile-based method.

2. Recurrent Neural Networks (RNNs)

Recurrent Neural Networks (RNNs) are neural networks with feedback connections specifically designed to model sequences. They are computationally more powerful and biologically more reasonable than feed-forward networks (no internal states). The feedback connections provide a RNN the memory of past activations, which allows it to learn the temporal dynamics of sequential data. A RNN is powerful because it uses contextual information when mapping between input and output sequences. However, the traditional RNNs have a problem called vanishing or exploding gradient. To handle this problem, Hochreiter and Schmidhuber [23] proposed the Long Short-Term Memory (LSTM) algorithm.

In LSTM, the hidden units are replaced by memory blocks, which contain one or more self-connected memory cells and three multiplicative units (input, output, forget gates). These gates allow writing, reading, and resetting operations within a memory block, and they control the overall behavior internally. A representation of a single LSTM unit is shown in Figure 1. Let c_t be the sum of inputs at time step t, then LSTM updates for time step i at given inputs x_t, h_{t-1}, and c_{t-1} are [24]:

$$i_t = \sigma(W_{xi}.x_t + W_{hi}.h_{t-1} + W_{ci}.c_{t-1} + b_i) \tag{1}$$

$$f_t = \sigma\left(W_{xf}.x_t + W_{hf}.h_{t-1} + W_{cf}.c_{t-1} + b_f\right) \tag{2}$$

$$c_t = i_t.tanh(W_{xc}.x_t + W_{hc}.h_{t-1} + b_c) + f_t.c_{t-1} \tag{3}$$

$$o_t = \sigma(W_{xo}.x_t + W_{ho}.h_{t-1} + W_{co}.c_t + b_o) \tag{4}$$

$$h_t = o_t.tanh(c_t) \tag{5}$$

where σ is an element-wised non-linearity such as a sigmoid function, W is the weight matrix, x_t is the input at time step t, h_{t-1} is the hidden state vector of the previous time step and b_i denotes the input bias vector.

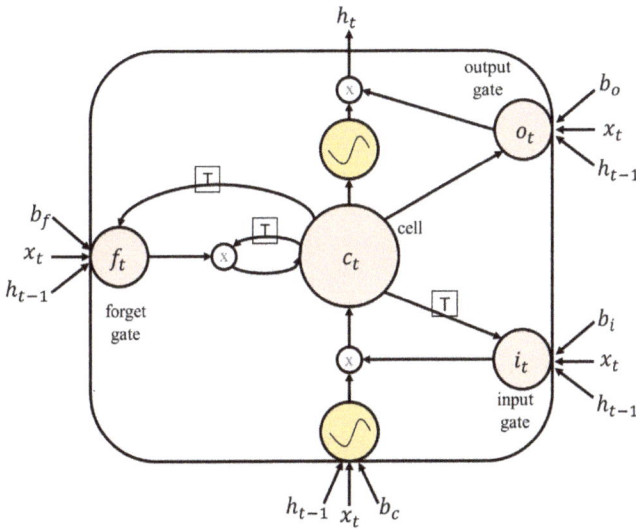

Figure 1. The structure of a memory cell in the Long Short-Term Memory–Recurrent Neural Network (LSTM–RNN).

3. The Proposed Network Framework

3.1. Network Architecture

Figure 2 shows the high-level architecture of the proposed severity prediction model based on deep learning. The advantage of LSTM-RNN over the traditional neural networks is that the traffic accidents are correlated to the historical and future incidents [25]. The RNN model comprises of five main layers which include input, LSTM, two dense layers and a Softmax layer.

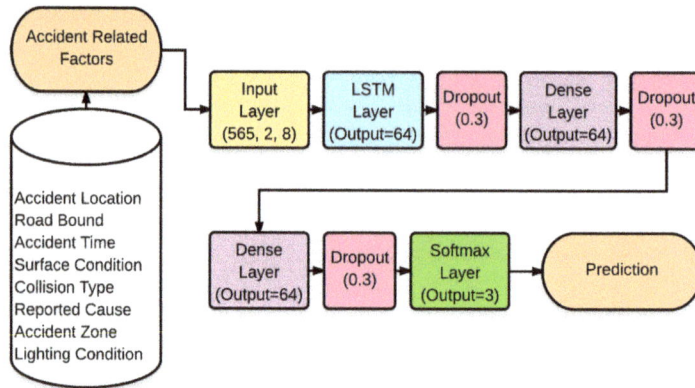

Figure 2. The high-level architecture of the proposed RNN model used in this study.

The main input of the network is a set of traffic accident-related factors, and the output is the corresponding accident severity classes (i.e., property damage only, possible/evident injury, disabling/fatality). The input dimension of the LSTM layer is equal to eight (number of input factors) with 64 nodes. The output of each node in the LSTM layer is results from the Rectified Linear Unit (ReLU) activation function applied to a weighted sum of both inputs from the previous layer and the previous outputs of the layer. The ReLU computes the function $f(x) = \max(0, x)$. This activation function accelerates (e.g., a factor of 6 in [26]) the convergence of Stochastic Gradient Descent (SGD) compared to the sigmoid/tanh functions. In addition, it can be implemented without expensive operations as in the case of sigmoid/tanh functions. Then, two fully connected layers were trained on top of the LSTM layers. They were added to match the output of the LSTM layer and the size of the accident severity classes. The output layer was a fully connected feed-forward layer and used to directly map the learned features to the three accident severity classes. A Softmax function is used to activate the output layer. In addition, three dropout layers with probability of 0.3 were used to reduce the complexity of the model to prevent over-fitting [27].

3.2. Training Methodology

The network was trained with Backpropagation Through Time (BPTT) [28] and Stochastic Gradient Descent (SGD) algorithms in Tensorflow on a personal CPU system (Core i7 with 16 GB RAM). As a RNN can be seen as a normal feedforward NN with shared weights, the BPTT begins by unfolding the RNN through several steps in time. The network training then proceeds in a manner similar to training a feed-forward neural network with backpropagation, except that the training patterns were visited in a sequential order. It determines the gradients of the objective function (Equation (6)) with respect to a parameter at each time step. Finally, the BPTT calculates the gradient outputs by taking the average of the individual step-dependent gradients.

$$H_{\hat{y}}(y) = -\sum_i y_l \log(y_l) \tag{6}$$

where y is the predicted probability distribution and $\hat{y}$ is the actual distribution (the one-hot vector with injury severity outcomes).

In addition, the SGD algorithm uses few examples from the input training vector and computes the outputs and the errors, and then adjusts the weights. The process is recurrent for many small sets of examples until the average of the objective function stops. The SGD method could determine a good set of weights when compared with other optimization techniques [29,30]. The training was run on a batch size 32 with 100 epochs. The best model was achieved using SGD with the decay of 0.9 and a momentum = 0.8. The best learning rate was 0.01. In addition, gradient clipping [31] with threshold 2.0 was useful to stabilize the training and to avoid gradient explosion.

3.3. Mitigating Overfitting

Networks with more complicated functions can perform better generalization performance as they learn different representations in each of their layers. However, complicated networks can easily over-fit the training data, producing poor generalization capacity and testing performance. Over-fitting occurs when a network model with high capacity fits the noise in the data instead of the underlying relationship.

Therefore, to avoid over-fitting in the proposed RNN model, three standard techniques were used. These were Gaussian noise injection into the training data [32], using a ReLU activation function in the hidden layers [20], and subsequently applying the dropout technique [29]. Dropout leads to big improvements in the prediction performance of the model. When a dropout technique is applied, the generalization capacity of the model is improved because the network is forced to learn multiple independent representations of the data. A probability of 30% was used in the dropout technique. This is because low probability has minimal effect and a high probability results in under-learning by the network.

3.4. Hyperparameter Tuning

The hyper-parameters of the RNN model were selected by performing a systematic grid search implemented in scikit-learn [33] using 100 epochs. Even though the systematic grid search requires high computational cost, better results could be obtained by systematically tuning the hyper-parameter values. Models with various combination of parameters were constructed, and a 3-fold cross-validation was used to evaluate each model. The parameters of the model with the highest validation accuracy were found to be the best parameters among the evaluated ones. Table 1 shows the optimized parameters used in the network.

Table 1. The optimized hyperparameters of the proposed RNN model.

Hyper-Parameter	Best Value	Description
Minibatch size	32	Number of training cases over which SGD update is computed.
Loss function	Categorical crossentropy	The objective function or optimization score function is also called as multiclass logloss which is appropriate for categorical targets.
Optimizer	SGD	Stochastic gradient descent optimizer.
Learning rate	0.01	The learning rate used by SGD optimizer
Gradient momentum	0.80	Gradient momentum used by SGD optimizer.
Weight decay	0.9	Learning rate decay over each update.

4. Experimental Results and Discussion

The proposed RNN model was implemented in Python using the open source TensorFlow deep learning framework developed by Google [34]. TensorFlow has automatic differentiation and parameter sharing capabilities, which allows a wide range of architectures to be easily defined and executed [34]. The proposed network was trained with 791 samples and validated with 339 samples using the Tensorflow framework. The SGD optimization algorithm was applied, with a batch size of 32 and a learning rate of 0.01.

4.1. Data

The traffic accident data for the period 2009–2015 from the North-South Expressway (NSE, Petaling Jaya, Malaysia), Malaysia were used in this study. The NSE is the longest expressway

(772 km) operated by Projek Lebuhraya Usaha Sama (PLUS) Berhad (the largest expressway operator in Malaysia, Petaling Jaya, Malaysia) and links many major cities and towns in Peninsular Malaysia. The data were obtained from the PLUS accident databases. The files used in this study were accident frequency and accident severity files in the form of an Excel spreadsheet. The accident frequency file contains the positional and descriptive accident location and the number of accidents in each road segment of 100 m. The accident records were separated according to the road bound (south, north). In contrast, the accident severity file contains the general accident characteristics such as accident time, road surface and lighting conditions, collision type, and the reported accident cause. To link the two files, the unique identity field (accident number) was used.

According to the vehicle type information in the accident data, three scenarios were found: (1) single-vehicle with object accidents, (2) two-vehicle accidents, (3) and multiple vehicle accidents (mostly three vehicles). Training a deep learning model requires many samples to capture the data structure and to avoid model overfitting. Therefore, the analysis in this study did not focus just on a single scenario, but instead, all scenarios were included in training the proposed RNN model. In total, 1130 accident records were reported during 2009–2015. Of these, 740 (approximately 65.4%) resulted in drivers damaging property only. On the other hand, 172 (15.2%) drivers were involved in possible/evident injury, and 218 (19.4%) in disabling injury.

The section of the NSE used in this study has a length of 15 km running from Ayer Keroh (210 km) to Pedas Linggi (225 km) (Figure 3). The accident severity data showed that the last section (220–225) of the NSE experienced several accidents resulting in serious injury (82) than the other sections (Table 2). Most accidents have occurred on the main route and southbound of the expressway. During the accident events, the actual accident causes were documented. The data showed that lost control, brake failure, and obstacles were the main accident causes on the NSE. With respect to lighting and surface conditions, most accidents occurred in daylight conditions and with a dry road surface. The main collision types in the accident records were out of control and rear collision. In addition, the accident time factor showed that 91.68% of the accidents occurred during the daytime. Additionally, the data also revealed that two-car accidents, single heavy car with an object and motorcycle with an object were the most recorded crashes on the NSE.

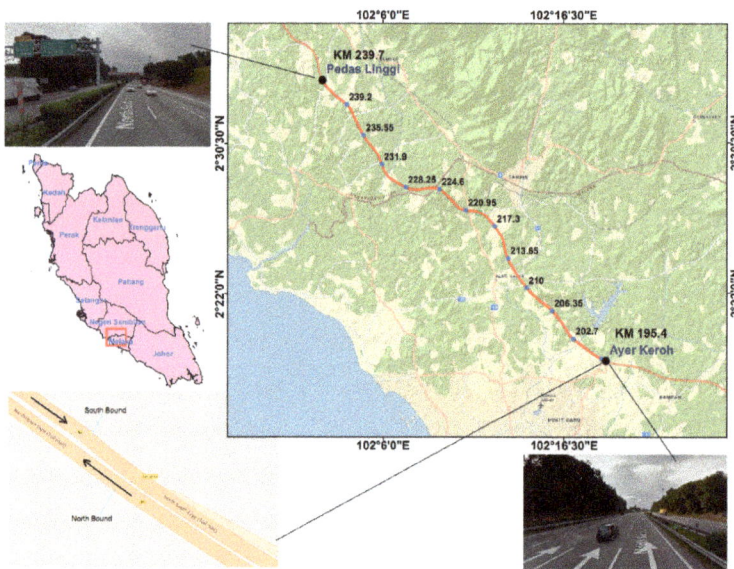

Figure 3. Location of the North-South Expressway (NSE) section analyzed in this study.

Table 2. Driver injury severity distribution according to accident related factors.

Factor	Property Damage Only	Evident Injury	Disabling Injury	Total
Location				
210–214	185	172	58	415
215–219	234	47	56	337
220–225	238	58	82	378
Road-bound				
South	453	99	139	691
North	287	73	79	439
Accident zone				
Interchange	14	3	0	17
Junction				
Lay-by	2	0	1	3
Main Route	666	155	209	1030
North Bound Entry Ramp	8	2	0	10
North Bound Exit Ramp	4	2	0	6
Rest and Service Area	21	4	2	27
South Bound Entry Ramp	2	0	1	3
South Bound Exit Ramp	7	1	3	11
Toll Plaza	16	5	2	23
Accident reported cause				
Bad Pavement Condition	0	1	0	1
Brake Failure	6	2	1	9
Bump-bump	37	12	27	76
Dangerous Pedestrian Behaviour	0	0	1	1
Drunk	0	0	1	1
Loss of Wheel	1	0	2	3
Lost control	75	18	22	115
Mechanical	5	1	0	6
Mechanical/Electrical Failure	11	0	1	12
Obstacle	43	12	6	61
Other Bad Driving	15	1	4	20
Other Human Factor/Over Load/Over Height	3	0	0	3
Over speeding	345	61	91	497
Parked Vehicle	4	4	10	18
Skidding	1	0	0	1
Sleepy Driver	134	44	42	220
Stray Animal	13	1	2	16
Tire burst	47	15	8	70
Lighting condition				
Dark with Street Light	47	6	8	61
Dark without Street Light	225	74	89	388
Dawn/Dusk	35	9	9	53
Day Light	433	83	112	628
Surface condition				
Dry	460	146	190	796
Wet	280	26	28	334
Collision type				
Angular Collision	9	2	0	11
Broken Windscreen	2	0	0	2
Cross direction	2	0	1	3
Head-on Collision	0	1	4	5
Hitting Animal	12	1	2	15
Hitting Object On Road	44	12	7	63
Others	20	0	6	26
Out of Control	457	92	107	656
Overturned	33	11	7	51
Rear Collision	137	48	81	266
Right Angle Side Collision	11	1	1	13
Side Swipe	13	4	2	19
Accident time				
Day time	677	156	203	1036
Night time	63	16	15	94

Table 2. *Cont.*

Factor	Property Damage Only	Evident Injury	Disabling Injury	Total
Vehicle type				
Car-Bus	7	3	6	16
Car-Car	499	68	60	627
Car-Heavy Car	51	11	14	76
Car-Motorcycle	4	7	22	33
Heavy Car	131	23	25	179
Heavy Car-Bus	2	3	3	8
Heavy Car-Heavy Car	24	9	15	48
Heavy Car-Motorcycle	0	1	6	7
Heavy Car-Taxi	2	0	0	2
Motorcycle	11	42	60	113
Motorcycle-Taxi	0	1	1	2
Motorcycle-Van	0	0	2	2
Taxi	1	0	1	2
Van	8	4	3	15

Before feeding the accident data into the recurrent neural network, the data must be preprocessed. The steps included were: removing missing data, data transformation, and detection of highly correlated factors. Some data were missing in the accident records; therefore, the complete raw data in which a missing data is found was removed. The data transformation included one-hot encoding for the categorical factors. In addition, correlation between predictors was assessed to detect any multicollinearity problem. First, the multiple R^2 was calculated for each factor. Second, the Variance Inflation Factor (*VIF*) was calculated from the multiple R^2 for each factor (Table 3). The highest correlation of 0.27 was found between lighting condition factor and surface condition factor. However, the highest multiple R^2 and *VIF* were found to be 0.135 and 1.156 for surface condition factor. There was no multicollinearity found among the factors if *VIF* = 1.0, however when the value exceeded 1.0, then moderate multicollinearity was found. In both cases, no high correlation was found during the model training and testing phase. Therefore, none of the factors was removed.

Table 3. Multicollinearity assessment among the accident related factors.

Factor	Multiple R^2	VIF
Accident location	0.062	1.066
Road bound	0.012	1.012
Accident zone	0.040	1.042
Accident reported cause	0.060	1.063
Lighting condition	0.095	1.105
Surface condition	0.135	1.156
Collision type	0.033	1.034
Accident time	0.009	1.009
Vehicle type	0.090	1.099

4.2. Results of RNN Model Performance

Figure 4 shows the accuracy performance and loss of the RNN model calculated for 100 epochs (iterations) using the training (80%) and validation (20%) datasets. In general, model accuracy on the training and validation datasets increases after each iteration with fluctuations. The fluctuations in accuracy are due to the use of the dropout technique in the model, which results in different training during every iteration. The dropout technique, which was used to prevent over-fitting, introduces some of the randomnesses to the network. In the first iteration, the accuracy was 61.28% and 66.37% for training and testing data, respectively. As the model trains during the first pass through the data, both training and validation losses decline, indicating that the model is learning the structure of the traffic accident data and possibly its temporal correlations. In the first and consecutive iterations the validation loss did not increase significantly and was always less than the training loss, indicating that the network did not overfit the training data and accurately generalized to the unseen validation data. After 100 epochs, the validation accuracy of the model was 71.77%.

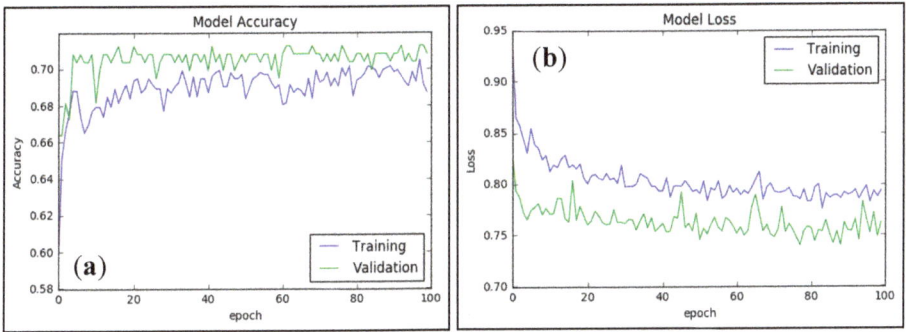

Figure 4. Accuracy performance and loss of the RNN model calculated for 100 epochs; (**a**) model accuracy; and (**b**) model loss.

4.3. Sensitivity Analysis of Optimization Algorithm

Several optimization algorithms such as SGD, Adagrad [35], RMSprop [36] and Adam [37] are available to adjust the weights and biases of a network by iterating through the training data. Each algorithm has its own advantages and disadvantages, and there are no clear guidelines for selecting an optimizer for a particular problem. Therefore, in this study, several optimization algorithms were evaluated as we searched for the best to train the proposed RNN model. The evaluated algorithms are SGD, RMSprop, Adagrad, Adadelta, Adam, Adamax, and Nadam [38]. The parameters of each algorithm were obtained from the aforementioned literature. Table 4 shows the performance of each optimization technique in predicting the severity of traffic accidents. The best validation accuracy (71.77) was achieved by the SGD method with a learning rate of 0.01. The SGD uses a small batch of the total data set to calculate the gradient of the loss function with respect to the weights and biases, using a backpropagation algorithm. The algorithm moves down the gradient by an amount proportional to its learning rate. Optimization methods such as Adam and Nadam have performed reasonably well.

Table 4. The performance of different optimization methods evaluated in this study.

Optimizer	Parameters	Training Accuracy (%)	Validation Accuracy (%)
SGD	lr = 0.01, momentum = 0.0, decay = 0.0, nesterov = False	71.79	71.77
RMSprop	lr = 0.001, rho = 0.9, epsilon = 1×10^{-8}, decay = 0.0	71.90	70.80
Adagrad	lr = 0.01, epsilon = 1×10^{-8}, decay = 0.0	71.35	71.24
Adadelta	lr = 1.0, rho = 0.95, epsilon = 1×10^{-8}, decay = 0.0	70.24	71.24
Adam	lr = 0.001, beta_1 = 0.9, beta_2 = 0.999, epsilon = 1×10^{-8}, decay = 0.0	73.12	71.68
Adamax	Lr = 0.002, beta_1 = 0.9, beta_2 = 0.999, epsilon = 1×10^{-8}, decay = 0.0	70.46	71.24
Nadam	Lr = 0.002, beta_1 = 0.9, beta_2 = 0.999, epsilon = 1×10^{-8}, schedule_decay = 0.004	74.67	71.68

4.4. Sensitivity Analysis of Learning Rate and RNN Sequence Length

Furthermore, the effect of the learning rate and RNN sequence length on the generalization ability of the proposed network in predicting the severity of traffic accidents was investigated. The learning rates varied by 0.5, 0.1, 0.05, 0.01, and 0.001. The sequence length varied by 2, 5, and 10. The choices of learning rate and sequence length were purely on an arbitrary basis. Figure 5 shows the results of these experiments. The highest validation accuracy was achieved when a learning rate of 0.01 was used. Reducing the learning rate up to 0.001 did not improve the validation accuracy, but reduced it to 70.8%. Similarly, large learning rates (e.g., 0.5) significantly reduced the performance of the network with a validation accuracy of 66.37%. On the other hand, the best validation accuracy (71.77%) was achieved with a sequence length of 2. In addition, the model performed better with a sequence length of 10 (70.87%) than 5 (70.43%). The unseen structured pattern in the model performance with different learning rates and sequence lengths indicated that the optimization of these parameters is critical to design a network that can best predict the severity of traffic accidents.

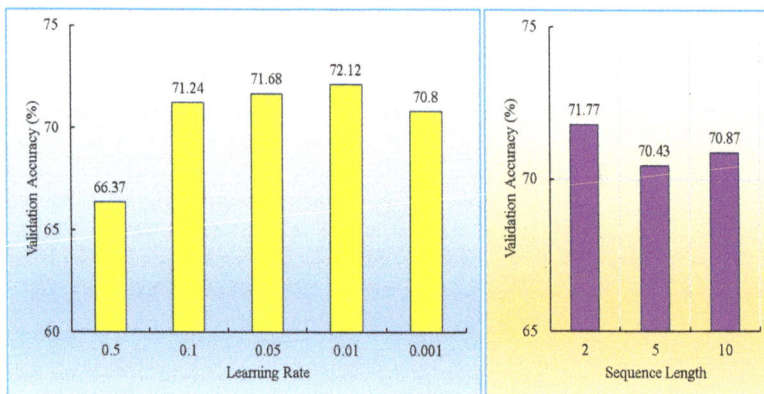

Figure 5. The sensitivity of the RNN model for different learning rate and sequence length configurations.

4.5. Network Depth Analysis

Generally, in a deep learning model, several modules and multilayers can be stacked on top of each other so that it is significant to analyze the network depth (number of hidden layers) to understand the network behavior. Table 5 shows the accuracy of the RNN model with various numbers of dense layers used on top of the LSTM layer. The best accuracy was achieved by using two dense (fully-connected) layers with 64 hidden units. This model achieved training and validation accuracies of 71.79% and 71.77% respectively. When another dense layer was added to the network, the training and validation accuracies were reduced to 70.09% and 71.24 respectively. When using five dense layers, the validation accuracy (70.35%) was found to be better than the training accuracy (67.26%); however, it could not perform better than using only two dense layers. The model started over-fitting when more than five dense layers were used. The validation accuracy of the network with eight dense layers was 56.37%, i.e., less by almost 10% than the training accuracy.

Table 5. The training and validation accuracy of the proposed RNN model with a different number of dense layers.

Number of Dense Layers	Training Accuracy (%)	Validation Accuracy (%)
2	71.79	71.77
3	70.09	71.24
5	67.26	70.35
8	65.27	56.37

In addition, the effect of the number of LSTM layers on model accuracy was assessed. Table 6 shows that the best training (71.79%) and validation (71.77%) accuracy could be achieved with a network of one LSTM layer. The accuracy was slightly reduced with another LSTM layer which was added to the network. Finally, when three LSTM layers were used, the accuracy of the model was gradually decreased and overfitted the training data.

Table 6. The training and validation accuracy of the proposed RNN model with a different number of LSTM layers.

Number of LSTM Layers	Training Accuracy (%)	Validation Accuracy (%)
1	71.79	71.77
2	71.30	71.58
3	67.11	65.34

4.6. Extraction Factor Contribution in the RNN Model

To compute the contribution of each factor in predicting the severity of traffic accidents, the profile method [39,40] was used. This technique examines the evolution of each factor with a scale of values, while the remaining factors keep their values fixed. Each factor x_i takes 11 values resulting from the division of the range, between its minimum and maximum value, into 10 equal intervals. All factors except one were initially fixed at their minimum value and then were fixed successively at their first quartile, median, third quartile and maximum values. Five values of the response variable were obtained for each 11 value and adopted by x_i, and the median of those five values was calculated. Finally, a curve with the profile of variation was obtained for every factor.

Table 7 shows the calculated weight of each factor by the profile method implemented in Python. The results indicate that the road bound and accident time have a significant effect on injury severity; the drivers had a greater risk of injuries during daytime along the southbound lanes of the NSE. Also, dry surface conditions were found to be more dangerous than a wet surface, as far as driver injury severity is concerned. During rainy times (surface is wet), drivers decrease their speed due to traffic jams, hence decreasing the severity level of driver injury. Lighting conditions, such as dark with and without street lights, increase the potential of possible injury and fatality, whereas daylight reduces the severity level of driver injury. The negative weight of the accident location factor indicates that severe accidents are most likely to happen in the section 220–225 km of the expressway. When vehicle types such as cars and motorcycles are involved in crashes, drivers are more prone to possible injury and fatality than for heavy vehicles and buses. Fatigue and speeding can give the driver a greater chance of experiencing a severe injury than other causes reported in the dataset. Accidents at the entry and exit ramps, toll plaza, and main route are more dangerous than accidents in other zones. Collision type also plays a significant role in increasing or decreasing the severity level of an accident. The analysis in this paper shows that collisions such as out of control and rear collisions increase the injury severity, whereas right angle side collision and sideswipe decrease the injury severity on the NSE.

Table 7. The calculated weights of accident related factors.

Factor	Weight
Accident location	−0.0074
Road bound	0.2899
Accident zone	0.0779
Accident reported cause	0.0469
Lighting condition	−0.0892
Surface condition	0.1785
Collision type	−0.0603
Accident time	0.3612
Vehicle type	−0.0468

4.7. Comparative Experiment

In this experiment, the proposed RNN model was compared with the traditional MLP model and the BLR approach. The advantages of MLP networks over Radial Basis Function (RBF) networks are that MLP networks are compact, less sensitive to including unnecessary inputs, and more effective for modeling data with categorical inputs.

The Grid Search algorithm was used to search the space of MLP network architectures and configurations. The number of units in the hidden layer, activation functions and the learning rate were optimized. The search space of the number of hidden units was 4 to 64 units. Five activation functions were tested such as Gaussian, Identity, Sigmoid, Softmax, and Exponential for the hidden and output layers. In addition, five values of learning rate (0.5, 0.1, 0.05, 0.01, and 0.001) were evaluated. The suboptimal network was then determined according to the best validation accuracy. The suboptimal network is the one that best represents the relationship between the input and output variables. In total, 100 network architectures with different combinations of selected parameters and network configurations were executed, and the validation accuracy on 20% of the data was computed for each created model. Then, the best network was retained. The best network had eight (8) hidden units with identity activation function, a Softmax activation in the output layer, and a learning rate of 0.01. The training and validation performance of the network was 68.90% and 65.48% respectively.

On the other hand, the BLR [41] and the computation of Markov chain Monte Carlo (MCMC) simulations were implemented in OpenBUGS software. BUGS modeling language (Bayesian Inference using Gibbs Sampling) is an effective and simplified platform to allow the computation using MCMC algorithms for all sorts of Bayesian models including BLR applied. The simulation of the posterior distribution of beta (β) allowed estimating the mean, standard deviation, and quartiles of the parameters of each explanatory variable. In the simulation stage, two MCMC chains were used to ensure convergence. The initial 100,000 iterations were discarded as burn-ins to achieve convergence and a further 20,000 iterations for each chain were performed and kept to calculate the posterior estimates of interested parameters.

In developing the Bayesian model, monitoring convergence is important because it ensures that the posterior distribution was achieved at the beginning of sampling of parameters. Convergence of multiple chains is assessed using the Brooks-Gelman-Rubin (BGR) statistic [42]. A value of less than 1.2 BGR statistic indicates convergence [42]. Convergence is also assessed by visual inspection of the MCMC trace plots for the model parameters and by monitoring the ratios of the Monte Carlo errors with respect to the corresponding standard deviation. The estimates of these ratios should be less than 0.05. In addition, MC error less than 0.05 also indicates that convergence may have been achieved [43]. The results of BLR showed training and validation accuracies of 70.30% and 58.30% respectively. In addition, the model had an average MC error of 0.047, Brooks-Gelman-Rubin (BGR) of 0.98, and model fitting performance (Deviance Information Criterion-DIC) of 322.

The comparative analysis showed that the proposed RNN model outperformed both MLP and BLR in terms of training and validation accuracies (Table 8). The BLR model performed better than MLP on the training dataset; however, its accuracy on the validation dataset was less than the MLP model. The MLP model uses only local contexts and therefore does not capture the spatial and temporal correlations in the dataset. The hidden units of the RNN model contain historical information from previous states, hence increasing the information about the data structure, which may be the main reason for its high validation accuracy over the other two methods. Building BLR models is more difficult than NN models, because they require expert domain knowledge and effective feature engineering processes. The NN model automatically captures the underlying structure of the dataset and extracts different levels of features abstractions. However, building deep learning models with various modules (e.g., fully connected networks, LSTM, convolutional layers) can be a challenging task. In addition, the BLR models are less prone to over-fitting of the training dataset than NN models, because they involve simpler relationships between the outcome and the explanatory variables. Complex networks with more hidden units and many modules often tend to over-fit more, because they detect almost any possible interaction so that the model becomes too specific to the training dataset. Therefore, optimization of network structures is very critical to avoid over-fitting and build practical prediction models. In computing the importance of an explanatory factor, the BLR models could easily calculate the factor importance and the confidence intervals of the predicted probabilities. In contrast, NN methods, which are not built primarily for statistical use, cannot easily calculate the factor importance or generate confidence intervals of the predicted probabilities unless extensive computations were done.

Table 8. Performance comparison of the proposed RNN model with Multilayer Perceptron (MLP) and Bayesian Logistic Regression (BLR) models.

Method	Training Accuracy (%)	Validation Accuracy (%)
MLP	68.90	65.48
BLR	70.30	58.30
RNN	71.79	71.77

In addition, based on the estimated weight of each accident related factor, the factors could be ranked by scaling the weights into the range of 1–100 and then giving them a rank from 1 to 9 as shown in Table 9. The three methods (MLP, BLR, and RNN) did not agree on the factors' ranking. For example, the RNN model ranked accident time as the most influential factor in injury severity, whereas accident time was ranked 2 and 7 by the BLR and MLP models, respectively. Both the RNN and BLR models agreed on several factors' ranking i.e., accident location (6), accident reported cause (5), and collision type (8). The correlation (R^2) between the RNN and BLR ranks was 0.72. In contrast, RNN and MLP did not agree on the ranking of the factors and their ranking correlation was the lowest (0.38).

Table 9. Calculated ranks of the accident related factors in the RNN, MLP, and BLR models.

Factor	MLP	BLR	RNN
Accident location	2	6	6
Road bound	8	4	2
Accident zone	5	3	4
Accident reported cause	3	5	5
Lighting condition	4	7	9
Surface condition	9	1	3
Collision type	6	8	8
Accident time	7	2	1
Vehicle type	1	9	7

BLR:MLP: $R^2 = 0.51$
RNN:MLP: $R^2 = 0.38$
MLP:BLR: $R^2 = 0.51$
RNN:BLR: $R^2 = 0.72$

4.8. Computational Complexity of the Model

The time complexity of the RNN model was measured in terms of training and testing time per iteration. Table 10 gives information about the average training time per iteration with batch size of 32 and the average testing time per prediction. It can be seen that the model on average spends around 150 milliseconds per iteration during training and only ~13 milliseconds per prediction for new unseen examples. Although this experiment shows the computational efficiency of the model, it is also worth to note that the training time can be increased by reducing the batch size or sequence length, and also by increasing the volume of the training data.

Table 10. Average training and testing time per iteration/ prediction of the proposed model.

Time (milliseconds per iteration)	RNN Model
Training Time	150.23
Testing Time	13.17

Additionally, the major computational problem of NNs is their scalability; sometimes they become unstable when applied to larger problems. However, recent advancements in hardware and computing performance such as Graphics Processing Units (GPUs), parallel computing, and cloud computing have decreased most of the limitations of NN models including RNN-based computations.

4.9. Applicability and Limitations of the Proposed Method

The proposed RNN model is a promising traffic accident forecasting tool that has several applications in practice. First, identifying the most risky road sections (i.e., site ranking) is a daily practice of transportation agencies in many cities around the world. Accident prediction models, such as the one we proposed in this research, are often an effective solution for identifying risky road sections and helping to conduct investigations into methods for improving the safety performance of the road systems. Second, the RNN model is able to explain the underlying relationships between several accidents' related factors, such as accident time and collision type, and the injury severity outcomes. Information about the effects of accident factors on the injury severity outcomes provides huge amount of information to the transportation agencies and stakeholders. Finally, estimating the expected number of traffic accidents in a road section can help road designers to optimize the geometric alignments of the roads based on the accident scenarios.

However, the proposed RNN model has some constraints and limitations. The major limitation of the model is that the input factors are prerequisite and if any of them is missing, the output probabilities cannot be accurately estimated. Another constraint of the model is the sequence length of the RNN model, which mainly depends on the number of accident records in the training dataset. To handle this limitation, the future works should develop RNN models that operate on input sequences of variable lengths via Tensorflow dynamic calculation.

5. Conclusions

In this paper, a Recurrent Neural Network (RNN) model was developed to predict the injury severity of traffic accidents on the NSE, Malaysia. An optimized network architecture was determined through a systematic grid search for the suboptimal network hyper-parameters. Several hyper-parameters of the RNN model were critical to achieve the highest validation accuracy. The best optimization algorithm was determined to be the SGD with learning rate, momentum, and weight decay of 0.01, 0.80, and 0.9, respectively. In addition, the dropout technique helped us to reduce the complexity of the model and also the chance of over-fitting the training dataset. The RNN model achieved the best validation accuracy of 71.77% when compared to the MLP and BLR models. This indicates that additional information about temporal and contextual correlations among accident records could help the RNN model to perform better than the other models. In the sensitivity analysis of RNN sequence length, the best accuracy was achieved with a sequence length of 2, whereas the accuracy was decreased when a sequence length of 10 or 5 was used. This means that the accident data has temporal and contextual structures that could not be used by the MLP and BLR models. The impact of each factor in the RNN model was calculated using the profile method. The results showed that the most influential factors in predicting injury severity of traffic accidents are accident time and road bound. In the study area, the model predicted that the southbound section is more dangerous than the northbound, with respct to injury severity. The model also predicted drivers had a greater risk of injury on a dry surface and with lighting conditions such as dark with and without street lights.

While the proposed RNN model outperformed the MLP and BLR models and could estimate the factors' impacts, further studies should focus on developing more flexible and optimized network structures to predict accident frequency and injury severity. Future theoretical studies are encouraged to focus on improving the ability of the RNN model to represent variables and data structures, and to store data over long timescales. In addition, more applied research is recommended to develop new techniques to make use of additional information in accident datasets, such as contextual structures, spatial and temporal interactions and underlying relationships between accident factors and injury severity outcomes.

Author Contributions: Maher Ibrahim Sameen and Biswajeet Pradhan conceived and designed the experiments; Maher Ibrahim Sameen performed the experiments and analyzed the data; Biswajeet Pradhan and Maher Ibrahim Sameen contributed reagents/materials/analysis tools; Maher Ibrahim Sameen and Biswajeet Pradhan wrote the paper.

Conflicts of Interest: The authors declare no conflict of interest.

References

1. Sameen, M.I.; Pradhan, B. Assessment of the effects of expressway geometric design features on the frequency of accident crash rates using high-resolution laser scanning data and gis. *Geomat. Nat. Hazards Risk* **2016**, 1–15. [CrossRef]
2. Pei, X.; Wong, S.; Sze, N.-N. A joint-probability approach to crash prediction models. *Accid. Anal. Prev.* **2011**, *43*, 1160–1166. [CrossRef] [PubMed]
3. Fogue, M.; Garrido, P.; Martinez, F.J.; Cano, J.-C.; Calafate, C.T.; Manzoni, P. A system for automatic notification and severity estimation of automotive accidents. *IEEE Trans. Mob. Comput.* **2014**, *13*, 948–963. [CrossRef]
4. Pawlus, W.; Reza, H.; Robbersmyr, K.G. Application of viscoelastic hybrid models to vehicle crash simulation. *Int. J. Crashworthiness* **2011**, *16*, 195–205. [CrossRef]
5. Pawlus, W.; Robbersmyr, K.G.; Karimi, H.R. Mathematical modeling and parameters estimation of a car crash using data-based regressive model approach. *Appl. Math. Model.* **2011**, *35*, 5091–5107. [CrossRef]
6. Pawlus, W.; Karimi, H.R.; Robbersmyr, K.G. Mathematical modeling of a vehicle crash test based on elasto-plastic unloading scenarios of spring-mass models. *Int. J. Adv. Manuf. Technol.* **2011**, *55*, 369–378. [CrossRef]
7. Karimi, H.R.; Pawlus, W.; Robbersmyr, K.G. Signal reconstruction, modeling and simulation of a vehicle full-scale crash test based on morlet wavelets. *Neurocomputing* **2012**, *93*, 88–99. [CrossRef]
8. Abdelwahab, H.; Abdel-Aty, M. Development of artificial neural network models to predict driver injury severity in traffic accidents at signalized intersections. *Transp. Res. Rec. J. Transp. Res. Board* **2001**, *1746*, 6–13. [CrossRef]
9. Lv, Y.; Duan, Y.; Kang, W.; Li, Z.; Wang, F.-Y. Traffic flow prediction with big data: A deep learning approach. *IEEE Trans. Intell. Transp. Syst.* **2015**, *16*, 865–873. [CrossRef]
10. Duan, Y.; Lv, Y.; Liu, Y.-L.; Wang, F.-Y. An efficient realization of deep learning for traffic data imputation. *Transp. Res. C* **2016**, *72*, 168–181. [CrossRef]
11. Alkheder, S.; Taamneh, M.; Taamneh, S. Severity prediction of traffic accident using an artificial neural network. *J. Forecast.* **2017**, *36*, 100–108. [CrossRef]
12. Yu, R.; Li, Y.; Shahabi, C.; Demiryurek, U.; Liu, Y. Deep learning: A generic approach for extreme condition traffic forecasting. In Proceedings of the 2017 SIAM International Conference on Data Mining (SDM), Houston, TE, USA, 27–29 April 2017.
13. Delen, D.; Sharda, R.; Bessonov, M. Identifying significant predictors of injury severity in traffic accidents using a series of artificial neural networks. *Accid. Anal. Prev.* **2006**, *38*, 434–444. [CrossRef] [PubMed]
14. Hashmienejad, S.H.-A.; Hasheminejad, S.M.H. Traffic accident severity prediction using a novel multi-objective genetic algorithm. *Int. J. Crashworthiness* **2017**, 1–16. [CrossRef]
15. Zeng, Q.; Huang, H. A stable and optimized neural network model for crash injury severity prediction. *Accid. Anal. Prev.* **2014**, *73*, 351–358. [CrossRef] [PubMed]
16. Dean, J.; Corrado, G.; Monga, R.; Chen, K.; Devin, M.; Mao, M.; Senior, A.; Tucker, P.; Yang, K.; Le, Q.V. Advances in neural information processing systems. In *Large Scale Distributed Deep Networks*; Association for Computing Machinery: New York, NY, USA, 2012; pp. 1223–1231.
17. Collobert, R.; Weston, J. A unified architecture for natural language processing: Deep neural networks with multitask learning. In Proceedings of the 25th International Conference on Machine learning, New York, NY, USA, 5–9 July 2008; pp. 160–167.
18. Huang, W.; Song, G.; Hong, H.; Xie, K. Deep architecture for traffic flow prediction: Deep belief networks with multitask learning. *IEEE Trans. Intell. Transp. Syst.* **2014**, *15*, 2191–2201. [CrossRef]

19. Xie, Z.; Yan, J. Detecting traffic accident clusters with network kernel density estimation and local spatial statistics: An integrated approach. *J. Transp. Geogr.* **2013**, *31*, 64–71. [CrossRef]
20. LeCun, Y.; Bengio, Y.; Hinton, G. Deep learning. *Nature* **2015**, *521*, 436–444. [CrossRef] [PubMed]
21. Sutskever, I.; Martens, J.; Hinton, G.E. Generating text with recurrent neural networks. In Proceedings of the 28th International Conference on Machine Learning (ICML-11), Bellevue, WA, USA, 28 June–2 July 2011; pp. 1017–1024.
22. Graves, A.; Mohamed, A.-r.; Hinton, G. Speech recognition with deep recurrent neural networks. In Proceedings of the 2013 IEEE International Conference on Acoustics, Speech and Signal Processing, Vancouver, BC, Canada, 26–31 May 2013; pp. 6645–6649.
23. Hochreiter, S.; Schmidhuber, J. Long short-term memory. *Neural Comput.* **1997**, *9*, 1735–1780. [CrossRef] [PubMed]
24. Donahue, J.; Anne Hendricks, L.; Guadarrama, S.; Rohrbach, M.; Venugopalan, S.; Saenko, K.; Darrell, T. Long-Term Recurrent Convolutional Networks for Visual Recognition and Description. In Proceedings of the IEEE Conference on Computer Vision and Pattern Recognition, Boston, MA, USA, 7–15 June 2015; pp. 2625–2634.
25. Wang, X.; Abdel-Aty, M. Temporal and spatial analyses of rear-end crashes at signalized intersections. *Accid. Anal. Prev.* **2006**, *38*, 1137–1150. [CrossRef] [PubMed]
26. Krizhevsky, A.; Sutskever, I.; Hinton, G.E. Imagenet classification with deep convolutional neural networks. In Proceedings of the Advances in Neural Information Processing Systems, Lake Tahoe, NV, USA, 3–6 December 2012; pp. 1097–1105.
27. Srivastava, N.; Hinton, G.E.; Krizhevsky, A.; Sutskever, I.; Salakhutdinov, R. Dropout: A simple way to prevent neural networks from overfitting. *J. Mach. Learn. Res.* **2014**, *15*, 1929–1958.
28. Hinton, G.; Rumelhart, D.; Williams, R. Learning internal representations by back-propagating errors. In *Parallel Distributed Processing: Explorations in the Microstructure of Cognition*; MIT Press: London, UK, 1985; Volume 1.
29. Hinton, G.E.; Srivastava, N.; Krizhevsky, A.; Sutskever, I.; Salakhutdinov, R.R. Improving neural networks by preventing co-adaptation of feature detectors. *arXiv* **2012**, arXiv:1207.0580.
30. Bousquet, O.; Bottou, L. The tradeoffs of large scale learning. In Proceedings of the Advances in Neural Information Processing Systems, Vancouver, BC, Canada; 2008; pp. 161–168.
31. Pascanu, R.; Mikolov, T.; Bengio, Y. On the difficulty of training recurrent neural networks. In Proceedings of the 30th International Conference on International Conference on Machine Learning, Atalnta, GA, USA, 16–21 June 2013; pp. 1310–1318.
32. Zur, R.M.; Jiang, Y.; Pesce, L.L.; Drukker, K. Noise injection for training artificial neural networks: A comparison with weight decay and early stopping. *Med. Phys.* **2009**, *36*, 4810–4818. [CrossRef] [PubMed]
33. Pedregosa, F.; Varoquaux, G.; Gramfort, A.; Michel, V.; Thirion, B.; Grisel, O.; Blondel, M.; Prettenhofer, P.; Weiss, R.; Dubourg, V. Scikit-learn: Machine learning in python. *J. Mach. Learn. Res.* **2011**, *12*, 2825–2830.
34. Abadi, M.; Agarwal, A.; Barham, P.; Brevdo, E.; Chen, Z.; Citro, C.; Corrado, G.S.; Davis, A.; Dean, J.; Devin, M. Tensorflow: Large-scale machine learning on heterogeneous distributed systems. *arXiv* **2016**, arXiv:1603.04467.
35. Duchi, J.; Hazan, E.; Singer, Y. Adaptive subgradient methods for online learning and stochastic optimization. *J. Mach. Learn. Res.* **2011**, *12*, 2121–2159.
36. Tieleman, T.; Hinton, G. Lecture 6.5-rmsprop: Divide the gradient by a running average of its recent magnitude. *COURSERA Neural Netw. Mach. Learn.* **2012**, *4*, 26–31.
37. Kingma, D.; Ba, J. Adam: A method for stochastic optimization. *arXiv* **2014**, arXiv:1412.6980.
38. Ruder, S. An overview of gradient descent optimization algorithms. *arXiv* **2016**, arXiv:1609.04747.
39. Lek, S.; Belaud, A.; Dimopoulos, I.; Lauga, J.; Moreau, J. Improved estimation, using neural networks, of the food consumption of fish populations. *Oceanogr. Lit. Rev.* **1996**, *9*, 929. [CrossRef]
40. Lek, S.; Belaud, A.; Dimopoulos, I.; Lauga, J.; Moreau, J. Improved estimation, using neural networks, of the food consumption of fish populations. *Mar. Freshw. Res.* **1995**, *46*, 1229–1236. [CrossRef]
41. Xu, C.; Wang, W.; Liu, P.; Guo, R.; Li, Z. Using the bayesian updating approach to improve the spatial and temporal transferability of real-time crash risk prediction models. *Transp. Res. C* **2014**, *38*, 167–176. [CrossRef]

42. El-Basyouny, K.; Sayed, T. Application of generalized link functions in developing accident prediction models. *Saf. Sci.* **2010**, *48*, 410–416. [CrossRef]

43. Wang, C.; Quddus, M.A.; Ison, S.G. Predicting accident frequency at their severity levels and its application in site ranking using a two-stage mixed multivariate model. *Accid. Anal. Prev.* **2011**, *43*, 1979–1990. [CrossRef] [PubMed]

![applied sciences logo] *applied sciences*

MDPI

Article

Characterization of Surface Ozone Behavior at Different Regimes

Nádia F. Afonso and José C. M. Pires *

Laboratório de Engenharia de Processos, Ambiente Biotecnologia e Energia (LEPABE), Departamento de Engenharia Química, Faculdade de Engenharia, Universidade do Porto, Rua Dr. Roberto Frias, 4200-465 Porto, Portugal; nadia.afonso@iol.pt
* Correspondence: jcpires@fe.up.pt; Tel.: +351-225-082-262

Received: 25 July 2017; Accepted: 12 September 2017; Published: 14 September 2017

Abstract: Previous studies showed that the influence of meteorological variables and concentrations of other air pollutants on O_3 concentrations changes at different O_3 concentration levels. In this study, threshold models with artificial neural networks (ANNs) were applied to characterize the O_3 behavior at an urban site (Porto, Portugal), describing the effect of environmental and meteorological variables on O_3 concentrations. ANN characteristics, and the threshold variable and value, were defined by genetic algorithms (GAs). The considered predictors were hourly average concentrations of NO, NO_2, and O_3, and meteorological variables (temperature, relative humidity, and wind speed) measured from January 2012 to December 2013. Seven simulations were performed and the achieved models considered wind speed (at 4.9 m·s^{-1}), temperature (at 17.5 °C) and NO_2 (at 26.6 µg·m^{-3}) as the variables that determine the change of O_3 behavior. All the achieved models presented a similar fitting performance: R^2 = 0.71–0.72, $RMSE$ = 14.5–14.7 µg·m^{-3}, and the index of agreement of the second order of 0.91. The combined effect of these variables on O_3 concentration was also analyzed. This statistical model was shown to be a powerful tool for interpreting O_3 behavior, which is useful for defining policy strategies for human health protection concerning this air pollutant.

Keywords: air pollution; artificial neural network; genetic algorithms; surface ozone; threshold models

1. Introduction

Surface ozone (O_3) is considered one of the most concerning air pollutants in Europe. It is a secondary pollutant (it is not directly emitted), generated by chemical reactions that occur in the atmosphere between primary air pollutants (nitrogen oxides—NO_x—and volatile organic compounds—VOCs) catalyzed by sunlight [1]. The impact of this air pollutant has been studied in different areas [2–5]. Concerning human health, O_3 can cause injuries to airway epithelial cells (and lung diseases such as asthma), hyperplasia, headaches, and nausea, particularly in sensitive people, such as children and elderly [6–10]. Regarding vegetation, O_3 can damage plant leaves (decreases in both leaf photosynthesis and leaf area), reducing crop yields associated with a high negative economic impact [11–13]. Moreover, as a strong oxidant, it is responsible for the degradation of material via corrosion [14].

As already mentioned, O_3 is produced by chemical reactions between primary pollutants present in the atmosphere. It can also be transported from other locations by the wind (horizontal transport) and from the stratosphere (vertical transport) [15,16]. Thus, the atmosphere works as an open chemical reactor, in which reaction kinetics depend on the concentrations of reactants (primary air pollutants), mixture (wind speed and direction), temperature (influencing exponentially kinetic reaction constants—Arrhenius equation), and solar radiation. Thus, the O_3 behavior is highly dependent on the environment (urban, rural, or background) and meteorological variables.

In urban areas, O_3 concentrations are usually lower than the values observed in rural areas [17–20]. This phenomenon occurs mainly due to high NO_x concentrations. Photochemical equilibrium is defined by the following equations [21–23]:

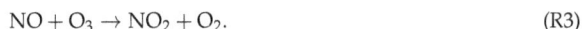

$$NO_2 + h\nu \; (\lambda < 420 \text{ nm}) \rightarrow NO + O \tag{R1}$$

$$O + O_2 + M \rightarrow O_3 + M \tag{R2}$$

$$NO + O_3 \rightarrow NO_2 + O_2. \tag{R3}$$

The photochemical reaction of NO_2 decomposition (chemical reaction R1) leads to the production of NO molecules and oxygen atoms that combine with molecular oxygen to produce ozone (chemical reaction R2). O_3 can also react with NO, forming NO_2 (chemical reaction R3). The complexity of phenomena associated with O_3 formation makes it hard to understand and predict its concentration in ambient air. Moreover, there are insufficient data (e.g., inventories of air pollutant emissions) to develop consistent phenomenological models to describe O_3 behavior.

Alternatively, statistical models have the capability of characterizing the relationship between variables using collected data and they involve mere pattern-recognition using mathematical operations. One of the most applied statistical models is the artificial neural network (ANN). ANNs are nonlinear models, which are inspired in the biological neural processing system [24,25]. These models are composed by artificial neurons (grouped in layers; three layers—input, hidden, and output—are often applied) that receive an input value and converts to an output through a selected function (activation function). Additionally, ANNs are characterized by a high fitting performance. The rapid development of computer hardware has increased the processing capabilities, which have led to achievement of ANN models with less computation time [26]. Therefore, these models have been used in a wide range of applications, including classification, regression, and mapping [27–29]. However, there are too many variables that need to be defined before the model parameters can be determined, including (i) the number of processing neurons in the hidden layer, and (ii) the activation function for each neuron. In recent years, genetic algorithms (GAs) have been applied to help in the definition of these variables. GAs are commonly applied to generate high-quality solutions for optimization and search problems, based on bio-inspired operators, such as mutation, crossover, and selection [30,31]. In GAs, a set of candidate solutions (called population—a group of individuals) are iteratively modified though the mentioned genetic operators in order to find a group of better solutions for the next generation (new iteration). GAs present the following advantages: (i) continuous or discrete variables can be optimized; (ii) a derivative function is not required; (iii) multivariable problems can be optimized; (iv) extremely complex cost surfaces can be dealt with; and (v) a list of optimal solutions (and not just a single one) is provided.

In Porto (selected area in this study—Portugal), an increasing trend of O_3 concentrations (147% higher) has been observed since the 19th century due to the photochemical production of this pollutant, associated with the increase in anthropogenic emissions mainly due to traffic [32]. Additionally, in the north of Portugal, Lamas d'Olo is a rural site where the highest O_3 concentrations are usually measured. Consequently, this site is often selected to evaluate O_3 behavior. Russo et al. [33] mentioned that high-ozone episodes can be explained by several factors: (i) atmospheric stagnation; (ii) horizontal transport by the wind of ozone-rich air masses; (iii) high solar radiation and temperature; and (iv) the influence of local winds (sea breezes and valley winds). Carvalho et al. [19] observed a positive correlation between O_3 concentrations with temperature and a negative correlation with relative humidity. Regarding the effect of wind field, the northeast flow from Spain (Galicia and Asturias) was observed, and this can be associated with the long-range transport of atmospheric pollutants to Portugal. Fernández-Guisuraga et al. [34] compared O_3 trends at urban and rural sites. At the rural site, O_3 concentrations were mainly influenced by the wind (transport), showing low variability with the concentrations of other pollutants. On the other hand, at the urban site, most of the variance was explained by the NO_2/NO_x ratio. Several research studies can be found where

ANN models were applied to determine O_3 trends and to predict its concentration (to provide early warning to the population when high O_3 concentration episodes occur). Comrie [35] compared the performance of an ANN model with a multiple linear regression (MLR) model to predict daily average O_3 concentrations in different cities with distinct climate and O_3 regimes. ANN models presented slightly better performance than MLR. Abdul-Wahab and Al-Alawi [36] developed ANN models to predict O_3 concentrations through meteorological and environmental data. The contribution of the meteorological data was defined between 33% and 41%, while the remaining variation was attributed to chemical pollutants. NO, SO_2, relative humidity (the highest contribution), non-methane hydrocarbon, and NO_2 were the variables that most influenced the O_3 concentrations. Additionally, temperature also presents an important role, while solar radiation had a lower effect than expected. Pires et al. [37] compared threshold autoregressive (TAR) models, autoregressive (AR) models, and ANN in the prediction of the next day hourly average O_3 concentrations. In the training period, ANN presented a higher performance. However, in the test period, TAR models presented more accurate results and the distinction became greater when the evaluation was performed for the prediction of extreme values.

In recent studies, O_3 concentrations have shown different behaviors regarding certain explanatory variables [25,37], which can be classified as O_3 regimes. This observation can be justified by the chemical reactions associated with O_3 formation/destruction that are influenced by certain variables, such as temperature, solar radiation, and wind speed [32]. To take these regimes account, threshold regression models were considered in this study [38]. Thus, GAs were used to define the threshold variable and value (the value of the explanatory variable corresponding to the change of the regime; two regimes were selected), the number of hidden neurons, and the activation function in the hidden and output layers. In this study, hourly average O_3 concentrations were modeled using threshold models with an ANN, whose structure was iteratively optimized by GAs. The achieved models enable the characterization of O_3 variability with selected meteorological and environmental variables in different regimes.

2. Materials and Methods

2.1. Data

Air quality data were obtained from an urban background site (*Sobreiras—Lordelo do Ouro*, see Figure 1) of the Air Quality Monitoring Network (AQMN) of Porto, Portugal. The AQMN is managed by the Regional Commission of Coordination and Development of Northern Portugal (*Comissão de Coordenação e Desenvolvimento Regional do Norte*), under the responsibility of the Ministry of Environment. Hourly average concentrations of NO, NO_2, and O_3 from the period from January 2012 to December 2013 (8760 hourly average values in 2012 and 7481 in 2013) were used to develop the proposed models. NO and NO_2 were obtained through the chemiluminescence method according to European Union (EU) Directive 1999/30/EC (European Community). According to EU Directive 2002/3/EC, O_3 measurements were performed through UV-absorption photometry using the equipment 41 M UV Photometric Ozone Analyzer (Environment S.A., Poissy, France). This monitoring equipment was subject to a rigid maintenance program, calibrated every 4 weeks. Measurements were continuously registered, and hourly average concentrations (in $\mu g \cdot m^{-3}$) were recorded.

The meteorological data were collected in a meteorological station located at *Pedras Rubras*, which is managed by *Instituto Português do Mar e da Atmosfera* (IPMA, I.P.); these values are considered representative for the entire Metropolitan Area of Porto. In this study, hourly averages of temperature (T, °C), relative humidity (RH, %), and wind speed (WS, m·s^{-1}) were used to analyze the influence of meteorological conditions on O_3 concentrations.

Figure 1. Monitoring site of *Sobreiras—Lordelo do Ouro* (from http://qualar.apambiente.pt/).

2.2. Statistical Model

In this study, threshold models with ANNs were defined with GAs, aiming to evaluate the effect of environmental and meteorological variables in O_3 concentrations. The applied model is defined as the following:

$$y = \begin{cases} \text{net}_1(x_i), \text{ if } x_d \leq r \\ \text{net}_2(x_i), \text{ if } x_d > r \end{cases} \tag{1}$$

where y is the output variable, net_1 and net_2 are ANN models, x_i are the exploratory variables, x_d is the threshold variable, and r is the threshold value. Applied feedforward ANN models had three layers (input, hidden, and output) and considered eight input variables (hourly average data): NO concentration, NO_2 concentration (due to the chemical reactions R1 and R3), the ratio NO_2/NO (due to the equilibrium constant of the chemical reaction R3), T, RH, $1/RH$ (as RH usually shows a negative effect on O_3 levels), WS, and $1/WS$ (the same as the RH effect). The output variable was the hourly average O_3 concentrations measured at the same time of the input data to infer the direct influence of these variables on O_3 chemistry. Regarding the activation functions, the linear function was considered for the output neuron and four functions were selected by GAs: sigmoid, hyperbolic tangent, inverse, and radial basis. The data were divided in training (75%) and validation (25%) sets and the early stopping method (ANN training procedure is stopped when an increase in validation error is observed) was applied to avoid overfitting. The division of the data was performed by time: 75% for training (January 2012 to 25 May 2013); 25% for validation (25 May 2013 to 19 December 2013). In the training set, O_3 concentrations ranged from 0 to 161 $\mu g \cdot m^{-3}$, while O_3 concentrations ranged from 0 to 170 $\mu g \cdot m^{-3}$ in the validation set.

GAs are a search and optimization technique based on Darwin principles of evolution and natural genetics [30,31]. This procedure begins with a set of individuals (population) that is randomly generated. Each individual (also called chromosome) is a binary code string and contains information about a set of parameters, which is a potential solution to a given problem. To evaluate the quality of the proposed solution (to rank the individuals in the population), a fitness function should be defined. To create new chromosomes for the next generation, the fittest chromosomes are submitted to the genetic operations [30]: (i) selection; (ii) crossover; (iii) mutation. These new chromosomes are then evaluated according to the fitness function, and the ones with the highest performance were selected. The repetition of this procedure generates a sequence of populations containing better solutions. The termination criteria can be (i) stop after a previously defined maximum number of generations is achieved, or (ii) stop when a desired fitness value is achieved. In this study, GAs were used to define the threshold variable and value, the number of hidden neurons, and the activation function in the hidden layer, and to select the explanatory variables to be used in each ANN model. The determination of the models was coded by the authors with MATLAB® software (R2014a, MathWorks, Natick, MA, USA, 2014) using the following specifications:

- a population size of 100;
- a selection probability of 0.20 (proportion of the individuals of the new generation obtained by selection operator);
- a selection criterion based on elitism (a small proportion of the fittest candidates is copied unchanged into the next generation);
- a crossover probability of 0.70 (proportion of the individuals of the new generation obtained by crossover operator);
- a mutation probability of 0.1 (proportion of the individuals of the new generation obtained by mutation operator);
- an evaluation of root mean squared error (*RMSE*) in training and validation sets;
- a stopping criterion based on the maximum number of generations.

Figure 2 shows an example of chromosome (37 bits). It is divided in 8 sets of bits (SB_i). SB_1 (3 bits) defines the threshold variable (from the explanatory variables; the maximum number of 8) through the conversion from binary to decimal numbers (MATLAB function *bin2dec*). SB_2 (8 bits) defines the threshold value. With the threshold variable already defined, the maximum (x_{max}) and minimum (x_{min}) values of this variable are determined. Threshold value is calculated based on Equation (2).

$$r = \frac{bin2dec(SB_2)}{255} \times (x_{max} - x_{min}) + x_{min}. \tag{2}$$

SB_3 and SB_6 (2 bits) define the activation function for the hidden layer of each ANN: 00—log-sigmoid (*logsig*); 01—hyperbolic tangent sigmoid (*tansig*); 10—inverse (*netinv*); 11—radial basis (*radbas*). SB_4 and SB_7 (3 bits) define the number of neurons in the hidden layer through the conversion from binary to decimal number (1 to 8). SB_5 and SB_8 (8 bits) define the explanatory variables that are used in each ANN (1 bit for each explanatory variable): 0—not selected; 1—selected.

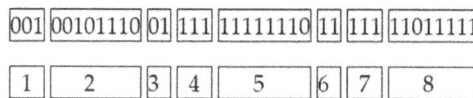

001	00101110	01	111	11111110	11	111	11011111
1	2	3	4	5	6	7	8

Figure 2. Example of a chromosome.

3. Results and Discussion

3.1. Air Quality and Meteorological Data Characterization

During the analyzed period, the hourly average O_3 concentrations were between 0 and 170 $\mu g \cdot m^{-3}$ (not exceeding the information neither the alert threshold—180 and 240 $\mu g \cdot m^{-3}$, respectively). Regarding O_3 exceedances to EU limits for the protection of human health, the 8 h average O_3 concentrations were higher than 120 $\mu g \cdot m^{-3}$ twice in September 2012, twice in July 2013, and once in August 2013. Figure 3 shows the average daily profile of O_3 concentrations. As a photochemical pollutant, its concentration increases during the daylight period, presenting a maximum between 14 and 15 h and a minimum at night time. The observed profile is characteristic of an urban site, as it does not present a high amplitude of concentrations (due to the presence of high NO_x concentrations). Figure 4 shows the monthly average values of NO, NO_2, and O_3 concentrations, as well as the analyzed meteorological variables (temperature, relative humidity, and wind speed). High O_3 concentrations were observed in April 2012 (63.7 $\mu g \cdot m^{-3}$) and from March to July 2013 (59.9–71.3 $\mu g \cdot m^{-3}$). In this period, low concentrations of NO and NO_2 were also measured.

High temperatures and low relative humidity were also observed. On the other hand, lower O_3 concentrations were measured for periods with high NO and NO_2 concentrations and lower temperatures. These observations are in agreement with other research studies in which the behavior of O_3 was analyzed [18,39–41]. Pires et al. [40] compared several linear models to predict O_3 concentrations at an urban site in Porto. The correlation analysis performed between O_3 concentrations and meteorological variables showed also negative correlations with NO, NO_2, and RH and a positive correlation with T. In another study focusing on the same region [41], O_3 concentrations were negatively correlated with NO, NO_2, and RH and positively correlated with T and WS. Zhang, Wang, Park, and Deng [18] analyzed high O_3 concentration episodes and related them with meteorological variables. O_3 concentrations were highly correlated with maximum temperature and minimum relative humidity. The effect of minimum WS was also analyzed at urban, suburban, and rural sites. O_3 concentrations were positively (negatively) correlated with minimum WS at urban (suburban and rural) sites. Shan, Yin, Zhang, Ji, and Deng [39] analyzed the effect of meteorological variables on O_3 concentrations at an urban site in China. Daily average O_3 concentrations were negatively correlated with pressure and RH, and positively correlated with temperature, solar radiation, sunshine duration, and wind speed.

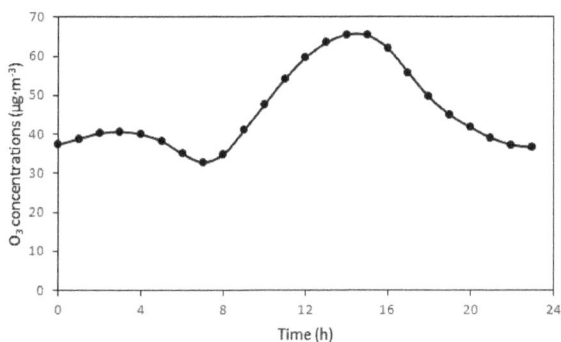

Figure 3. Daily average profile of O_3 concentrations at the monitoring site.

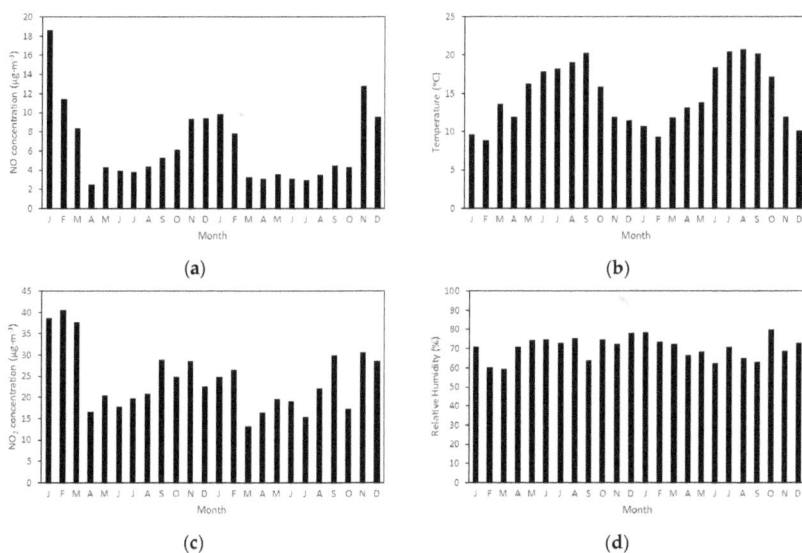

(a)

(b)

(c)

(d)

Figure 4. *Cont.*

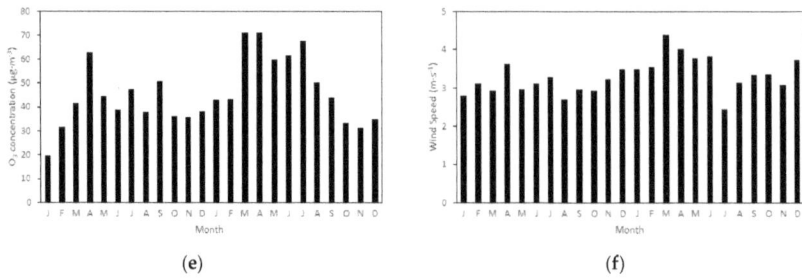

(e) (f)

Figure 4. Monthly average values of (**a**) NO concentrations, (**b**) temperature, (**c**) NO$_2$ concentrations, (**d**) relative humidity, (**e**) O$_3$ concentrations, and (**f**) wind speed.

3.2. Linear Correlation Analysis

Figure 5 shows the variation in linear correlation between O$_3$ and meteorological parameters on a monthly basis. Negative correlations were observed for NO (−0.547 to −0.296) and NO$_2$ (−0.807 to −0.276) concentrations. The effect of these air pollutants was more significant in winter periods than in summer periods. Regarding the effect of meteorological variables, temperature was usually positively correlated with O$_3$, which was in agreement with what was expected. The highest value ($R = 0.661$) was determined in September 2013 and an unusual negative correlation ($R = -0.376$) was determined in July 2013. *RH* was negatively correlated in almost all periods. The highest impact was also observed in September 2013 ($R = -0.685$) and an unusual positive correlation was determined in July 2013 ($R = 0.419$). Chen et al. [42] demonstrated that *RH* favors O$_3$ decomposition, justifying the associated negative effect. Regarding *WS*, this variable can have two different effects on O$_3$ concentrations. Low *WS* can promote the accumulation of O$_3$ produced in the region (increasing its concentration), while high values reduce the levels of other air pollutants (such as NO$_x$) that influence the O$_3$ chemistry (in the case of NO$_x$, its concentration decrease leads to the increase in O$_3$ levels). Thus, the effect of *WS* on O$_3$ concentrations depends on the studied environment: urban or rural. In this study (urban environment), *WS* was positively correlated with O$_3$ concentrations, with the highest value ($R = 0.593$) in February 2013.

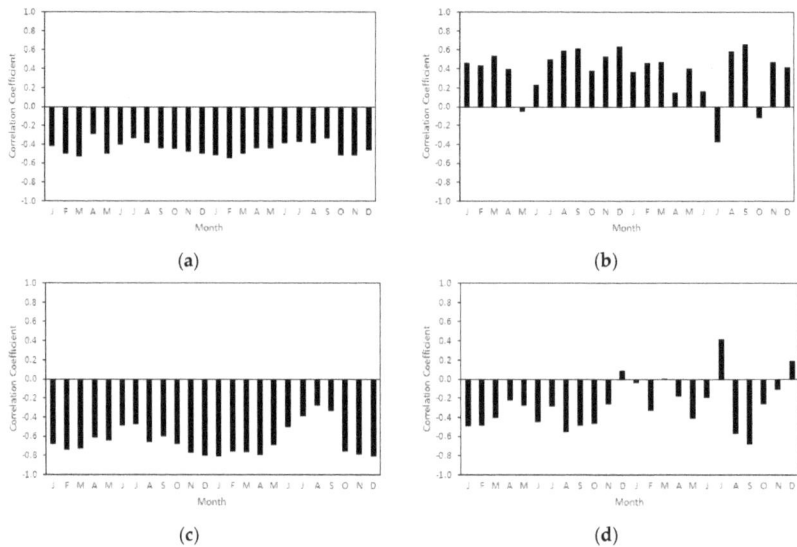

(a) (b)

(c) (d)

Figure 5. *Cont.*

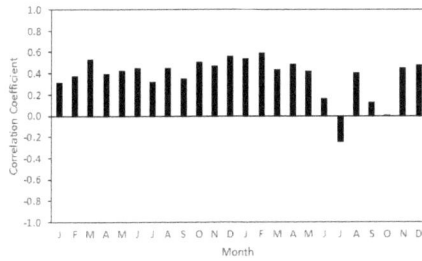

(e)

Figure 5. Temporal variation of linear correlation between O_3 concentrations and the following: (**a**) NO concentrations, (**b**) temperature, (**c**) NO_2 concentrations, (**d**) relative humidity, and (**e**) wind speed.

3.3. ANN Models and Interpretation

Seven simulations were performed to determine the models able to describe the relationship between O_3 concentrations with NO, NO_2, T, RH, and WS (measured at same time). These models are threshold models, considering two O_3 regimes where the relationship between output and input variables are different. The change from one regime to another depends on the value (threshold value) of a specific input variable (threshold variable). GAs were used to optimize the ANN characteristics (the number of hidden neurons, the activation function in the hidden layer, and the input variables), and the threshold variable and value. The models were evaluated according their fitting performance in training and validation sets. Table 1 shows the best models for the seven simulations. All of them presented similar fitting performance: (i) $R^2 = 0.71$–0.72; (ii) $RMSE$ between 14.5 and 14.7 $\mu g \cdot m^{-3}$; and (iii) the index of agreement of the second order of 0.91. Three explanatory variables were selected, each one with a specific threshold value: (i) WS with 4.9 $m \cdot s^{-1}$; (ii) T with 17.5 °C; and (iii) NO_2 with 26.6 $\mu g \cdot m^{-3}$. Generally, hyperbolic tangent and radial basis were the functions selected for the hidden layer, composed by 7 or 8 neurons. In almost all models, all input variables were selected as ANN inputs in both O_3 regimes.

The analysis of the combined effect of input variables (two variables) was performed for the three threshold variables, considering the two regimes determined by the best models in Simulations I, II, and III. Figure 6 shows the combined effect of NO_2, T, and WS on O_3 concentrations for $WS \leq 4.9$ $m \cdot s^{-1}$ and for $WS > 4.9$ $m \cdot s^{-1}$. For $WS \leq 4.9$ $m \cdot s^{-1}$, O_3 concentrations (i) decreased with NO_2 except when $T > 17$ °C (without significant variation), (ii) increased with T except when $WS > 2.8$ $m \cdot s^{-1}$ (O_3 presented a maximum between 17 and 20 °C), and (iii) did not change significantly with WS except when $T > 26$ °C (presenting a decreasing tendency). For $WS > 4.9$ $m \cdot s^{-1}$, O_3 concentrations (i) decreased with NO_2 except when $T > 17$ °C (presenting a slight increase), (ii) presented high values for high T with all tested ranges of NO_2 concentrations (presenting a local maximum—≈ 87.4 $\mu g \cdot m^{-3}$—for $T \approx 11$ °C and $NO_2 \approx 6$ $\mu g \cdot m^{-3}$), (iii) increased with T for the tested range of WS, (iv) did not change significantly with WS, and (v) presented higher values than those where $WS \leq 4.9$ $m \cdot s^{-1}$. The combined effect of T-NO_2 is in agreement with what was concluded in linear correlation analysis. The effect of NO_2 is more significant in the winter period, in which temperatures are usually low and NO_2 concentrations are high (see Figure 4). With high NO_2 concentrations, the chemical equilibrium given by Equation (R3) limits the increase in O_3 concentrations. In addition, based on a comparison of the two regimes defined by WS, the combined effect of these two variables presented similar behavior; however, O_3 concentrations were higher when $WS > 4.9$ $m \cdot s^{-1}$ (49–104 $\mu g \cdot m^{-3}$) than they were when $WS \leq 4.9$ $m \cdot s^{-1}$ (11–41 $\mu g \cdot m^{-3}$). High values of WS are associated with the dispersion of air pollutants, reducing their concentration. As NO_2 concentrations decrease, O_3 concentrations can achieve higher values [17,19,20,22]. The combined effect of T-WS showed the highest variability of O_3 with T, showing a positive correlation between these variables. The O_3 variability with WS is almost insignificant.

Regarding the combined effect NO_2-WS, similar conclusions were drawn: O_3 concentrations presented a decreasing tendency with NO_2 concentrations, and their variability with WS is almost insignificant. Figures S1 and S2 present the effect of the same combinations of variables on O_3 concentrations considering T (simulation II) and NO_2 (simulation III) as threshold variables, respectively. Similar analysis can be performed through these figures.

Table 1. ANN models: their input variables, activation functions (AF), number of hidden neurons (HN), and performance indexes (R^2, RMSE (root mean squared error) and d_2) for each performed simulation (Sim).

Sim	Model	AF	HN	R^2/RMSE/d_2
I	$O_{3|t} = \begin{cases} net_1\left(NO, NO_2, \frac{NO_2}{NO}, T, RH, \frac{1}{RH}, WS, \frac{1}{WS}\right), \text{ if } WS \leq 4.9 \\ net_2\left(NO, NO_2, \frac{NO_2}{NO}, T, RH, \frac{1}{RH}, WS, \frac{1}{WS}\right), \text{ if } WS > 4.9 \end{cases}$	tansig radbas	8 8	0.71/14.7/0.91
II	$O_{3|t} = \begin{cases} net_1\left(NO, NO_2, \frac{NO_2}{NO}, T, RH, \frac{1}{RH}, WS, \frac{1}{WS}\right), \text{ if } T \leq 17.5 \\ net_2\left(NO, NO_2, \frac{NO_2}{NO}, T, RH, \frac{1}{RH}, WS, \frac{1}{WS}\right), \text{ if } T > 17.5 \end{cases}$	tansig tansig	7 7	0.72/14.5/0.91
III	$O_{3|t} = \begin{cases} net_1\left(NO, NO_2, \frac{NO_2}{NO}, T, RH, \frac{1}{RH}, WS\right), \text{ if } NO_2 \leq 26.6 \\ net_2\left(NO, NO_2, \frac{NO_2}{NO}, T, RH, \frac{1}{RH}, WS, \frac{1}{WS}\right), \text{ if } NO_2 > 26.6 \end{cases}$	tansig radbas	8 7	0.71/14.7/0.91
IV	$O_{3|t} = \begin{cases} net_1\left(NO, NO_2, \frac{NO_2}{NO}, T, RH, \frac{1}{RH}, WS, \frac{1}{WS}\right), \text{ if } WS \leq 4.9 \\ net_2\left(NO, NO_2, \frac{NO_2}{NO}, T, RH, \frac{1}{RH}, WS, \frac{1}{WS}\right), \text{ if } WS > 4.9 \end{cases}$	tansig radbas	8 8	0.71/14.7/0.91
V	$O_{3|t} = \begin{cases} net_1\left(NO, NO_2, \frac{NO_2}{NO}, T, RH, \frac{1}{RH}, WS, \frac{1}{WS}\right), \text{ if } T \leq 17.5 \\ net_2\left(NO, NO_2, \frac{NO_2}{NO}, T, RH, \frac{1}{RH}, WS, \frac{1}{WS}\right), \text{ if } T > 17.5 \end{cases}$	tansig tansig	7 7	0.72/14.5/0.91
VI	$O_{3|t} = \begin{cases} net_1\left(NO, NO_2, \frac{NO_2}{NO}, T, RH, \frac{1}{RH}, WS\right), \text{ if } NO_2 \leq 26.6 \\ net_2\left(NO, NO_2, T, RH, \frac{1}{RH}, WS, \frac{1}{WS}\right), \text{ if } NO_2 > 26.6 \end{cases}$	tansig radbas	8 8	0.71/14.7/0.91
VII	$O_{3|t} = \begin{cases} net_1\left(NO, NO_2, \frac{NO_2}{NO}, T, RH, WS\right), \text{ if } T \leq 17.5 \\ net_2\left(NO_2, \frac{NO_2}{NO}, T, RH, \frac{1}{RH}, WS, \frac{1}{WS}\right), \text{ if } T > 17.5 \end{cases}$	tansig tansig	7 7	0.72/14.5/0.91

The application of this statistical methodology allows for the determination of the influence of environmental and meteorological variables on O_3 concentration. Consequently, it is possible to develop more accurate predictive models for this secondary pollutant, which is important for the definition of policy measures for human health protection.

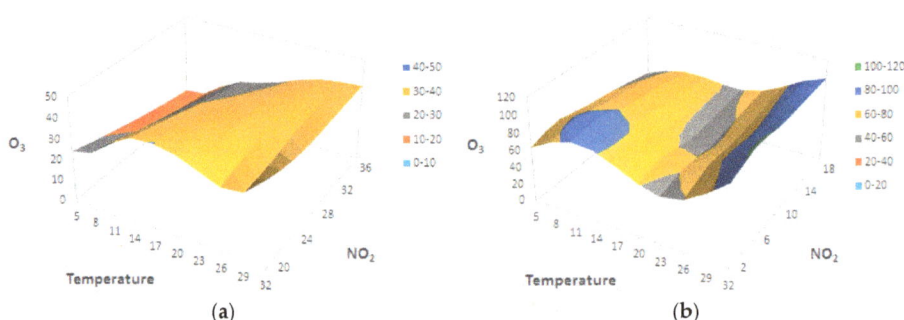

(a)   (b)

Figure 6. *Cont.*

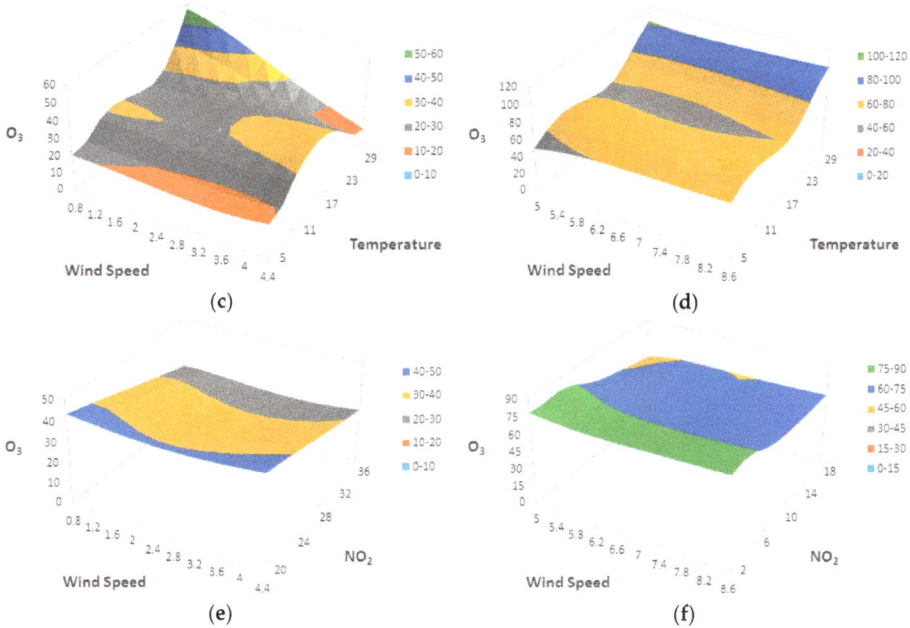

Figure 6. The combined effect of NO_2 concentrations, temperature, and wind speed on O_3 concentrations based on model determined in Simulation I where *WS* (wind speed) ≤ 4.9 m·s^{-1} (**a,c,e**) and where *WS* > 4.9 m·s^{-1} (**b,d,f**).

4. Conclusions

Linear correlation analysis showed a positive relationship between O_3 concentrations with *T* and *WS*, while NO, NO_2, and *RH* showed a negative effect. In the studied period, the highest O_3 concentrations were observed for low NO_x concentrations and high wind speed. Threshold models with ANNs and those defined by genetic algorithms define three important variables that could define different O_3 regimes: (i) a wind speed of 4.9 m·s^{-1}; (ii) a temperature of 17.5 °C; and (iii) an NO_2 concentration of 26.6 µg·m^{-3}. The achieved models enabled the evaluation of the combined effect of two input variables in different O_3 regimes. This information may be useful for defining policy strategies for human health protection concerning surface ozone.

Supplementary Materials: The following are available online at http://www.mdpi.com/2076-3417/7/9/944/s1. Figure S1: The combined effect of NO_2 concentrations, temperature, and wind speed on O_3 concentrations based on the model determined in Simulation II; Figure S2: The combined effect of NO_2 concentrations, temperature, and wind speed on O_3 concentrations based on the model determined in Simulation III.

Acknowledgments: This work was financially supported by Project POCI-01-0145-FEDER-006939 (LEPABE) funded by FEDER funds through COMPETE2020—Programa Operacional Competitividade e Internacionalização (POCI)—and by national funds through FCT—Fundação para a Ciência e a Tecnologia. J.C.M.P. acknowledges the FCT Investigator 2015 Programme (IF/01341/2015).

Author Contributions: N.F.A. collected the data, ran the simulations, analyzed the results, and wrote the paper; J.C.M.P. conceived the study, developed the model code, and revised the paper.

Conflicts of Interest: The authors declare no conflict of interest.

References

1. Lin, M.Y.; Horowitz, L.W.; Payton, R.; Fiore, A.M.; Tonnesen, G. Us surface ozone trends and extremes from 1980 to 2014: Quantifying the roles of rising asian emissions, domestic controls, wildfires, and climate. *Atmos. Chem. Phys.* **2017**, *17*, 2943–2970. [CrossRef]

2. Lefohn, A.S.; Malley, C.S.; Simon, H.; Wells, B.; Xu, X.; Zhang, L.; Wang, T. Responses of human health and vegetation exposure metrics to changes in ozone concentration distributions in the European Union, United States, and China. *Atmos. Environ.* **2017**, *152*, 123–145. [CrossRef]

3. Hollaway, M.J.; Arnold, S.R.; Collins, W.J.; Folberth, G.; Rap, A. Sensitivity of midnineteenth century tropospheric ozone to atmospheric chemistry-vegetation interactions. *J. Geophys. Res.-Atmos.* **2017**, *122*, 2452–2473. [CrossRef]

4. Proietti, C.; Anav, A.; De Marco, A.; Sicard, P.; Vitale, M. A multi-sites analysis on the ozone effects on gross primary production of European forests. *Sci. Total Environ.* **2016**, *556*, 1–11. [CrossRef] [PubMed]

5. Jiang, Z.; Miyazaki, K.; Worden, J.R.; Liu, J.J.; Jones, D.B.A.; Henze, D.K. Impacts of anthropogenic and natural sources on free tropospheric ozone over the Middle East. *Atmos. Chem. Phys.* **2016**, *16*, 6537–6546. [CrossRef]

6. Jorres, R.A.; Holz, O.; Zachgo, W.; Timm, P.; Koschyk, S.; Muller, B.; Grimminger, F.; Seeger, W.; Kelly, F.J.; Dunster, C.; et al. The effect of repeated ozone exposures on inflammatory markers in bronchoalveolar lavage fluid and mucosal biopsies. *Am. J. Respir. Crit. Care Med.* **2000**, *161*, 1855–1861. [CrossRef] [PubMed]

7. Frank, R.; Liu, M.C.; Spannhake, E.W.; Mlynarek, S.; Macri, K.; Weinmann, G.G. Repetitive ozone exposure of young adults—Evidence of persistent small airway dysfunction. *Am. J. Respir. Crit. Care Med.* **2001**, *164*, 1253–1260. [CrossRef] [PubMed]

8. Holz, O.; Mucke, M.; Paasch, K.; Bohme, S.; Timm, P.; Richter, K.; Magnussen, H.; Jorres, R.A. Repeated ozone exposures enhance bronchial allergen responses in subjects with rhinitis or asthma. *Clin. Exp. Allergy* **2002**, *32*, 681–689. [CrossRef] [PubMed]

9. McConnell, R.; Berhane, K.; Gilliland, F.; London, S.J.; Islam, T.; Gauderman, W.J.; Avol, E.; Margolis, H.G.; Peters, J.M. Asthma in exercising children exposed to ozone: A cohort study. *Lancet* **2002**, *359*, 386–391. [CrossRef]

10. Goldberg, M.S.; Burnett, R.T.; Brook, J.; Bailar, J.C.; Valois, M.F.; Vincent, R. Associations between daily cause-specific mortality and concentrations of ground-level ozone in Montreal, Quebec. *Am. J. Epidemiol.* **2001**, *154*, 817–826. [CrossRef] [PubMed]

11. Carter, C.A.; Cui, X.; Ding, A.; Ghanem, D.; Jiang, F.; Yi, F.; Zhong, F. Stage-specific, nonlinear surface ozone damage to rice production in China. *Sci. Rep.* **2017**, *7*, 44224. [CrossRef] [PubMed]

12. Morgan, P.B.; Ainsworth, E.A.; Long, S.P. How does elevated ozone impact soybean? A meta-analysis of photosynthesis, growth and yield. *Plant Cell Environ.* **2003**, *26*, 1317–1328. [CrossRef]

13. Sadiq, M.; Tai, A.P.K.; Lombardozzi, D.; Val Martin, M. Effects of ozone-vegetation coupling on surface ozone air quality via biogeochemical and meteorological feedbacks. *Atmos. Chem. Phys.* **2017**, *17*, 3055–3066. [CrossRef]

14. Christodoulakis, J.; Tzanis, C.G.; Varotsos, C.A.; Ferm, M.; Tidblad, J. Impacts of air pollution and climate on materials in Athens, Greece. *Atmos. Chem. Phys.* **2017**, *17*, 439–448. [CrossRef]

15. Banta, R.M.; Senff, C.J.; White, A.B.; Trainer, M.; McNider, R.T.; Valente, R.J.; Mayor, S.D.; Alvarez, R.J.; Hardesty, R.M.; Parrish, D.; et al. Daytime buildup and nighttime transport of urban ozone in the boundary layer during a stagnation episode. *J. Geophys. Res.-Atmos.* **1998**, *103*, 22519–22544. [CrossRef]

16. Monier, E.; Weare, B.C. Climatology and trends in the forcing of the stratospheric ozone transport. *Atmos. Chem. Phys.* **2011**, *11*, 6311–6323. [CrossRef]

17. Tong, L.; Zhang, H.L.; Yu, J.; He, M.M.; Xu, N.B.; Zhang, J.J.; Qian, F.Z.; Feng, J.Y.; Xiao, H. Characteristics of surface ozone and nitrogen oxides at urban, suburban and rural sites in Ningbo, China. *Atmos. Res.* **2017**, *187*, 57–68. [CrossRef]

18. Zhang, H.; Wang, Y.H.; Park, T.W.; Deng, Y. Quantifying the relationship between extreme air pollution events and extreme weather events. *Atmos. Res.* **2017**, *188*, 64–79. [CrossRef]

19. Carvalho, A.; Monteiro, A.; Ribeiro, I.; Tchepel, O.; Miranda, A.I.; Borrego, C.; Saavedra, S.; Souto, J.A.; Casares, J.J. High ozone levels in the northeast of Portugal: Analysis and characterization. *Atmo. Environ.* **2010**, *44*, 1020–1031. [CrossRef]

20. Pires, J.C.M.; Alvim-Ferraz, M.C.M.; Martins, F.G. Surface ozone behaviour at rural sites in Portugal. *Atmos. Res.* **2012**, *104*, 164–171. [CrossRef]

21. Sanchez, B.; Santiago, J.L.; Martilli, A.; Palacios, M.; Kirchner, F. Cfd modeling of reactive pollutant dispersion in simplified urban configurations with different chemical mechanisms. *Atmos. Chem. Phys.* **2016**, *16*, 12143–12157. [CrossRef]

22. Pires, J.C.M. Ozone weekend effect analysis in three European urban areas. *Clean-Soil Air Water* **2012**, *40*, 790–797. [CrossRef]

23. Gressent, A.; Sauvage, B.; Cariolle, D.; Evans, M.; Leriche, M.; Mari, C.; Thouret, V. Modeling lightning-nox chemistry on a sub-grid scale in a global chemical transport model. *Atmos. Chem. Phys.* **2016**, *16*, 5867–5889. [CrossRef]

24. Agatonovic-Kustrin, S.; Beresford, R. Basic concepts of artificial neural network (ann) modeling and its application in pharmaceutical research. *J. Pharm. Biomed.* **2000**, *22*, 717–727. [CrossRef]

25. Pires, J.C.M.; Goncalves, B.; Azevedo, F.G.; Carneiro, A.P.; Rego, N.; Assembleia, A.J.B.; Lima, J.F.B.; Silva, P.A.; Alves, C.; Martins, F.G. Optimization of artificial neural network models through genetic algorithms for surface ozone concentration forecasting. *Environ. Sci. Pollut. Res.* **2012**, *19*, 3228–3234. [CrossRef] [PubMed]

26. Tawadrous, A.S.; Katsabanis, P.D. Prediction of surface crown pillar stability using artificial neural networks. *Int. J. Numer. Anal. Methods* **2007**, *31*, 917–931. [CrossRef]

27. Pasini, A. Artificial neural networks for small dataset analysis. *J. Thorac. Dis.* **2015**, *7*, 953–960. [PubMed]

28. Keshavarzi, A.; Sarmadian, F.; Omran, E.-S.E.; Iqbal, M. A neural network model for estimating soil phosphorus using terrain analysis. *Egypt. J. Remote Sens. Space Sci.* **2015**, *18*, 127–135. [CrossRef]

29. Zare, M.; Pourghasemi, H.R.; Vafakhah, M.; Pradhan, B. Landslide susceptibility mapping at vaz watershed (iran) using an artificial neural network model: A comparison between multilayer perceptron (mlp) and radial basic function (rbf) algorithms. *Arab. J. Geosci.* **2013**, *6*, 2873–2888. [CrossRef]

30. Goldberg, D.E. *Genetic Algorithms in Search, Optimization and Machine Learning*, 1st ed.; Addison-Wesley Professional: Boston, MA, USA, 1989.

31. Holland, J.H. *Adaptation in Natural and Artificial Systems*; University of Michigan Press: Ann Arbor, MI, USA, 1975.

32. Alvim-Ferraz, M.C.M.; Sousa, S.I.V.; Pereira, M.C.; Martins, F.G. Contribution of anthropogenic pollutants to the increase of tropospheric ozone levels in the oporto metropolitan area, Portugal since the 19th century. *Environ. Pollut.* **2006**, *140*, 516–524. [CrossRef] [PubMed]

33. Russo, A.; Trigo, R.M.; Martins, H.; Mendes, M.T. NO$_2$, PM10 and O$_3$ urban concentrations and its association with circulation weather types in Portugal. *Atmos. Environ.* **2014**, *89*, 768–785. [CrossRef]

34. Fernandez-Guisuraga, J.M.; Castro, A.; Alves, C.; Calvo, A.; Alonso-Blanco, E.; Blanco-Alegre, C.; Rocha, A.; Fraile, R. Nitrogen oxides and ozone in Portugal: Trends and ozone estimation in an urban and a rural site. *Environ. Sci. Pollut. Res.* **2016**, *23*, 17171–17182. [CrossRef] [PubMed]

35. Comrie, A.C. Comparing neural networks and regression models for ozone forecasting. *J. Air Waste Manag.* **1997**, *47*, 653–663. [CrossRef]

36. Abdul-Wahab, S.A.; Al-Alawi, S.M. Assessment and prediction of tropospheric ozone concentration levels using artificial neural networks. *Environ. Model. Softw.* **2002**, *17*, 219–228. [CrossRef]

37. Pires, J.C.M.; Alvim-Ferraz, M.C.M.; Pereira, M.C.; Martins, F.G. Evolutionary procedure based model to predict ground-level ozone concentrations. *Atmos. Pollut. Res.* **2010**, *1*, 215–219. [CrossRef]

38. Terui, N.; van Dijk, H.K. Combined forecasts from linear and nonlinear time series models. *Int. J. Forecast.* **2002**, *18*, 421–438. [CrossRef]

39. Shan, W.P.; Yin, Y.Q.; Zhang, J.D.; Ji, X.; Deng, X.Y. Surface ozone and meteorological condition in a single year at an urban site in central-eastern China. *Environ. Monit. Assess.* **2009**, *151*, 127–141. [CrossRef] [PubMed]

40. Pires, J.C.M.; Alvim-Ferraz, M.C.M.; Pereira, M.C.; Martins, F.G. Comparison of several linear statistical models to predict tropospheric ozone concentrations. *J. Stat. Comput. Simul.* **2012**, *82*, 183–192. [CrossRef]

41. Pires, J.C.M.; Martins, F.G.; Sousa, S.I.V.; Alvim-Ferraz, M.C.M.; Pereira, M.C. Selection and validation of parameters in multiple linear and principal component regressions. *Environ. Model. Softw.* **2008**, *23*, 50–55. [CrossRef]

42. Chen, H.H.; Stanier, C.O.; Young, M.A.; Grassian, V.H. A kinetic study of ozone decomposition on illuminated oxide surfaces. *J. Phys. Chem. A* **2011**, *115*, 11979–11987. [CrossRef] [PubMed]

MDPI AG

St. Alban-Anlage 66

4052 Basel, Switzerland

Tel. +41 61 683 77 34

Fax +41 61 302 89 18

http://www.mdpi.com

Applied Sciences Editorial Office

E-mail: applsci@mdpi.com

http://www.mdpi.com/journal/applsci

www.ingramcontent.com/pod-product-compliance
Lightning Source LLC
Chambersburg PA
CBHW051839210326

41597CB00033B/5714